Introduction to Protein Structure

John Kendrew (*left*) and Max Perutz (*right*) in front of the wire model of myoglobin in 1962, framed by a present-day computer graphics workstation.

Introduction to Protein Structure

Carl Branden

Uppsala Biomedical Center
Swedish University of Agricultural Sciences
Uppsala, Sweden

John Tooze

European Molecular Biology Organization
Heidelberg, Germany

Garland Publishing, Inc.
New York and London, 1991

THE COVER

Front: The background photograph of the cover is of a Laue x-ray diffraction pattern produced by a crystal of the plant enzyme ribulose bisphosphate carboxylase. This technique is described in Chapter 17. Information derived from such x-ray patterns, together with a knowledge of the amino acid sequence, enabled the three-dimensional arrangement of atoms in the protein to be determined. A simplified representation of this protein structure is shown in color, superimposed on the diffraction pattern. The enzyme, which is involved in the fixation of carbon dioxide, is a member of the large class of α/β barrel protein structures. This class of structures is discussed in detail in Chapter 4.

Back: Tomato bushy stunt virus is a spherical virus made from 180 protein subunits. Arms extending from sixty of these subunits contribute to an internal framework that determines the size of the correctly assembled virus particle. The interdigitated arms from three subunits meet at each of the twenty icosahedral threefold axes of the virus. One such axis is shown here with the β strands from three subunits shown in different shades of green. Virus structure is described in more detail in Chapter 11.

Library of Congress Cataloging-in-Publication Data

Branden, Carl.
   Introduction to protein structure / Carl Branden, John Tooze.
   p.   cm.
   Includes index.
   ISBN 0-8153-0344-0 — ISBN 0-8153-0270-3 (pbk.)
   1. Proteins—Structure. I. Tooze, John. II. Title.
QP551.B7635 1991
574.19'245—dc20   91-11788
                  CIP

Published by Garland Publishing, Inc.
136 Madison Ave., New York, New York, 10016

Printed in the United States of America

15 14 13 12 11 10 9 8 7 6 5 4 3 2 1

Preface

The ultimate goal of molecular biology is to understand biological processes in terms of the chemistry and physics of the macromolecules that participate in them. One of the essential differences between the chemistry of living systems and that of the nonliving is the great structural complexity of biological macromolecules. We shall not unravel the chemistry of life in molecular detail without knowing at atomic or close to atomic resolution the structure of biological macromolecules, especially the proteins.

The importance of molecular structure for an understanding of function is best exemplified, of course, by DNA. The simple and beautiful double-helical, base-paired structure of DNA immediately made genetics intelligible in chemical terms. Genes, the previously mysterious factors that controlled inheritance of particular traits, were segments of the DNA molecules that could be spooled out of solution at the end of a rotating glass rod, like cotton candy on a stick. Understanding DNA structure explained the two cardinal properties of genes: their ability to replicate and their ability to determine the structure of proteins. Molecular genetics came into being and flourished as the genetic code was deciphered and patterns of gene organization and expression were elucidated.

By contrast, the determination of the structures of myoglobin and hemoglobin, less than a decade after the structure of the DNA, and of many more proteins since, has not yielded a simple and all-embracing explanation of protein structure and hence function. Despite knowing today the three-dimensional structures of some 300 different proteins, we are still unable to formulate a set of general rules that allows us to predict a protein's three-dimensional structure from the amino acid sequence of its polypeptide chain. With hindsight it is perhaps not surprising that protein structures are so much more complex than that of DNA. Proteins are built up from twenty different amino acids compared with the four nucleotides of DNA. Moreover, proteins fulfill a much wider range of biological functions than does DNA, and functional diversity has dictated structural diversity.

By comparison with molecular genetics, progress in research on protein structure has been painfully slow, not only because of the great diversity that protein structures have proven to have, but also because of the simple technical problem of obtaining protein crystals that are large enough to use for crystallographic analyses and also diffract x-rays well enough to allow the structure to be determined to high, atomic resolution. Crystallizing protein is a hit-and-miss business with no theory. Some proteins crystallize readily, others not at all; some investigators seem to have "green thumbs," like good gardeners, and can grow crystals where others fail. Inescapably, therefore, crystallographers have been forced to work on those proteins of which crystals can be obtained rather than those proteins with the most interesting biological properties. Since, by biochemical standards, large quantities of a protein are needed before its crystallization can be attempted, the work until recently also has been confined to proteins naturally present in large amounts.

Recent technical developments have removed the latter limitation, and we believe that it is a particularly opportune moment to attempt to write a book on protein structure aimed at a wider audience than the professional crystallographer. With over 300 proteins solved certain basic themes or principles of protein structure can be discerned. Second, protein crystallography is currently undergoing something of a renaissance brought about by technical advances. On the one hand, recombinant DNA techniques now make it feasible to clone and overexpress genes whose protein products in nature are present in extremely low amounts. Through recombinant DNA it is possible not only to produce amounts of naturally very rare proteins that are sufficient to allow their crystallization but also to determine the amino acid sequences of the proteins, without which their x-ray structures cannot be solved. Thus, one of the crippling limitations on the choice of protein to study has been removed. The other, whether or not a protein can be persuaded to crystallize, remains, but again recombinant DNA techniques may provide at least a partial remedy. So far almost all attempts to crystallize overexpressed proteins have succeeded, probably because the gentle purification conditions do not damage the molecules and the large quantities of protein permit many more experiments.

Contemporaneously with the development of gene cloning there have been rapid increases in the power of computers, improvements in relevant software, the development of computer graphics devices, and the development of detectors to replace x-ray film as the means of collecting x-ray diffraction data. In combination these have speeded up the mathematical manipulations that convert a diffraction pattern into a structural model. Computer graphics has rendered model building with space-filling and skeletal models obsolescent. Structures can now be manipulated and analyzed in simulated three-dimensional space on the screen of a graphics terminal; substrates and inhibitors can be fitted into enzymes, DNA-binding proteins docked into DNA, all by the manipulation of a joystick.

In addition to these developments in x-ray crystallography and computer graphics, nuclear magnetic resonance has been developed into a tool for rapid determination of the three-dimensional structures of small protein molecules. However, it should always be kept in mind that the structural models obtained by these methods are based on interpretations of experimental data and might contain errors. In a few cases, preliminary protein structures based on x-ray data have been shown to be incorrect. Such models can, however, be corrected by subsequent refinement based on additional experimental data.

Molecular geneticists using site-directed mutagenesis and other techniques of reverse genetics are now beginning to redesign proteins both to investigate their biological function and to tailor enzymes for biotechnological processes. The realization of the full potential of protein engineering, however, will depend critically on the genetic engineer having a thorough grasp of the three-dimensional structure of the protein that he or she is redesigning. Without an understanding of protein structure, protein engineering degenerates into tinkering.

Our aim in writing this book, therefore, has been to bring out the structural and functional logic that has emerged from the accumulated data on protein structures. In the first part we identify recurring structural motifs and show how proteins with quite unrelated functions are built up from combinations of these. Although we cannot yet predict structure from sequence, it is possible to recognize certain structural themes that during the course of evolution have been utilized time and again. In the second part we consider particular biological functions and discuss the different protein structural solutions that have evolved to fulfill them. In short, we discuss how common basic structures can be elaborated to fulfill different functions and how the same biological function has more than one structural solution.

There is one insoluble problem inherent in writing a book about the structure of proteins, namely, the impossibility of illustrating satisfactorily on a flat sheet of paper complex three-dimensional structures. An illustration on a page can never convey properly the third dimension; the conventional schematic drawings, invaluable though they are, do scant justice to the intricacies of real structure. Short of providing everyone with computer graphics terminals there is no really satisfactory solution to this problem. We purposely refrained from using stereopair images, the results of which do not, in our view, justify their expense and inconvenience. We hope that readers will be stimulated by this book to seek out structural biologist colleagues with graphics terminals and see the molecules for themselves.

Molecular biology began some 40 years ago with the realization that structure was crucial for a proper understanding of function. Paradoxically, the dazzling achievements of molecular genetics and biochemistry led to the eclipse of structural studies. We believe the wheel has now come full circle, and those very achievements have increased the need for structural analysis at the same time that they have provided the means for it. If this book helps in any way to foster a wider interest in protein structure, it will have fulfilled its purpose.

Carl Branden
John Tooze

Acknowledgments

This book was conceived and the writing begun while one of us (C. Branden) was on sabbatical leave at the European Molecular Biology Laboratory in Heidelberg, and we wish first to thank Lennart Philipson, the Director General of EMBL, for making that sabbatical leave possible. We have benefitted from the advice of many colleagues, and we are particularly grateful to the following for their comments and corrections on individual chapters of the book—David Davies, NIH; Steve Fuller, EMBL; Stephen Harrison, Harvard; Carl-Henrik Heldin, Uppsala; Michael James, Edmonton; Alwyn Jones, Uppsala; Per Kraulis, Uppsala; Arthur Lesk, EMBL; Jane Richardson, Duke; and Simon Phillips, Leeds.

We also wish to thank the following, who sent us figures from, and pre-prints of, their work—Tom Alber, Utah; Hans Eklund, Uppsala; Richard Feldman, NIH; Richard Hendersson, Cambridge; John Kendrew, Cambridge; Kaspar Kirschner, Basel; Ylva Lindqvist, Uppsala; Brian Matthews, Eugene; Art Olsen, Scripps; Michael Rossmann, Purdue; Tom Steitz, Yale; Dmitri Tsernoglou, EMBL; Don Wiley, Harvard; and Kurt Wüthrich, Zürich.

We are deeply indebted to Keith Roberts for his hospitality and for working with Nigel Orme to convert simple drawings and sketches into the magnificent figures that illustrate the text. We thank Miranda Robertson for reading the entire manuscript and weeding out our infelicities. Finally, we are grateful to Gavin Borden, our publisher, for his faith in this project.

Contents

Basic Structural Principles

Part 1

The Building Blocks

Recombinant DNA techniques have provided tools for the rapid determination of DNA sequences and, by inference, the amino acid sequences of proteins from structural genes. The number of such sequences is now increasing almost exponentially, but by themselves these sequences tell little more about the biology of the system than a New York City telephone directory tells about the function and marvels of that city.

The proteins we observe in nature have evolved, through selection pressure, to perform specific functions. The functional properties of proteins depend upon their three-dimensional structures. The three-dimensional structure arises because particular sequences of amino acids in polypeptide chains fold to generate, from linear chains, compact domains with specific three-dimensional structures (Figure 1.1). The folded domains either can serve as modules for building up large assemblies such as virus particles or muscle fibers or can provide specific catalytic or binding sites as found in enzymes or proteins that carry oxygen or that regulate the function of DNA.

To understand the biological function of proteins we would therefore like to be able to deduce or predict the three-dimensional structure from the amino acid sequence. This we cannot do. In spite of considerable efforts over the last 25 years, this folding problem is still unsolved and remains one of the most basic intellectual challenges in molecular biology.

The fundamental reason why the folding problem remains unsolved lies in the fact that there are 20 different amino acids and therefore a vast number of ways in which similar structural domains can be generated in proteins by

Figure 1.1 The amino acid sequence of a protein's polypeptide chain is called its **primary** structure. Different regions of the sequence form local regular **secondary** structure, such as alpha (α) helices or beta (β) strands. The **tertiary** structure is formed by packing such structural elements into one or several compact globular units called **domains**. The final protein may contain several polypeptide chains arranged in a **quaternary** structure. By formation of such tertiary and quaternary structure amino acids far apart in the sequence are brought close together in three dimensions to form a functional region, an **active site**.

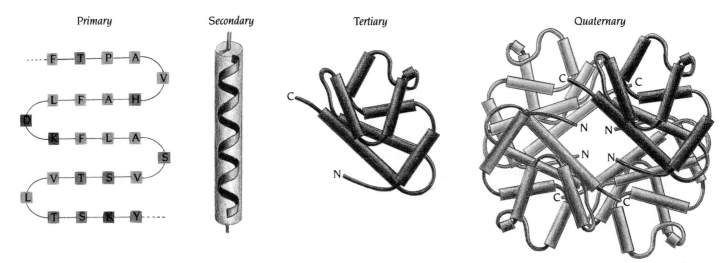

| Primary | Secondary | Tertiary | Quaternary |

different amino acid sequences. By contrast, the structure of DNA, made up of only four different nucleotide building blocks that occur in two pairs, is relatively simple, regular, and predictable.

Since the three-dimensional structures of individual proteins cannot be predicted, they have instead to be determined experimentally by x-ray crystallography or NMR techniques. Over the past 30 years the structures of around 500 proteins have been solved by x-ray methods. This has generated a body of information from which a set of basic principles of protein structure has emerged. These principles make it easier for us to understand how protein structure is generated, to identify common structural themes, to relate structure to function, and to see fundamental relationships between different proteins. The science of protein structure is at the stage of taxonomy where we can begin to discern patterns and motifs amongst the relatively small number of proteins whose three-dimensional structure is known.

The first five chapters of this book deal with the basic principles of protein structure as far as we understand them today, with examples given of the different major classes of protein structures. Chapter 6 contains a brief discussion on DNA structures with emphasis on recognition by proteins of specific nucleotide sequences. The remaining chapters illustrate how during evolution different structural solutions have been selected to fulfill particular functions.

Proteins are polypeptide chains

All of the 20 amino acids have in common a central carbon atom (C_α) to which are attached a hydrogen atom, an amino group (NH_2), and a carboxy group (COOH) (Figure 1.2a). What distinguishes one amino acid from another is the side chain attached to the C_α through its fourth valency. There are 20 different side chains specified by the genetic code; others occur, in rare cases, as the products of enzymatic modifications after translation.

Amino acids are joined end to end during protein synthesis by the formation of peptide bonds. The carboxy group of the first amino acid condenses with the amino group of the next to eliminate water and yield a peptide bond (see Figure 1.2b). This process is repeated as the chain elongates. One consequence is that the amino group of the first amino acid of a polypeptide chain and the carboxy group of the last amino acid remain intact, and the chain is said to run from its amino terminus to its carboxy terminus. The formation of a succession of peptide bonds generates a "main chain" or "backbone" from which project the various side chains.

The main-chain atoms are a carbon atom C_α to which the side chain is attached, an NH group bound to C_α, and a carbonyl group C´=O, where the carbon atom C´ is attached to C_α. These units, residues, are linked into a polypeptide by a peptide bond between the C´ atom of one residue and the nitrogen atom of the next (see Figure 1.2b). The basic repeating unit along the main chain from a biochemical or genetic viewpoint is thus (NH– C_αH– C´O), which is the residue of the common parts of amino acids after peptide bonds have been formed (Figure 1.2).

The genetic code specifies 20 different amino acid side chains

The 20 different side chains that occur in proteins are shown in Panel 1.1. Their names are abbreviated with both a three-letter and a one-letter code, which are also given in the figure. It is worthwhile trying to memorize the one-letter code, which is widely used in the literature. A mnemonic device to remember this one-letter code is given on page 7.

Depending on the chemical nature of the side chain, the amino acids are usually divided into three different classes. The first class comprises those with strictly hydrophobic side chains Ala (A), Val (V), Leu (L), Ile (I), Phe (F), Pro (P), and Met (M). The four charged residues Asp (D), Glu (E), Lys (K), and Arg (R) form the second class. The third class comprises those with polar side chains Ser (S), Thr (T), Cys (C), Asn (N), Gln (Q), His (H), Tyr (Y), and Trp (W). The amino acid

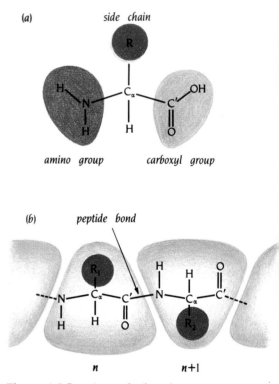

Figure 1.2 Proteins are built up by amino acids that are linked by peptide bonds into a polypeptide chain. (a) Schematic diagram of an amino acid, illustrating the nomenclature used in this book. A central carbon atom C_α is attached to an amino group, NH_2, a carboxy group C´OOH, a hydrogen atom, H, and a side chain, R. (b) In a polypeptide chain the carboxy group of amino acid n has formed a peptide bond, C–N, to the amino group of amino acid n+1. One water molecule is eliminated in this process. The repeating units, which are called residues, are divided into main-chain atoms and side chains. The main-chain part, which is identical in all residues, contains a central C_α atom attached to an NH group, a C´=O group, and an H atom. The side-chain R, which is different for different residues, is bound to the C_α atom.

glycine (G), which only has a hydrogen atom as a side chain and so is the simplest of the 20 acids, has special properties and is usually considered either to form a fourth class or to belong to the first class.

The four groups attached to the C_α atom are chemically different for all the amino acids except glycine, where two H atoms bind to C_α. All amino acids except glycine are thus chiral molecules which can exist in two different forms with different hands, L- or D-form (see Figure 1.3).

Biological systems depend on specific detailed recognition of molecules involving differentiation between chiral forms. The translation machinery for protein synthesis has evolved to utilize only one of the chiral forms of amino acids, the L-form. All amino acids that occur in proteins therefore have the L-form. There is, however, no obvious reason why the L-form was chosen during evolution and not the D-form.

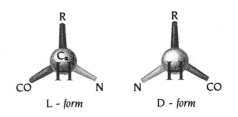

Figure 1.3 The handedness of amino acids. Looking down the H–C_α bond from the hydrogen atom, the L-form has CO, R, and N substituents from C_α going in a clockwise direction. There is a mnemonic to remember this; for the L-form the groups read CORN in clockwise direction.

Cysteines can form disulfide bridges

Two cysteine residues in different parts of the polypeptide chain but adjacent in the three-dimensional structure of a protein can be oxidized to form a disulfide bridge (Figure 1.4). This reaction requires an oxidative environment, and such disulfide bridges are usually not found in intracellular proteins, which spend their lifetime in an essentially reductive environment. Disulfide bridges do, however, occur quite frequently among extracellular proteins that are secreted from cells, and in eucaryotes formation of these bridges occurs within the lumen of the endoplasmic reticulum, the first compartment of the secretory pathway.

Disulfide bridges stabilize three-dimentional structure. In some proteins these bridges hold together different polypeptide chains; for example, the A and B chains of insulin are linked by two disulfide bridges between the chains. More frequently intramolecular disulfide bridges stabilize the folding of a single polypeptide chain, making the protein less susceptible to degradation. There are many examples of this, including snake venom toxins and protease inhibitors. Much effort is currently spent on introducing extra intramolecular disulfide bridges into enzymes by site-directed mutagenesis in order to make them more thermostable and hence more useful for industrial applications as catalysts as described in Chapter 16.

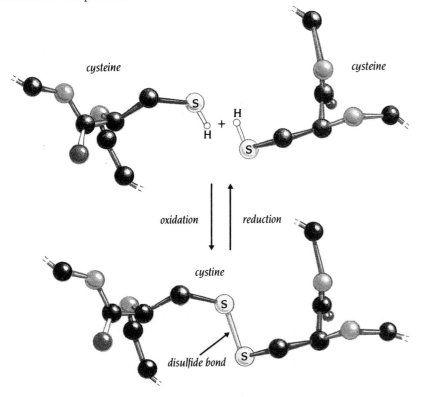

Figure 1.4 Disulfide bonds form between the side chains of two cysteine residues. Two SH groups from cysteine residues, which may be in different parts of the amino acid sequence but adjacent in the three-dimensional structure, are oxidized to form one S–S (disulfide) group.

5

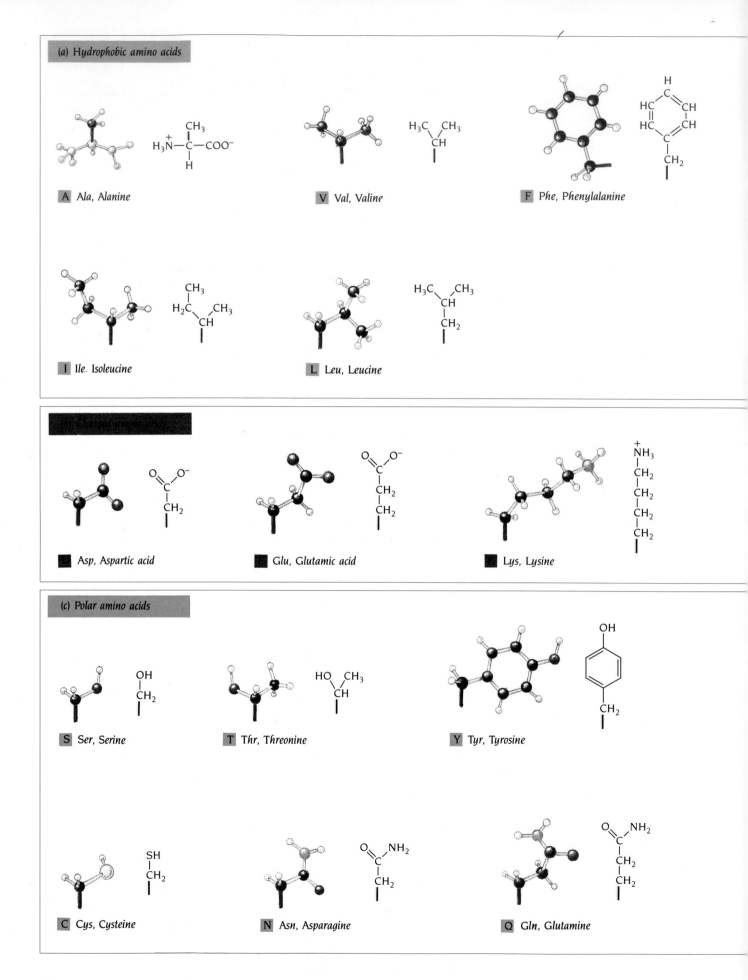

(a) Hydrophobic amino acids

A Ala, Alanine

$H_3\overset{+}{N}-\overset{\overset{\displaystyle CH_3}{|}}{\underset{\overset{\displaystyle |}{H}}{C}}-COO^-$

V Val, Valine

$\overset{\displaystyle H_3C}{}\overset{\displaystyle CH_3}{}$ CH

F Phe, Phenylalanine

I Ile. Isoleucine

$\overset{\displaystyle CH_3}{}$ $H_2C \quad CH_3$ CH

L Leu, Leucine

$H_3C \quad CH_3$ CH CH$_2$

(b) Charged amino acids

Asp, Aspartic acid

$\overset{O}{\parallel}\overset{}{C}\overset{O^-}{}$ CH$_2$

Glu, Glutamic acid

$\overset{O}{\parallel}\overset{}{C}\overset{O^-}{}$ CH$_2$ CH$_2$

K Lys, Lysine

$\overset{+}{N}H_3$ CH$_2$ CH$_2$ CH$_2$ CH$_2$

(c) Polar amino acids

S Ser, Serine

OH CH$_2$

T Thr, Threonine

HO CH$_3$ CH

Y Tyr, Tyrosine

OH CH$_2$

C Cys, Cysteine

SH CH$_2$

N Asn, Asparagine

$\overset{O}{\parallel}\overset{}{C}\overset{NH_2}{}$ CH$_2$

Q Gln, Glutamine

$\overset{O}{\parallel}\overset{}{C}\overset{NH_2}{}$ CH$_2$ CH$_2$

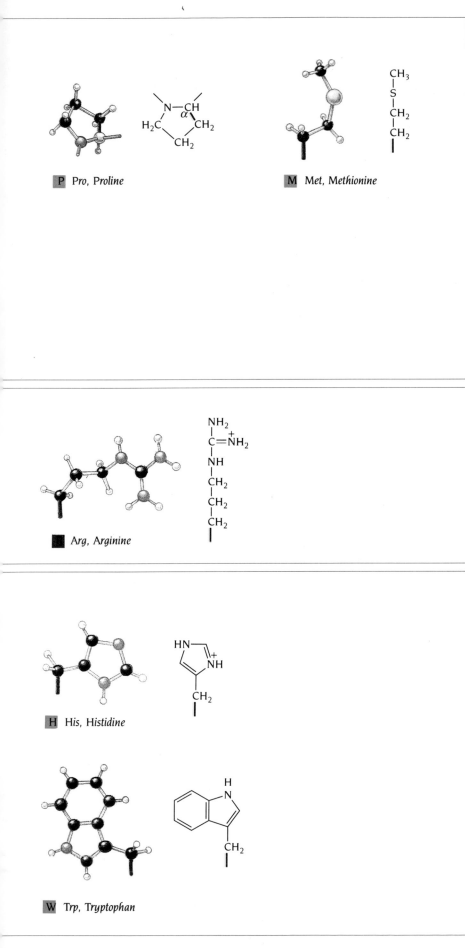

P Pro, Proline

M Met, Methionine

Arg, Arginine

H His, Histidine

W Trp, Tryptophan

Panel 1.1 The 20 different amino acids that occur in proteins. Only side chains are shown except for the first amino acid, alanine, where all atoms are shown. The bond from the side chain to C$_\alpha$ is red. A ball-and-stick model, the chemical formula, the full name as well as the three-letter and one-letter codes are given for each amino acid.

There are some easy ways of remembering the one-letter code for amino acids. If only one amino acid begins with a certain letter, that letter is used:

C = Cys = Cysteine
H = His = Histidine
I = Ile = Isoleucine
M = Met = Methionine
S = Ser = Serine
V = Val = Valine

If more than one amino acid begins with a certain letter, that letter is assigned to the most commonly occurring amino acid:

A = Ala = Alanine
G = Gly = Glycine
L = Leu = Leucine
P = Pro = Proline
T = Thr = Threonine

Some of the others are phonetically suggestive:

F = Phe = Phenylalanine ("Fenylalanine")
R = Arg = Arginine ("aRginine")
Y = Tyr = Tyrosine ("tYrosine")
W = Trp = Tryptophan (double ring in the molecule)

In other cases a letter close to the initial is used. Amides have letters from the middle of the alphabet. The smaller molecules (D, N, B) are earlier in the alphabet than the larger (E, Q, Z).

D = Asp = Aspartic acid (near A)
N = Asn = Asparagine (contains N)
B = Asx = Either of D or N
E = Glu = Glutamic acid (near G)
Q = Gln = Glutamine ("Q-tamine")
Z = Glx = Either of E or Q
K = Lys = Lysine (near L)
X = X = Undetermined amino acid

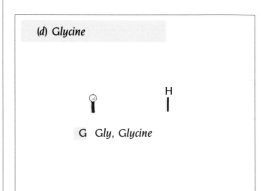

(d) Glycine

G Gly, Glycine

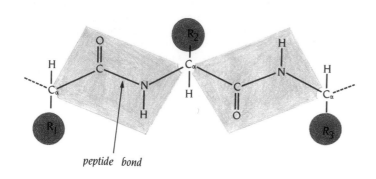

peptide bond

Figure 1.5 Part of a polypeptide chain that is divided into peptide units, represented as blocks in the diagram. Each peptide unit contains the C_α atom and the $C'=O$ group of residue n as well as the NH group and C_α atom of residue n+1. Each such unit is a planar rigid group with known bond distances and bond angles. R1, R2, and R3 are the side chains attached to the C_α atoms that link the peptide units in the polypeptide chain.

Peptide units are building blocks of protein structures

Figure 1.2 shows one way of dividing up a polypeptide chain, the biochemist's way. There is, however, a different way to divide the main chain into repeating units that is better to use when we want to describe the structural properties of proteins. We divide the polypeptide chain into peptide units that go from one C_α atom to the next C_α (see Figure 1.5). Each C_α atom except the first and the last thus belongs to two such units. The reason for dividing the chain in this way is that all the atoms in such a unit are fixed in a plane with the bond lengths and bond angles very nearly the same in all units in all proteins. Note that the peptide units of the main chain do not involve the different side chains (Figure 1.5). We will use both of these alternative descriptions of polypeptide chains— the biochemical and the structural—and discuss proteins in terms of the sequence of different amino acids and the sequence of planar peptide units.

Since the peptide units are effectively rigid groups that are linked into a chain by covalent bonds at the C_α atoms, the only degrees of freedom they have are rotations around these bonds. Each unit can rotate around two such bonds: the C_α–C' and the N–C_α bonds (see Figure 1.6). A convention has been adopted to call the angle of rotation around the N–C_α-bond **phi** (ϕ) and the angle around the C_α–C' bond from the same C_α-atom **psi** (ψ).

In this way each amino acid residue is associated with two conformational angles ϕ and ψ. Since these are the only degrees of freedom, the conformation of the whole main chain of the polypeptide is completely determined when the ϕ and ψ angles for each amino acid are defined.

Glycine residues can adopt many different conformations

Most combinations of ϕ and ψ angles for an amino acid are not allowed because of steric collisions between the side chains and main chain. It is reasonably straightforward to calculate those combinations that are allowed. Since the D- and L-forms of the amino acids have their side chain oriented differently with respect to the CO group, they have different allowed ϕ, ψ angles. If they existed, proteins built from D amino acids would thus have different conformations than those found in nature that are exclusively made of L amino acids.

The angle pairs ϕ and ψ are usually plotted against each other in a diagram called a **Ramachandran plot** after the Indian biophysicist G. N. Ramachandran who first made calculations of sterically allowed regions. Figure 1.7 shows the results of such calculations and also a plot for all amino acids except glycine from a number of accurately determined protein structures. It is apparent that the observed values are clustered in the sterically allowed regions. There is one important exception. Glycine with only a hydrogen atom as a side chain can adopt a much wider range of conformations than the other residues as seen in Figure 1.7c. Glycine thus plays a structurally very important role; it allows unusual main chain conformations in proteins. This is one of the main reasons why a high proportion of glycine residues are conserved among homologous protein sequences.

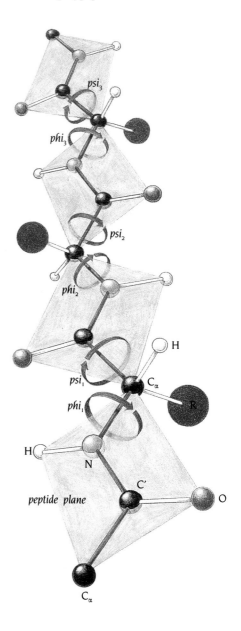

Figure 1.6 Diagram showing a polypeptide chain where the main chain atoms are represented as rigid peptide units, linked through the C_α atoms. Each unit has two degrees of freedom; it can rotate around two bonds, its C_α–C' bond and its N–C_α bond. The angle of rotation around the N–C_α bond is called phi (ϕ) and that around the C_α–C' bond psi (ψ). The conformation of the main chain atoms is therefore determined by the values of these two angles for each amino acid.

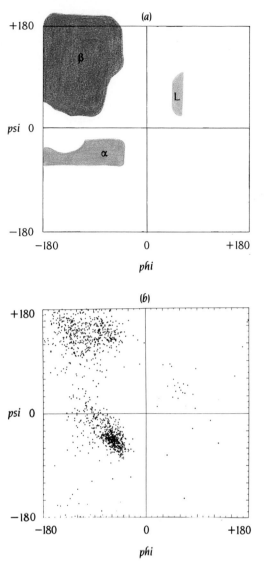

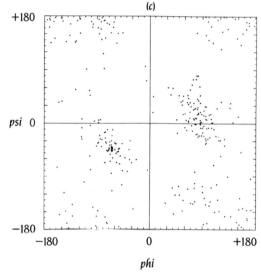

Figure 1.7 Ramachandran plots showing allowed combinations of the conformational angles phi and psi defined in Figure 1.6. Since phi (ϕ) and psi (ψ) refer to rotations of two rigid peptide units around the same C_α atom, most combinations produce steric collisions either between atoms in different peptide groups or between a peptide unit and the side chain attached to C_α. These combinations are therefore not allowed. (a) Colored areas show sterically allowed regions. The areas labeled α, β, and L correspond approximately to conformational angles found for the usual right-handed α helices, β strands, and left-handed α helices, respectively. (b) Observed values for all residue types except glycine. Each point represents ϕ and ψ values for an amino acid residue in a well-refined x-ray structure to high resolution. (c) Observed values for glycine. Notice that the values include combinations of ϕ and ψ that are not allowed for other amino acids. (From J. Richardson, *Adv. Prot. Chem.* 34: 174–175, 1981.)

Conclusion

All protein molecules are polymers built up from 20 different amino acids linked end to end by peptide bonds. The function of every protein molecule depends on its three-dimensional structure which in turn is determined by its amino acid sequence which in turn is determined by the nucleotide sequence of the structural gene.

Each amino acid has some atoms in common that form the main chain. The remaining atoms form side chains that are either hydrophobic, polar, or charged.

The conformation of the whole main chain of a protein is determined by two conformational angles, phi (ϕ) and psi (ψ), for each amino acid. Only certain combinations of these angles are allowed due to steric hindrance between main chain and side chain atoms, except for glycine.

Selected readings

General

Alberts, B., et al. Molecular Biology of the Cell, 2nd ed. New York: Garland, 1989.

Creighton, T.E. Proteins: Structures and Molecular Properties. New York: W.H. Freeman, 1983.

Fletterick, R.J.; Schroer, T.; Matela, R.J. Molecular Structure: Macromolecules in Three Dimensions. Oxford, UK: Blackwell Scientific, 1985.

Judson, H.F. The Eighth Day of Creation: Makers of the Revolution in Biology. New York: Simon & Schuster, 1979.

Mathews, C.K.; van Holde, K.E. Biochemistry. Menlo Park, CA: Benjamin/Cummings, 1990.

Ramachandran, G.N.; Sassiekharan, V. Conformation of polypeptides and proteins. *Adv. Prot. Chem.* 28: 283–437, 1968.

Schulz, G.E.; Schirmer, R.H. Principles of Protein Structure. New York: Springer, 1979.

Stryer, L. Biochemistry, 3rd ed. New York: W.H. Freeman, 1988.

Zubay, G. Biochemistry, 2nd ed. New York: Macmillan, 1988.

Motifs of Protein Structure

2

Few general principles emerged from the first protein structure

X-ray structural studies have played a major role in transforming inorganic and organic chemistry from a descriptive science at the beginning of this century to the present state where theoretical concepts are used to predict properties of novel compounds. When W. L. Bragg solved the very first crystal structure, that of rock salt, NaCl, the results completely changed prevalent concepts of bonding forces in ionic compounds.

The first x-ray crystallographic structural results on a globular protein molecule, myoglobin (see Figure 2.1), reported in 1958, came as a shock to those who had believed that they would reveal general simple principles of how proteins are folded and function, analogous to the simple and beautiful double-stranded DNA structure that had been determined five years before by James Watson and Francis Crick. John Kendrew at the Medical Research Council Laboratory of Molecular Biology, Cambridge, who determined the myoglobin structure to low resolution in 1958, expressed this disappointment in the following words: "Perhaps the most remarkable features of the molecule are its complexity and its lack of symmetry. The arrangement seems to be almost totally lacking in the kind of regularities which one instinctively anticipates, and it is more complicated than has been predicted by any theory of protein structure."

In retrospect it is easy to realize that such irregularity in structure is actually required for proteins to fulfill their diverse functions. Information storage and transfer from DNA is essentially linear and therefore the structures of DNA molecules can be similar to each other independent of their information content. In contrast, proteins must recognize many thousands of different molecules in the cell by detailed three-dimensional interactions, which require

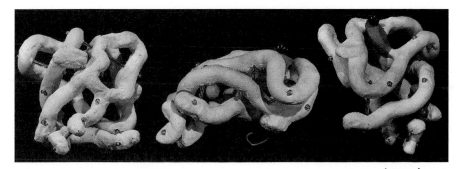

Figure 2.1 Kendrew's model of the low-resolution structure of myoglobin shown in three different views. The sausage-shaped regions represent α helices, which are arranged in a seemingly irregular manner to form a compact globular molecule.
(Courtesy of J. C. Kendrew.)

10Å

11

diverse and irregular structures of the protein molecules. In spite of these requirements, there are regular features present in protein structures, the most important of which is their **secondary structure**.

The interior of proteins is hydrophobic

Kendrew noticed from the subsequent high-resolution studies of myoglobin that the amino acids in the interior of the protein had almost exclusively hydrophobic side chains. This was one of the first important general principles that emerged from studies of protein structures. The main driving force for folding water-soluble globular protein molecules is to pack hydrophobic side chains into the interior of the molecule, thus creating a **hydrophobic core** and a hydrophilic surface.

There is a major problem, however, to create such a hydrophobic core from a protein chain. To bring the side chains into this core, the main chain must also fold into the interior. The main chain is highly polar and therefore hydrophilic, with one hydrogen bond donor, NH, and one hydrogen bond acceptor, C'=O, for each peptide unit. These main chain polar groups must be neutralized by hydrogen bond formation in a hydrophobic environment. This problem is solved in a very elegant way by the formation of regular secondary structure within the interior of the protein molecule. Such secondary structure is usually one of two types: **alpha helices** or **beta sheets**. Both types are characterized by having the main chain NH and C'O groups participating in hydrogen bonds to each other. They are formed when a number of consecutive residues have the same phi (ϕ), psi (ψ) angles. We will now have a closer look at these important structural elements.

The alpha (α) helix is an important element of secondary structure

The α helix is the classic element of protein structure. It was first described by Linus Pauling working at the California Institute of Technology in 1951. He predicted it as a structure that would be stable and favorable in proteins. He made this remarkable prediction on the basis of accurate geometrical parameters that he had derived for the peptide unit from the results of crystallographic analyses of the structures of a range of small molecules. This prediction almost immediately received strong experimental support from diffraction patterns obtained by Max Perutz in Cambridge from hemoglobin crystals and fibers of keratin. It was completely verified from John Kendrew's high-resolution structure of myoglobin, where all secondary structure was helical.

α helices in proteins are found when a stretch of consecutive residues all have the phi, psi angle pair approximately –60° and –50°, corresponding to the allowed region in the bottom left quadrant of the Ramachandran plot (see Figure 1.7a). The α helix has 3.6 residues per turn with hydrogen bonds between C'=O of residue n and NH of residue n+4 (see Figure 2.2). Thus all NH and C'O groups are joined with hydrogen bonds except the first NH groups and the last C'O groups at the ends of the α helix. As a consequence, the ends of α helices are polar and are almost always at the surface of protein molecules.

α helices vary considerably in length in globular proteins ranging from four or five amino acids to over 40 residues. The average length is around 10 residues, corresponding to three turns. Since the rise per residue of an α helix is 1.5 Å along the helical axis, this corresponds to about 15 Å from one end to the other of an average α helix.

An α helix can in theory be either right-handed or left-handed depending on the screw direction of the chain. A left-handed α helix is not, however, allowed for L amino acids due to the close approach of the side chains and the C'O group. Thus the α helix that is observed in proteins is almost always right-handed. Short regions (3–5 residues) of left-handed α helices occur occasionally.

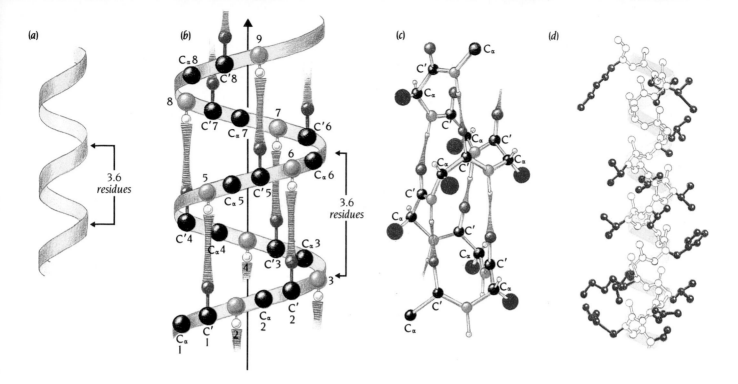

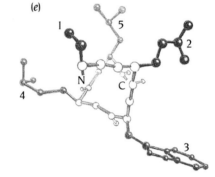

Figure 2.2 The α helix is one of the major elements of secondary structure in proteins. Main chain N and O atoms are hydrogen bonded to each other within α helices. (a) Idealized diagram of the path of the main chain in an α helix. α helices are frequently illustrated in this way. There are 3.6 residues per turn in an α helix, which corresponds to 5.4 Å (1.5 Å per residue). (b) The same as (a) but with approximate positions for main chain atoms and hydrogen bonds included. The arrow denotes the direction from N terminal to C terminal. (c) Schematic diagram of an α helix. Oxygen atoms are red, and N atoms are blue. Hydrogen bonds between O and N are red and striated. The side chains are represented as purple circles. (d) A stick-and-ball model of one α helix in myoglobin. The path of the main chain is outlined in yellow; side chains are purple. Main chain atoms are not colored. (e) One turn of an α helix viewed down the helical axis. The purple side chains project out from the α helix.

The α helix has a dipole moment

All the hydrogen bonds in an α helix point in the same direction so the peptide units are aligned in the same orientation along the helical axis. Since a peptide unit has a dipole moment arising from the different polarity of NH and C'O groups, these dipole moments are also aligned along the helical axis (Figure 2.3). The overall effect is a significant net dipole for the α helix that gives a partial positive charge at the amino end and a partial negative charge at the carboxy end of the α helix. These charges would be expected to attract ligands of opposite charge. Negatively charged ligands, especially when they contain phosphate groups, frequently bind at the N terminii of α helices, but, in contrast, positively charged ligands rarely bind at the C terminus. This may be due to the fact that in addition to the dipole effect the N terminus of an α helix has free NH groups with favorable geometry to position phosphate groups by specific hydrogen bonds (Figure 2.3). Such ligand binding occurs frequently in proteins; it provides examples of specific binding through main chain conformation in which side chains are not involved.

Some amino acids are preferred in α helices

The side chains project out from the α helix (Figure 2.2e) and do not interfere with it except for proline where the last atom of the side chain is bonded to the main chain N atom forming a ring structure, C_α–CH2–CH2–CH2–N (see

13

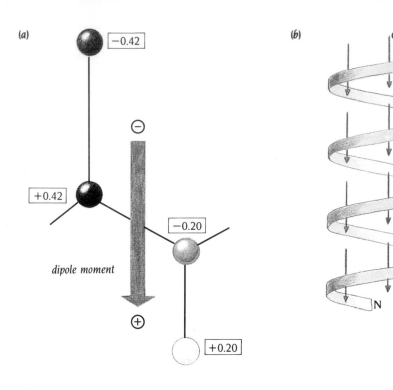

(a)

−0.42

⊖

+0.42

−0.20

dipole moment

⊕

+0.20

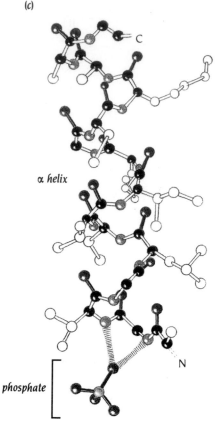

(b)

C

N

(c)

α *helix*

phosphate

C

N

Panel 1.1 on pages 6–7). This prevents the N atom from participating in hydrogen bonding and also provides some steric hindrance to the α-helical conformation. Proline fits very well in the first turn of an α helix but anywhere further on it usually produces a significant bend in the helix. Such bends occur in many α helices, not just in those few that contain a proline in the middle. So although we can predict that a proline residue may cause a bend in an α helix, it does not follow that all bends are the result of the presence of proline.

Different side chains have been found to have weak but definite preferences either for or against being in α helices. Thus Ala (A), Glu (E), Leu (L), and Met (M) are good α helix formers, while Pro (P), Gly (G), Tyr (Y), and Ser (S) are very poor. Such preferences were central to all early attempts to predict secondary structure from amino acid sequence, but they are not strong enough to give accurate predictions.

The most common location for an α helix in a protein structure is along the outside of the protein with one side of the helix facing the solution and the other side toward the hydrophobic interior of the protein. With 3.6 residues per turn there is therefore a tendency for side chains to change from hydrophobic to hydrophilic with a periodicity of three to four residues. This trend can sometimes be seen in the amino acid sequence but it is not strong enough for reliable structural prediction by itself, because residues that face the solution can be hydrophobic and furthermore α helices can be either completely buried or completely exposed within the protein. Table 2.1 shows examples of the amino acid sequences of a totally buried, a partially buried, and a completely exposed α helix.

Figure 2.3 Negatively charged groups such as phosphate ions frequently bind to the amino ends of α helices. The dipole moment of an α helix as well as the possibility to form hydrogen bonds to free NH groups at the end of the helix favors such binding. (a) The dipole of a peptide unit. Numbers in boxes give the approximate fractional charges of the atoms of the peptide unit. (b) The dipoles of peptide units are aligned along the α helical axis, which creates an overall dipole moment of the α helix, positive at the amino end and negative at the carboxy end. (c) A phosphate group hydrogen bonded to the NH end of an α helix. Nitrogen atoms are blue; oxygen atoms are red; main-chain carbon atoms are black; and phosphorous is green.

Table 2.1 Amino acid sequences of three α helices

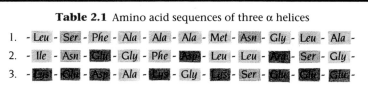

1. - Leu - Ser - Phe - Ala - Ala - Ala - Met - Asn - Gly - Leu - Ala -
2. - Ile - Asn - Glu - Gly - Phe - Asp - Leu - Leu - Arg - Ser - Gly -
3. - Lys - Glu - Asp - Ala - Lys - Gly - Lys - Ser - Glu - Glu - Glu -

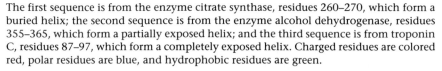

The first sequence is from the enzyme citrate synthase, residues 260–270, which form a buried helix; the second sequence is from the enzyme alcohol dehydrogenase, residues 355–365, which form a partially exposed helix; and the third sequence is from troponin C, residues 87–97, which form a completely exposed helix. Charged residues are colored red, polar residues are blue, and hydrophobic residues are green.

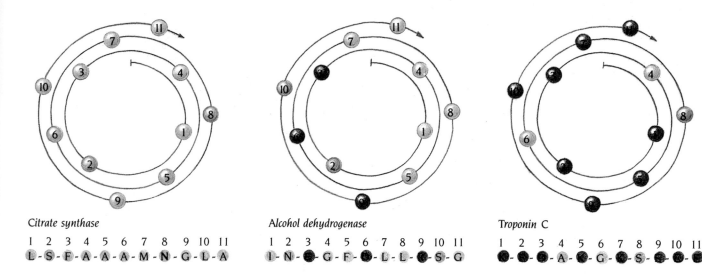

Citrate synthase

1 2 3 4 5 6 7 8 9 10 11
L - S - F - A - A - A - M - N - G - L - A

Alcohol dehydrogenase

1 2 3 4 5 6 7 8 9 10 11
I - N - G - G - F - G - L - L - G - S - G

Troponin C

1 2 3 4 5 6 7 8 9 10 11
G - G - G - A - G - G - G - S - G - G - G

A convenient way to illustrate the amino acid sequences in helices is the **helical wheel** or spiral. Since one turn in an α helix is 3.6 residues long, each residue can be plotted every 360/3.6 = 100° around a circle or a spiral as shown in Figure 2.4. Such a plot shows the projection of the position of the residues on to a plane perpendicular to the helical axis. Residues on one side of the helix are plotted on one side of the spiral. The three helices whose sequences are given in Table 2.1 are plotted in this way in Figure 2.4, using a color code for hydrophobic, polar, and charged residues. It is immediately obvious that one side of the helix from alcohol dehydrogenase is hydrophilic and the other side hydrophobic.

α helices that cross membranes are in a hydrophobic environment. Therefore, most of their side chains are hydrophobic. Long regions of hydrophobic residues in the amino acid sequence of a protein that is membrane bound can therefore be predicted to be a transmembrane helix with a high degree of confidence, as will be discussed in Chapter 13.

Figure 2.4 The helical wheel or spiral. Amino acid residues are plotted every 100° consecutive around the spiral, following the sequences given in Table 2.1. The following color code is used: green is an amino acid with a hydrophobic side chain; blue is a polar side chain; and red is a charged side chain. The first helix is all hydrophobic; the second is polar on one side and hydrophobic on the other side; and the third helix is all polar.

Beta (β) sheets usually have their beta (β) strands either parallel or antiparallel

The second major structural element found in globular proteins is the β sheet. This structure is built up from a combination of several regions of the polypeptide chain, in contrast to the α helix, which is built up from one continous region. These regions, β strands, are usually from 5 to 10 residues long and are in an almost fully extended conformation with ϕ, ψ angles within the broad structurally allowed region in the upper left quadrant of the Ramachandran plot (see Figure 1.7). These β strands are aligned adjacent to each other (see Figures 2.5 and 2.6) such that hydrogen bonds can form between C'O groups of one β strand and NH groups on an adjacent β strand and vice versa. The β sheets that are formed from several such β strands are **"pleated"** with C_α atoms successively a little above and below the plane of the β sheet. The side chains follow this pattern such that within a β strand they also point alternatively above and below the β sheet.

β strands can interact in two ways to form a pleated sheet. Either the amino acids in the aligned β strands can all run in the same biochemical direction, amino terminal to carboxy terminal, in which case the sheet is described as **parallel**, or the amino acids in successive strands can have alternating directions, amino terminal to carboxy terminal followed by carboxy terminal to amino terminal, followed by amino terminal to carboxy terminal, and so on, in which case the sheet is called **antiparallel**. Each of the two forms has a distinctive pattern of hydrogen bonding. The antiparallel β sheet (Figure 2.5) has narrowly spaced hydrogen bond pairs that alternate with widely spaced pairs. Parallel β sheets (Figure 2.6) have evenly spaced hydrogen bonds that angle

15

(a)

(b)

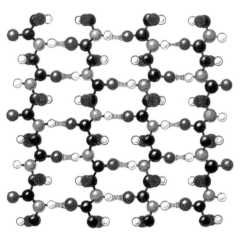

(c)

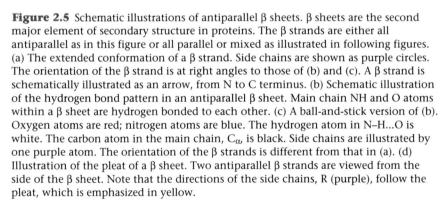

(d)

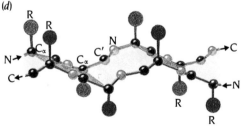

Figure 2.5 Schematic illustrations of antiparallel β sheets. β sheets are the second major element of secondary structure in proteins. The β strands are either all antiparallel as in this figure or all parallel or mixed as illustrated in following figures. (a) The extended conformation of a β strand. Side chains are shown as purple circles. The orientation of the β strand is at right angles to those of (b) and (c). A β strand is schematically illustrated as an arrow, from N to C terminus. (b) Schematic illustration of the hydrogen bond pattern in an antiparallel β sheet. Main chain NH and O atoms within a β sheet are hydrogen bonded to each other. (c) A ball-and-stick version of (b). Oxygen atoms are red; nitrogen atoms are blue. The hydrogen atom in N–H...O is white. The carbon atom in the main chain, C_α, is black. Side chains are illustrated by one purple atom. The orientation of the β strands is different from that in (a). (d) Illustration of the pleat of a β sheet. Two antiparallel β strands are viewed from the side of the β sheet. Note that the directions of the side chains, R (purple), follow the pleat, which is emphasized in yellow.

across between the β strands. Within both types of β sheets all possible main chain hydrogen bonds are formed, except for the two flanking strands of the β sheet that only have one neighboring β strand.

β strands can also combine into mixed β sheets with some β strand pairs parallel and some antiparallel. There is a strong bias against mixed β sheets; only about 20% of the strands inside the β sheets of known protein structures have parallel bonding on one side and antiparallel on the other. Figure 2.7 illustrates how the hydrogen bonds between the β strands are arranged in a mixed β sheet.

Almost all β sheets, parallel, antiparallel, and mixed, as they occur in known protein structures have their strands twisted. This twist always has the same handedness as shown in Figure 2.7, which is defined as a right-handed twist.

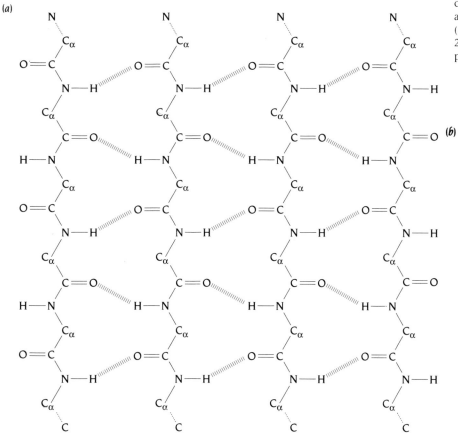

Figure 2.6 Parallel β sheet. (a) Schematic diagram showing the hydrogen bond pattern in a parallel β sheet. (b) Ball-and-stick version of (a). The same color scheme is used as in Figure 2.5c. (c) Schematic diagram illustrating the pleat of a parallel β sheet.

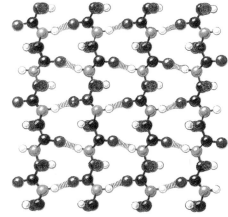

(b)

(c)

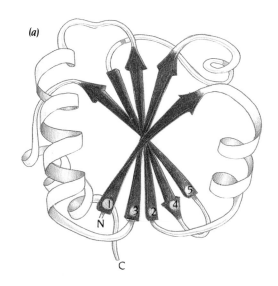

(a)

Figure 2.7 (a) Illustration of the twist of β sheets. β strands are drawn as arrows from the amino end to the carboxy end of the β strand in this schematic drawing of the protein thioredoxin from *E. coli*, the structure of which was determined in the laboratory of Carl Branden, Uppsala, Sweden, to 2.8 Å resolution. The mixed β sheet is viewed from one of its ends. (Adapted from B. Furugren.) (b) The hydrogen bonds between the β strands in the mixed β sheet of the same protein (page 18).

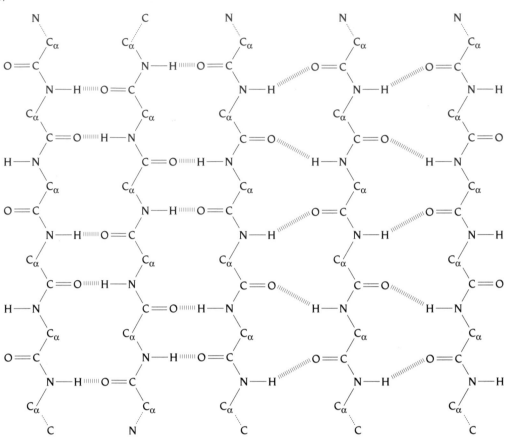

Loop regions are at the surface of protein molecules

Most protein structures are built up from combinations of secondary structure elements, α helices and β strands, which are connected by **loop regions** of various lengths and irregular shape. A combination of secondary structure elements forms the stable hydrophobic core of the molecule. The loop regions are at the surface of the molecule. The main chain C′O and NH groups of these loop regions, which in general do not form hydrogen bonds to each other, are exposed to the solvent and can form hydrogen bonds to water molecules.

Loop regions exposed to solvent are rich in charged and polar hydrophilic residues. This has been used in several prediction schemes, and it has proved possible to predict loop regions from an amino acid sequence with a higher degree of confidence than α helices or β strands, which is ironic since the loops have irregular structures.

When homologous amino acid sequences from different species are compared, it is found that insertions and deletions of a few residues occur almost exclusively in the loop regions. During evolution cores are much more stable than loops. Intron positions are also often found at sites in structural genes that correspond to loop regions in the protein structure. Since proteins that exhibit sequence homology in general have similar core structures, it is apparent that the specific arrangement of secondary structure elements in the core is rather insensitive to the lengths of the loop regions. In addition to their function as connecting units between secondary structure elements, loop regions frequently participate in forming binding sites and enzyme active sites. Thus antigen binding sites in antibodies are built up from six loop regions, which vary both in length and in amino acid sequence between different antibodies. Modeling

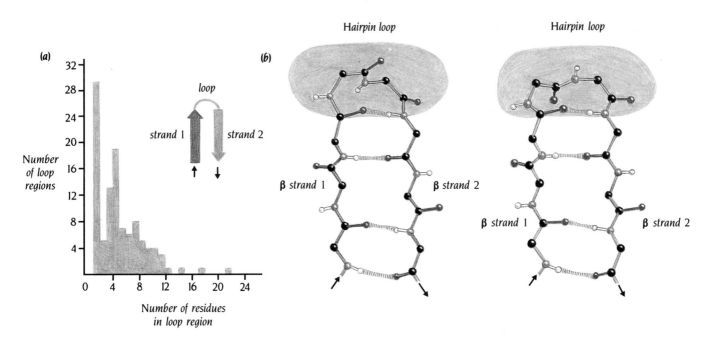

Hairpin loop

Hairpin loop

Number of loop regions

loop

strand 1   *strand 2*

β *strand 1*   β *strand 2*

β *strand 1*   β *strand 2*

Number of residues in loop region

antigen binding sites from a known antibody sequence is thus essentially a problem of modeling the three-dimensional structures of loop regions since the core structures of all antibodies are very similar. Such model building has been facilitated by the recent findings that loop regions have preferred structures. Surveys of known three-dimensional structures of loops have shown that they fall into a rather limited set of structures and are not a random collection of possible structures. Loop regions that connect two adjacent antiparallel β strands are called **hairpin** loops. Such loops have been particularly well analyzed (Figure 2.8) partly because they form the antigen binding sites of antibodies.

Figure 2.8 Adjacent antiparallel β strands are joined by hairpin loops. Such loops are frequently short and do not have regular secondary structure. Nevertheless, many loop regions in different proteins have similar structures. (a) Histogram showing the frequency of hairpin loops of different lengths in 62 different proteins. (b) The two most frequently occurring two-residue hairpin loops. Bonds within the hairpin loop are green. [(a) Adapted from B. L. Sibanda and J. M. Thornton, *Nature* 316: 170–174, 1985.]

Schematic pictures of proteins highlight secondary structure

Anyone who looks at a model of a protein structure where all atoms are displayed is confused by the sheer amount of information present (Figure 2.9a).

A two-dimensional picture of such a model is impossible to interpret. Even if the side chain atoms are stripped off, it is still difficult to extract meaningful information from a flat picture of such a model, mainly because it is difficult to see the relation between the secondary structural elements. Since these elements dominate the structure, the picture becomes more clear if they are simplified and highlighted in some way. This is usually done by representing the path of the polypeptide chain by three different symbols: cylinders for α helices, arrows for β strands, which give the direction of the strands from amino to carboxy end, and ribbons for the remaining parts. Such schematic diagrams give good and very useful overall views of protein structures but, of course, give no detailed information. Details are best studied on graphic display systems where one can manipulate the computer-generated model on the screen; in this way the power of the graphics device makes it possible for the viewer to study the much greater detail intelligibly.

Jane Richardson at Duke University has made a very popular collection of **schematic diagrams** of various protein molecules with an artistic touch that gives an aesthetic impression without losing too much accuracy. Arthur Lesk at the M.R.C. Laboratory of Molecular Biology in Cambridge, UK, and Karl Hardman at IBM, USA, pioneered the use of computer programs to generate schematic diagrams on a computer display from a list of atomic coordinates of the main chain atoms. Figures 2.9b–d and 2.10 show representative examples of both Lesk-type and Richardson-type diagrams.

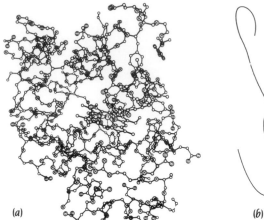

(a)

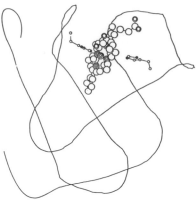

(b)

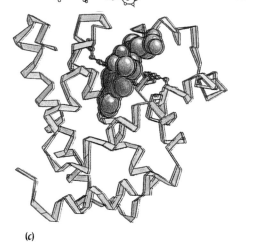

(c)

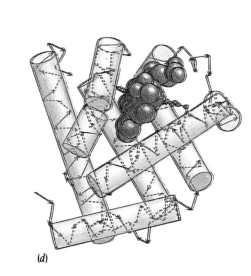

(d)

Figure 2.9 (a) The structure of myoglobin displaying all atoms as small circles connected by straight lines. Even though only side chains at the surface of the molecule are shown, the picture contains so many atoms that such a two-dimensional representation is very confusing and very little information can be gained from it. (The picture is one-half of a stereo diagram by H.C. Watson, Progress in Stereochemistry Vol. 4, 299–333, 1969.) (b–d) Computer-generated schematic diagrams at different degrees of simplification of the structure of myoglobin. (Courtesy of Arthur Lesk.)

Topology diagrams are useful for classification of protein structures

It is very convenient for some purposes to have an even more simplified schematic representation of the secondary structure elements, especially for β sheets. For example, when the structure of the enzyme alcohol dehydrogenase was solved in 1973, one important question we wanted to answer was whether or not the NAD-binding domain of this enzyme was similar to that of lactate dehydrogenase. This was by no means obvious from comparing a stereodiagram of one structure in one orientation with the electron density map of the second structure in a different orientation. However, simple **topology diagrams** of

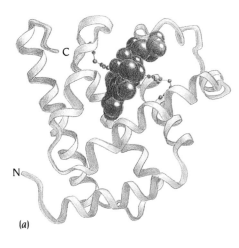

(a)

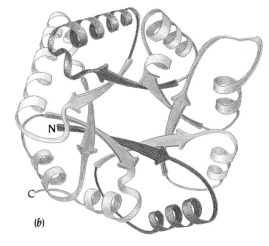

(b)

Figure 2.10 Examples of schematic diagrams of the type pioneered by Jane Richardson. Diagram (a) illustrates the structure of myoglobin in the same orientation as the computer-drawn diagrams of Figures 2.9 b–d. Diagram (b), which is adapted from J. Richardson, illustrates the structure of the enzyme triose phosphate isomerase, determined to 2.5 Å resolution in the laboratory of David Phillips, Oxford University.

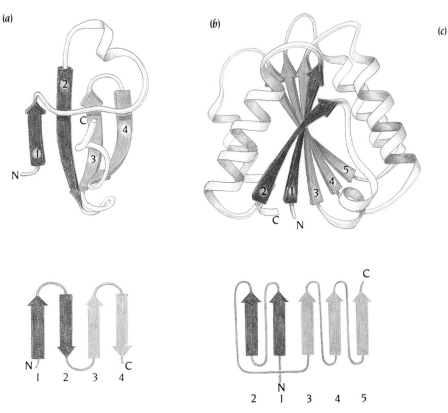

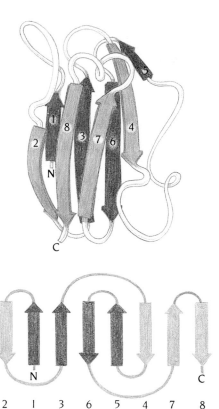

the two domains immediately revealed that they had the same fold, which is now called the nucleotide binding fold and which is discussed in Chapter 10. Topology diagrams were also used in the classic *Nature* paper published in 1976 by Michael Levitt and Cyrus Chothia of the M.R.C. Laboratory of Molecular Biology, Cambridge, which for the first time classified protein structures into four main classes.

The most characteristic features of a β sheet are the number of strands, their relative directions, parallel or antiparallel, and how the strands are connected along the polypeptide chain—in other words, the strand order. This information can be easily conveyed through simple diagrams of connected arrows like those in Figure 2.11, where such simple topology diagrams are compared to the more elaborate Richardson diagrams. The twist of the β sheet is not represented in these topology diagrams. They are, nevertheless, very useful when comparing β structures and will be used frequently in this book.

Secondary structure elements are connected into simple motifs

Simple combinations of a few secondary structure elements with a specific geometric arrangement have been found to occur frequently in protein structures. These units have been called either supersecondary structures or **motifs**. We will use the term motif throughout this book. Some of these motifs can be associated with a particular function such as DNA binding; others have no specific biological function alone but are part of larger structural and functional assemblies.

The simplest motif with a specific function consists of two α helices joined by a loop region. Two such motifs (see Figure 2.12), each with its own characteristic geometry and amino acid sequence requirements, have been observed as parts of many protein structures.

One of these motifs is specific for DNA binding and is described in detail in Chapter 7. The second motif is specific for calcium binding and is present in parvalbumin, calmodulin, troponin-C, and other proteins that bind calcium and thereby regulate cellular activities. This calcium-binding motif was first

Figure 2.11 β sheets are usually represented simply by arrows connected into topology diagrams that show both the direction of each β strand and the way the strands are connected to each other along the polypeptide chain. Such topology diagrams are here compared with more elaborate schematic diagrams for different types of β sheets. (a) Four strands. Antiparallel β sheet in one domain of the enzyme aspartate transcarbamylase. The structure of this enzyme has been determined to 2.8 Å resolution in the laboratory of William Lipscomb, Harvard University. (b) Five strands. Parallel β sheet in the redox protein flavodoxin, the structure of which has been determined to 1.8 Å resolution in the laboratory of Martha Ludwig, University of Michigan. (c) Eight strands. Antiparallel barrel in the electron carrier plastocyanin. This is a closed barrel where the sheet is folded such that β strands 2 and 8 are adjacent. The structure has been determined to 1.6 Å resolution in the laboratory of Hans Freeman in Sydney, Australia. (Adapted from J. Richardson.)

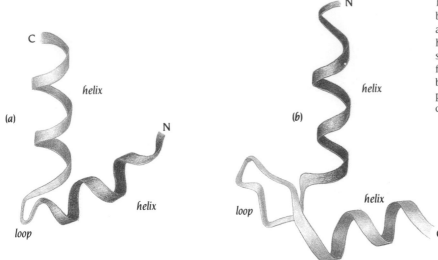

Figure 2.12 Two α helices that are connected by a short loop region in a specific geometric arrangement constitute a helix-loop-helix or helix-turn-helix motif. Two such motifs are shown: the DNA-binding motif (a), which is further discussed in Chapter 7, and the calcium-binding motif (b), which is present in many proteins whose function is regulated by calcium.

found in 1973 in the laboratory of Robert Kretsinger, University of Virginia, when he determined the structure of parvalbumin to 1.8 Å resolution.

Parvalbumin is a muscle protein with a single polypeptide chain of 109 amino acids. Its function is uncertain, but calcium binding to this protein probably plays a role in muscle relaxation. The **helix-loop-helix** motif appears three times in this structure, in two of the cases there is a calcium binding site. Figure 2.13 shows this motif, which is called an **EF hand** because the helices labeled E and F in the structure of parvalbumin are the parts of the structure that were originally used to illustrate calcium binding by this motif. The name has since remained in the literature.

The loop region between the two α helices binds the calcium atom. Carboxy side chains from Asp and Glu, main chain C'O and H_2O form the ligands to the metal atom (Figure 2.13b). Thus both the specific main chain conformation of the loop as well as specific side chains are required to provide the function of this motif. The helix-loop-helix motif provides a scaffold that holds the calcium ligands in the proper position to bind and release calcium.

This was one of the first functional motifs to be recognized in a protein structure. The amino acid sequences of a number of other calcium-binding proteins with important biological functions were known in 1973 but not their

Figure 2.13 Schematic diagrams of the calcium-binding motif. (a) The calcium-binding motif is symbolized by a right hand. Helix E (red) runs from the tip to the base of the forefinger. The flexed middle finger corresponds to the green loop region of 12 residues that binds calcium (pink). Helix F (blue) runs to the end of the thumb. (b) The calcium atom is bound to one of the motifs in the muscle protein troponin-C through six oxygen atoms: one each from the side chains of Asp (D) 9, Asn (N) 11, and Asp (D) 13; one from the main chain of residue 15; and two from the side chain of Glu (E) 20. In addition, a water molecule (W) is bound to calcium. (c) Schematic diagram illustrating that the structure of troponin-C is built up from four EF motifs—colored as in (a). Two of these bind Ca (pink balls) in the molecules that were used for the structure determination. (Adapted from a diagram by J. Richardson in O. Herzberg and M. James, *Nature* 313: 655, 1985.)

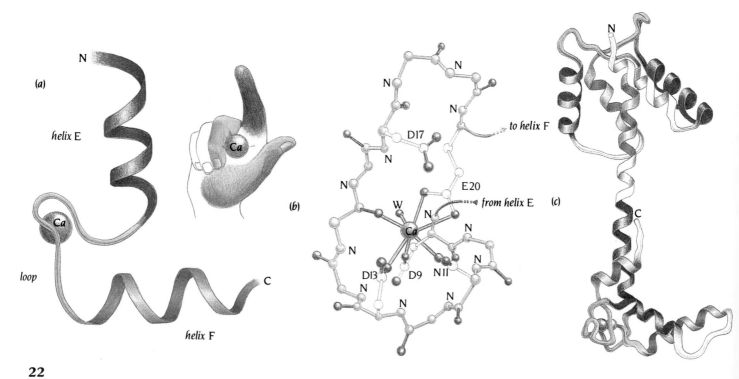

Table 2.2 Amino acid sequences of calcium-binding EF motifs
in three different proteins

Parvalbumin V K K A F A I I D Q D K S G F I E E D E L K L F L Q N F

Calmodulin F K E A F S L F D K D G D G T I T T K E L G T V M R S L

Troponin-C L A D C F R I F D K N A D G F I D I E E L G E I L R A T

E-helix	loop	F-helix

Calcium-binding residues are brown, and residues that form the hydrophobic core of the motif are light green. The helix-loop-helix region shown underneath is colored as in Figure 2.13.

three-dimensional structures. Kretsinger wanted to find out if these proteins could form a motif for calcium binding similar to that in parvalbumin. In order to do this, he analyzed in detail the structure of the calcium-binding motif in parvalbumin and deduced a set of constraints that an amino acid sequence must conform to in order to form such a motif. The motif comprises two α helices, E and F, that flank a loop of 12 contiguous residues. Five of the loop residues are calcium ligands, and their side chains should contain an oxygen atom and preferably be Asp or Glu (see Table 2.2). Residue 6 of the loop must be a glycine because the side chain of any other residue would disturb the structure of the motif. Finally, a number of side chains form a hydrophobic core between the α helices and must thus be hydrophobic. Using these constraints as a mask on their amino acid sequences, Kretsinger predicted that several different calcium-binding proteins could form this motif. Among these proteins were the important muscle protein troponin-C as well as calmodulin, which regulates a variety of cellular functions by calcium binding. Recent structure determinations of these two proteins have shown that this prediction was correct. The structure of one of these, troponin-C, which was determined by Osnat Herzberg in the laboratory of Michael James in Edmonton, Canada, to 2.0 Å resolution, is shown in Figure 2.13.

The hairpin β motif occurs frequently in protein structures

The simplest motif involving β strands, simpler than the calcium-binding motif, is two adjacent antiparallel strands joined by a loop. This motif, which is called either a hairpin or a β-β unit, occurs quite frequently; it is present in most antiparallel β structures both as an isolated ribbon and as part of more complex β sheets. There is a strong preference for β strands to be adjacent in β sheets when they are adjacent in the amino acid sequence and thus to form a **hairpin β motif** or β hairpin for short. The lengths of the loop regions between the β strands vary but are generally from two to five residues long (Figure 2.8). There is no specific function associated with this motif.

Figure 2.14 shows examples of both cases, an isolated ribbon and a β sheet. The isolated ribbon is illustrated by the structure of bovine trypsin inhibitor (Figure 2.14), a small, very stable polypeptide of 58 amino acids that inhibits the activity of the digestive protease trypsin. The structure has been determined to 1.0 Å resolution in the laboratory of Robert Huber in Munich, Germany. Hairpin motifs as parts of a β sheet are exemplified by the structure of a snake venom, erabutoxin (Figure 2.14), which binds to and inhibits the acetylcholine receptor in nerve cells. The structure has been determined to 1.4 Å resolution in the laboratory of Barbara Low at Columbia University, New York. The core of this structure is a β sheet of five strands that contains two hairpin motifs and one additional β strand.

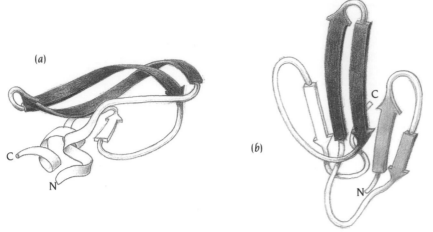

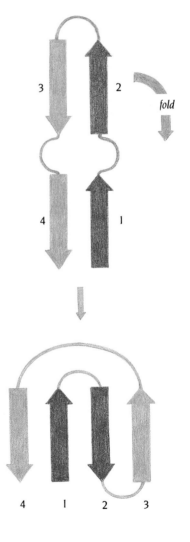

Figure 2.14 The hairpin motif is very frequent in β sheets and is built up from two adjacent β strands that are joined by a loop region. Two examples of such motifs are shown. (a) Schematic diagram of the structure of bovine trypsin inhibitor. The hairpin motif is colored red. (b) Schematic diagram of the structure of the snake venom erabutoxin. The two hairpin motifs within the β sheet are colored red and green. (Adapted from J. Richardson.)

The Greek key motif is found in antiparallel β sheets

Four adjacent antiparallel β strands are frequently arranged in a pattern similar to the repeating unit of one of the ornamental patterns or frets used in ancient Greece, which is now called a Greek key. The motif in proteins is therefore called a **Greek key motif**; Figure 2.15 shows an example in the structure of *Staphylococcus* nuclease, an enzyme that degrades DNA. The Greek key motif is not associated with any specific function, but it occurs frequently in protein structures.

Oleg Ptitsyn in Moscow has analyzed these motifs in detail and has suggested an explanation for their frequent occurrence compared to other arrangements of four antiparallel β strands. He proposed that they have folded from one long antiparallel structure with loops in the middle of both β strands as shown in Figure 2.16. By structural changes in the loop regions between β strands 1 and 2 as well as between β strands 3 and 4 the top part folds down, β strand 2 associates with β strand 1, and they form hydrogen bonds. Thus the Greek key motif has been made.

The β-α-β motif contains two parallel β strands

The hairpin motif is a simple and frequently used way to connect two anti-parallel β strands since the connected ends of the β strands are close together at

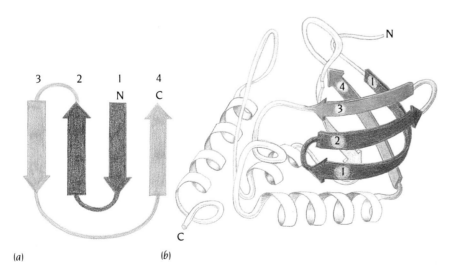

Figure 2.15 The Greek key motif is found in antiparallel β sheets when four adjacent β strands are arranged in the pattern shown as a topology diagram in (a). The motif occurs in many β sheets and is exemplified here by the enzyme *Staphylococcus* nuclease (b). The four β strands that form this motif are colored red and blue. The structure of this enzyme was determined to 1.5 Å resolution in the laboratory of Al Cotton at MIT. (Picture adapted from J. Richardson.)

Figure 2.16 Suggested folding pathway from a hairpinlike structure to the Greek key motif. β strands 2 and 3 fold over such that strand 2 is aligned adjacent and antiparallel to strand 1. The topology diagram of the Greek key shown here is the same as in Figure 2.15a but rotated 180° in the plane of the paper.

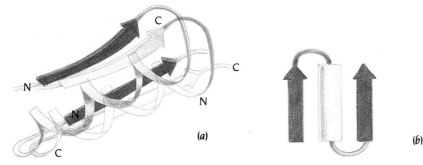

(a)

(b)

the same edge of the β sheet. How are parallel β strands connected? If two adjacent strands are consecutive in the amino acid sequence, the two ends that must be joined are at opposite edges of the β sheet. The polypeptide chain must cross the β sheet from one edge to the other and connect the next β strand close to the point where the first β strand started. It turns out that such crossover connections frequently are made by α helices. The polypeptide chain must turn twice using loop regions, and the motif that is formed is thus a β strand followed by a loop, an α helix, another loop, and, finally, the second β strand.

This motif is called a **beta-alpha-beta motif** (Figure 2.17) and is found as part of almost every protein structure that has a parallel β sheet. For example, the molecule shown in Figure 2.10b, triose phosphate isomerase, is entirely built up by repeated combinations of this motif, where two successive motifs share one β strand. Alternatively, it can be regarded as being built up from four consecutive β-α-β-α motifs.

The α helix in the β-α-β motif connects the carboxy end of one β strand with the amino end of the next β strand (Figure 2.17) and is usually oriented such that the helical axis is approximately parallel to the β strands. It packs against the β strands and thus shields the hydrophobic residues of the β strands from the solvent. The β-α-β motif thus consists of two parallel β strands, an α helix, and two loop regions. The loop regions can be of very different lengths from one or two residues to over a hundred. The two loops have different functions. The loop (dark green in Figure 2.17) that connects the carboxy end of the β strand with the amino end of the α helix is often involved in forming the functional binding site or active site of these structures. These loop regions thus usually have conserved amino acid sequences in homologous proteins. In contrast the other loop (light green in Figure 2.17) has not yet been found to contribute to an active site.

The β-α-β motif can be regarded as a loose helical turn, from one β strand, around the connection and into the next β strand. The motif can thus in principle have two different hands (Figure 2.18). Essentially every β-α-β motif in the known protein structures has been found to have the same hand as a right-handed α helix and therefore is called right-handed. No convincing explanation has been found for this regularity even though it is the only general rule that describes how three secondary structure elements are arranged relative to each other. This handedness has important structural and functional consequences when several of these motifs are linked into a domain structure, as will be described in Chapter 4.

Figure 2.17 Two adjacent parallel β strands are usually connected by an α helix from the C terminus of strand 1 to the N terminus of strand 2. Most protein structures that contain parallel β sheets are built up from combinations of such β-α-β motifs. β strands are red and α helices yellow. Arrows represent β strands and cylinders helices. (a) Schematic diagram of the path of the main chain. (b) Topological diagrams of the β-α-β motif.

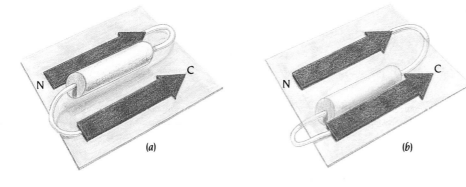

(a)

(b)

Figure 2.18 The β-α-β motif can in principle have two hands. (a) This connection with the helix above the sheet is found in almost all proteins and is called right-handed. (b) The left-handed connection with the helix below the sheet has so far only been found in one structure, subtilisin, where it serves a special function (Chapter 15).

Protein molecules are organized in a structural hierarchy

The Danish biochemist Kai Linderstrøm-Lang coined the terms primary, secondary, and tertiary structure to emphasize the structural hierarchy in proteins (see Figure 1.1). **Primary structure** is the linear amino acid sequence; in other words, the arrangement of amino acids along a polypeptide chain. When two different proteins have significant similarities in their primary structures, they are said to be homologous to each other. Since the corresponding DNA sequences also are significantly similar, it is generally assumed that they are evolutionarily related; they have evolved from a common ancestoral gene.

Secondary structure occurs mainly as α helices and β strands. The formation of secondary structure in a local region of the polypeptide chain is to some extent determined by the primary structure. Certain amino acid sequences favor either α helices or β strands; others favor formation of loop regions. Secondary structure elements usually arrange themselves in simple motifs, as described above. Motifs are formed by packing side chains from adjacent α helices or β strands close to each other.

Several motifs usually combine to form compact globular structures, which are called **domains**. In this book we will use **tertiary structure** as a common term both for the way motifs are arranged into domain structures and for the way a single polypeptide chain folds into one or several domains. Domains with homologous amino acid sequences in different proteins almost invariably have similar tertiary structures.

During the last decades the central dogma of protein folding has stated that the primary structure determines the tertiary structure; in other words, the final specific tertiary structure forms spontaneously from a polypeptide chain with a unique amino acid sequence. Recently, however, a set of proteins, called **chaperones**, have been identified that are required for formation of proper tertiary structure of many different proteins. They seem to act as catalysts by increasing the rates of the final steps in the folding process, and thus in principle they do not violate the central dogma of folding.

Many protein molecules have only one chain; these are monomeric proteins. But a fairly large number have several identical polypeptide chains that associate into a multimeric molecule with a specific **quaternary structure**. These subunits can function either independently of each other or cooperatively so that the function of one subunit is dependent on the functional state of other subunits. Other protein molecules are assembled from several different subunits with different functions, for example, RNA polymerase from *E. coli*, which contains five different polypeptide chains.

Large polypeptide chains fold into several domains

The fundamental unit of tertiary structure is the **domain**. A domain is defined as a polypeptide chain or a part of a polypeptide chain that can independently fold into a stable tertiary structure. Domains are also units of function. Often, the different domains of a protein are associated with different functions. For example, in the lambda repressor protein, discussed in Chapter 7, there is one domain at the N terminus of the polypeptide chain that binds DNA and a second C-terminal domain that holds two polypeptide chains together into a dimeric repressor molecule.

Proteins may comprise a single domain or as many as several dozen domains (Figure 2.19). There is no fundamental structural distinction between a domain and a subunit; there are many known examples where several biological functions that are carried out by separate polypeptide chains in one species are performed by domains of a single protein in another species. For example, synthesis of fatty acids requires catalysis of seven different chemical reactions. In plant chloroplasts these reactions are catalyzed by seven different proteins, whereas in mammals they are performed by one polypeptide chain arranged in seven domains. Such differences thus reflect the organization of the genome rather than the dictates of structure.

26

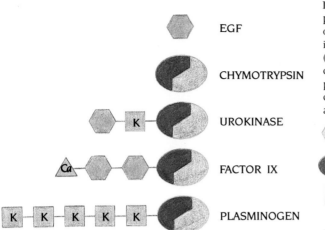

EGF

CHYMOTRYPSIN

UROKINASE

FACTOR IX

PLASMINOGEN

Figure 2.19 Organization of polypeptide chains into domains. Small protein molecules like the epidermal growth factor, EGF, comprise only one domain. Others like the serine proteinase chymotrypsin are arranged in two domains that are both required to form a functional unit (Chapter 15). Many of the proteins that are involved in blood coagulation and fibrinolysis, such as urokinase, factor IX, and plasminogen have long polypeptide chains that comprise different combinations of domains homologous to EGF and serine proteinases and, in addition, calcium-binding domains and Kringle domains.

Domains that are homologous to the epidermal growth factor, EGF, which is a small polypeptide chain of 53 amino acids;

Serine proteinase domains that are homologous to chymotrypsin, which has about 245 amino acids arranged in two domains;

K Kringle domains that have a characteristic pattern of three internal disulphide bridges within a region of about 85 amino acid residues;

Ca Calcium-binding domain (see Figure 2.13).

Domains are built from structural motifs

Domains are formed by different combinations of secondary structure elements and motifs. The α helices and β strands of the motifs are adjacent to each other in the three-dimensional structure and connected by loop regions. Sequentially adjacent motifs, in other words, motifs that are formed from consecutive regions of the primary structure of a polypeptide chain, are usually close together in the three-dimensional structure (Figure 2.20). Thus to a first approximation a polypeptide chain can be sequentially arranged in a number of these simple motifs. The number of such combinations found in proteins is limited, and some combinations seem to be structurally favored. Thus similar domain structures frequently occur in different proteins with different functions and with completely different amino acid sequences.

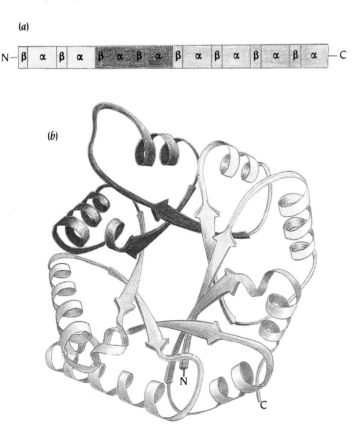

Figure 2.20 Motifs that are adjacent in the amino acid sequence are also usually adjacent in the three-dimensional structure. Triose phosphate isomerase (Figure 2.10) is built up from four β-α-β-α motifs that are consecutive both in the amino acid sequence (a) and in the three-dimensional structure (b).

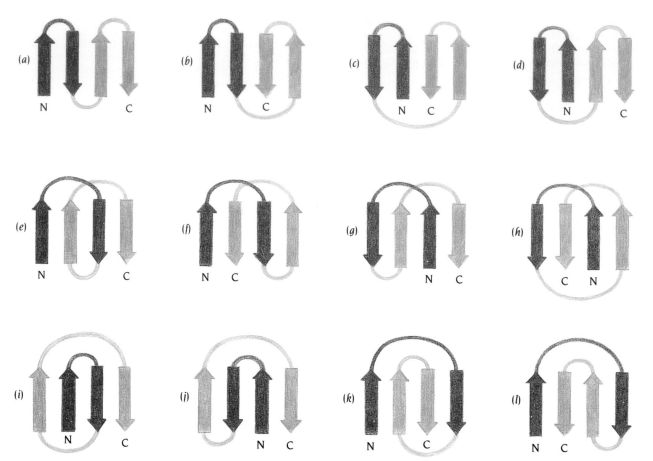

Simple motifs combine to form complex motifs

Figure 2.21 illustrates all the possible ways in which two adjacent β hairpin motifs, each consisting of two antiparallel β strands connected by a loop region, can be combined to make a more complex motif.

The simplest arrangement is that shown in Figure 2.21a. This occurs in many proteins and is illustrated by the structure of one domain of aspartate transcarbamylase (Figure 2.11a).

There are 11 additional ways to build up a β sheet of four strands from these two units. In a random sample of 10 β structures the arrangement in Figure 2.21a occurs 11 times out of a total number of 27 structural fragments of four adjacent β strands. In addition, arrangements 2.21j and 2.21l occur 6 times each. These are the Greek key motifs, previously discussed. Most of the theoretically possible arrangements do not occur within this sample of structures and some like 2.21b and 2.21d occur only once or twice.

Protein structures can be divided into three main classes

Based on simple considerations of connected motifs, Michael Levitt and Cyrus Chothia of the M.R.C. Laboratory of Molecular Biology in Cambridge derived a taxonomy of protein structures, and they could show that combinations of these motifs build up the core of most domain structures and also form the basis for a classification into three main groups: α domains, β domains, and α/β domains. In α structures the core is built up exclusively from α helices (Figure 2.9), and in β structures the core comprises antiparallel β sheets, usually two β sheets packed against each other (Figure 2.11c). The α/β structures are made from combinations of β-α-β motifs that form a predominantly parallel β sheet surrounded by α helices (Figures 2.10b and 2.11b).

Some proteins are built up from a combination of discrete α and β motifs and usually form one small antiparallel β sheet in one part of the domain packed against a number of α helices (Figure 2.15). These structures can be considered

Figure 2.21 Two sequentially adjacent hairpin motifs can be arranged in 12 different ways into a β sheet of four strands. (a) Topology diagram for the arrangement when sequentially adjacent β strands are adjacent in the structure. (b–l) are other possible arrangements, and of these j and l are the Greek key motifs. The topology diagram in j is the mirror image of that in Figure 2.15a, and the diagram in l is the same as that in Figure 2.16 but with the chain direction reversed. These four diagrams represent the Greek key motif.

to belong to a small fourth group. In addition to these groups, there are a number of small proteins that are rich in disulfide bonds or metals and form a special group. The structures of these proteins seem to be strongly influenced by the presence of these metals or disulfides and often look like distorted versions of more regular proteins.

The domains of the known protein structures are classified according to Levitt and Chothia's scheme in this book. The three main classes α, β, and α/β will be examined in more detail in Chapters 3–5.

Conclusion

The interiors of protein molecules contain mainly hydrophobic side chains. The main chain in the interior is arranged in secondary stuctures to neutralize its polar atoms through hydrogen bonds. There are two main types of secondary structure, α helices and β sheets. β sheets can have their strands parallel, antiparallel, or mixed.

Protein structures are built up by combinations of secondary structural elements, α helices, and β strands. These form the core regions—the interior of the molecule—and they are connected by loop regions at the surface. Schematic and simple topological diagrams where these secondary structure elements are highlighted are very useful and are frequently used. α helices or β strands that are adjacent in the amino acid sequence are also usually adjacent in the three-dimensional structure. Certain combinations are especially frequent and are called motifs, for example, the helix-loop-helix motif and the hairpin motif. Two helix-loop-helix motifs, each with its own specific geometry and amino acid sequence requirements, are used in many different proteins, one motif for DNA binding and one for calcium binding.

The β-α-β motif, which consists of two parallel β strands joined by an α helix, occurs in almost all structures that have a parallel β sheet. Four antiparallel β strands, arranged in a specific way, the Greek key motif, are frequently found in structures with antiparallel β sheets.

Polypeptide chains are folded into one or several discrete units, domains, which are the fundamental functional and three-dimensional structural units. The cores of domains are built up from combinations of small motifs of secondary structure, such as α-loop-α, β-loop-β or β-α-β motifs. Domains are classified into three main structural groups: α structures, where the core is built up exclusively from α helices; β structures, which comprise antiparallel β sheets; and α/β structures, where combinations of β-α-β motifs form a predominantly parallel β sheet surrounded by α helices.

Selected readings

General

Chothia, C. Principles that determine the structure of proteins. *Annu. Rev. Biochem.* 53: 537–572, 1984.

Doolittle, R.F. Proteins. *Sci. Am.* 253(4): 88–99, 1985.

Hardie, D.G.; Coggins, J.R. Multidomain Proteins: Structure and Evolution. Amsterdam: Elsevier, 1986.

Janin, J.; Chothia, C. Domains in proteins: definitions, location and structural principles. *Methods Enzymol.* 115: 420–430, 1985.

Klotz, I.M., et al. Quaternary structure of proteins. *Annu. Rev. Biochem.* 39: 25–62, 1970.

Lesk, A. M. Themes and contrasts in protein structures. *Trends Biochem. Sci.* 9: June V, 1984.

Levitt, M.; Chothia, C. Structural patterns in globular proteins. *Nature* 261: 552–558, 1976.

Matthews, B.W.; Bernhard, S.A. Structure and symmetry of oligomeric enzymes. *Annu. Rev. Biophys. Bioeng.* 2: 257–317, 1973.

Richardson, J.S. The anatomy and taxonomy of protein structure. *Adv. Prot. Chem.* 34: 167–339, 1981.

Richardson, J.S. Describing patterns of protein tertiary structure. *Methods Enzymol.* 115: 349–358, 1985.

Richardson, J.S. Schematic drawings of protein structures. *Methods Enzymol.* 115: 359–380, 1985.

Rossmann, M.G.; Argos, P. Protein folding. *Annu. Rev. Biochem.* 50: 497–532, 1981.

Schulz, G.E. Structural rules for globular proteins. *Angew. Chem.* Int. Ed. 16: 23–33, 1977.

Schulz, G.E. Protein differentiation: emergence of novel proteins during evolution. *Angew. Chem.* Int. Ed. 20: 143–151, 1981.

Strynadka, N.C.J.; James, M.N.G. Crystal structures of the helix-loop-helix calcium-binding proteins. *Annu. Rev. Biochem.* 58: 951–998, 1989.

Specific structures

Adams, M.J., et al. Structure of lactate dehydrogenase at 2.8 Å resolution. *Nature* 227: 1098–1103, 1970.

Baba, Y.S., et al. Three-dimensional structure of calmodulin. *Nature* 315: 37–40, 1985.

Banner, B.W., et al. Structure of chicken muscle triose phosphate isomerase determined crystallographically at 2.5 Å resolution using amino acid sequence data. *Nature* 255: 609–614, 1975.

Bourne, P.E., et al. Erabutoxin b. Initial protein refinement and sequence analysis at 0.140 nm resolution. *Eur. J. Biochem.* 153: 521–527, 1985.

Burnett, R.M., et al. The structure of oxidized form of clostridial flavodoxin at 1.9 Å resolution. *J. Biol. Chem.* 249: 4383–4392, 1974.

Chothia, C. Structural invariants in protein folding. *Nature* 254: 304–308, 1975.

Chothia, C.; Levitt, M.; Richardson, D. Structure of proteins: packing of α-helices and pleated sheets. *Proc. Natl. Acad. Sci. USA* 74: 4130–4134, 1977.

Chou, P.Y.; Fasman, G.D. β-turns in proteins. *J. Mol. Biol.* 115: 135–175, 1977.

Colman, P., et al. X-ray crystal structure analysis of plastocyanin at 2.7 Å resolution. *Nature* 272: 319–324, 1978.

Crawford, J.L.; Lipscomb, W.N.; Schellmann, C.G. The reverse turn as a polypeptide conformation in globular proteins. *Proc. Natl. Acad. Sci. USA* 70: 538–542, 1973.

Efimov, A.V. Stereochemistry of α-helices and β-sheet packing in compact globule. *J. Mol. Biol.* 134: 23–40, 1979.

Eklund, H., et al. Three-dimensional structure of horse liver alcohol dehydrogenase at 2.4 Å resolution. *J. Mol. Biol.* 102: 27–59, 1976.

Gouaux, J.E.; Lipscomb, W.N. Crystal structures of phosphonoacetamide ligated T and phosphonoacetamide and malonate ligated R states of aspartate carbamoyltransferase at 2.8 Å resolution and neutral pH. *Biochemistry* 29: 389–402, 1990.

Herzberg, O.; James, M.N.G. Structure of the calcium regulatory muscle protein troponin-C at 2.8 Å resolution. *Nature* 313: 653–659, 1985.

Hol, W.G.J.; van Duijnen, P.T.; Berendsen, H.J.C. The α-helix dipole and the properties of proteins. *Nature* 273: 443–446, 1978.

Holmgren, A., et al. Three-dimensional structure of *E. coli* thioredoxin-S_2 to 2.8 Å resolution. *Proc. Natl. Acad. Sci. USA* 72: 2305–2309, 1975.

Jones, A.; Thirup, S. Using known substructures in protein model building and crystallography. *EMBO J.* 5: 819–822, 1986.

Kendrew, J.C. The three-dimensional structure of a protein molecule. *Sci. Am.* 205: 96–110, 1961.

Kendrew, J.C., et al. A three-dimensional model of the myoglobin molecule obtained by x-ray analysis. *Nature* 181: 662–666, 1958.

Kendrew, J.C., et al. Structure of myoglobin. *Nature* 185: 422–427, 1960.

Kretsinger, R.H. Structure and evolution of calcium-modulated proteins. *CRC Crit. Rev. Biochem.* 8: 119–174, 1980.

Lesk, A.M.; Hardman, K.D. Computer-generated pictures of proteins. *Methods Enzymol.* 115: 381–390, 1985.

Levitt, M. Conformational preferences of amino acids in globular proteins. *Biochemistry* 17: 4277–4285, 1978.

Matthews, B.W.; Rossmann, M.G. Comparison of protein structures. *Methods Enzymol.* 115: 397–420, 1985.

Milner-White, E.J.; Poet, R. Loops, bulges, turns and hairpins in proteins. *Trends Biochem. Sci.* 12: 189–192, 1987.

Moews, P.C.; Kretsinger, R.H. Refinement of the structure of carp muscle calcium-binding parvalbumin by model building and difference Fourier analysis. *J. Mol. Biol.* 91: 201–228, 1975.

Park, C.H.; Tulinsky, A. Three-dimensional structure of the Kringle sequence: structure of prothrombin fragment 1. *Biochemistry* 25: 3977–3982, 1986.

Pauling, L.; Corey, R.B. Configurations of polypeptide chains with favored orientations around single bonds: two new pleated sheets. *Proc. Natl. Acad. Sci. USA* 37: 729–740, 1951.

Pauling, L.; Corey, R.B.; Branson, H.R. The structure of proteins: two hydrogen-bonded helical configurations of the polypeptide chain. *Proc. Natl. Acad. Sci. USA* 37: 205–211, 1951.

Perutz, M.F. New x-ray evidence on the configuration of polypeptide chains. Polypeptide chains in poly-g-benzyl-t-glutamate, keratin and haemoglobin. *Nature* 167: 1053–1054, 1951.

Perutz, M.F. Electrostatic effects in proteins. *Science* 201: 1187–1191, 1978.

Perutz, M.F., et al. Structure of haemoglobin. A three-dimensional Fourier synthesis at 5.5 Å resolution, obtained by x-ray analysis. *Nature* 185: 416–422, 1960.

Rao, S.T.; Rossmann, M.G. Comparison of super-secondary structures in proteins. *J. Mol. Biol.* 76: 241–256, 1973.

Remington, S.J.; Matthews, B.W. A systematic approach to the comparison of protein structures. *J. Mol. Biol.* 140: 77–99, 1980.

Richards, F.M. Calculation of molecular volumes and areas for structures of known geometry. *Methods Enzymol.* 115: 440–464, 1985.

Rose, G.D. Prediction of chain turns in globular proteins on a hydrophobic basis. *Nature* 272: 586–590, 1978.

Rose, G.D. Automatic recognition of domains in globular proteins. *Methods Enzymol.* 115: 430–440, 1985.

Rose, G.D.; Roy, S. Hydrophobic basis of packing in globular proteins. *Proc. Natl. Acad. Sci. USA* 77: 4643–4647, 1980.

Rose, G.D.; Young, W.B.; Gierasch, L.M. Interior turns in globular proteins. *Nature* 304: 654–657, 1983.

Sibanda, B.L.; Thornton, J.M. β-hairpin families in globular proteins. *Nature*, 316: 170–174, 1985.

Tucker, P.W.; Hazen, E.E.; Cotton, F.A. Staphylococcal nuclease reviewed: a prototypic study in contemporary enzymology. III. Correlation of the three-dimensional structure with the mechanisms of enzymatic action. *Mol. Cell. Biochem.* 23: 67–86, 1979.

Venkatachalam, C.M. Stereochemical criteria for polypeptides and proteins. V. Conformation of a system of three linked peptide units. *Biopolymers* 6: 1425–1436, 1968.

Wiegand, G., et al. Crystal structure analysis and molecular model of a complex of citrate synthase with oxaloacetate and S-acetonyl-coenzyme A. *J. Mol. Biol.* 174: 205–219, 1984.

Wlodawer, A.; Deisenhofer, J.; Huber, R. Comparison of two highly refined structures of bovine pancreatic trypsin inhibitor. *J. Mol. Biol.* 193: 145–156, 1987.

Wright, C.S.; Alden, R.; Kraut, J. Structure of subtilisin BPN' at 2.5 Å resolution. *Nature* 221: 235–242, 1969.

Alpha-Domain Structures

3

The first globular protein structure that was determined, myoglobin, belongs to the class of alpha- (α-) domain structures. As we described in Chapter 2, hydrophobic residues occupy the interior of the myoglobin structure. We now know from examination of a large number of structures that polypeptides fold such that they bury hydrophobic groups and expose hydrophilic groups. The hydrophobic side chains in the interior of the proteins form a hydrophobic core that fully occupies the space, leaving very few holes. The chains are as closely packed as hydrocarbon molecules are when packed into a crystal.

α-domain structures consist of a bundle of α helices that are connected by loop regions on the surface of the domains. The α helices are packed pairwise against each other so that one side of each α helix provides the hydrophobic side chains for packing interactions in the interior core and the other side faces the solution. α helices pack well against each other in several different arrangements, of which the two most frequently encountered to date in solved protein structures are the four-helix bundle and the globin fold. We will here describe these two folds, how they are derived from simple rules for packing α helices, and the way the globin fold accommodates mutational changes without losing its function. Of the three main classes of protein structures—α, β, and α/β—the α structures are the fewest in number.

Two adjacent α helices are usually antiparallel

In theory, the simplest way to pack a pair of adjacent helices against each other is to arrange them in an antiparallel fashion with a short connecting loop. This simple arrangement is, in fact, often observed. A particularly simple and illustrative example is the Rop protein, a small RNA-binding protein that is coded by certain plasmids and is involved in plasmid replication. The monomeric subunit of Rop is a polypeptide chain of 63 amino acids built up from two antiparallel α helices joined by a short loop of 3 amino acids to give a very stable structure (Figure 3.1a), which was determined to 1.7 Å resolution by David Banner at EMBL, Heidelberg, Germany.

The four-helix bundle is a common domain structure in α proteins

The Rop protein molecule is a dimer in which the two subunits, each with the structure shown in Figure 3.1a, are arranged such that a bundle of four α helices with their long axes aligned is formed. The four helices pack against each other

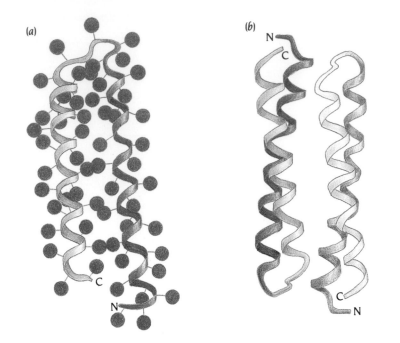

(a)

(b)

Figure 3.1 Two adjacent α helices are usually arranged antiparallel to each other. They pack closely together by intercalating hydrophobic side chains in the interaction area between them. Two such units are frequently arranged in a four-helix bundle, where hydrophobic side chains from all four α helices pack into a hydrophobic core in the center of the bundle. (a) The polypeptide chain of the RNA-binding protein ROP folds into two antiparallel helices joined by a short loop region. The main chain is represented as a ribbon, side chains as purple circles. (b) The ROP molecule consists of two polypeptide chains, where the two α helices of each chain are arranged into a four-helix bundle.

forming a hydrophobic core (Figure 3.1b). A similar arrangement of four α helices can be formed within a single polypeptide chain if the α helices are organized so that each pair of sequentially adjacent α helices is joined in an antiparallel fashion by relatively short loop regions.

A schematic representation of this domain structure is given in Figure 3.2. It occurs in several widely different proteins, such as myohemerythrin (Figure 3.3), a non-haem iron containing oxygen transport protein in marine worms; cytochrome c´ and cytochrome b_{562} (Figure 3.3), which are haem-containing electron carriers; ferritin, which is a storage molecule for Fe atoms in eucaryotic cells; and the coat protein of tobacco mosaic virus.

The side chains of each helix in the four-helix bundle are arranged such that hydrophobic side chains are buried between the helices and hydrophilic side chains are on the surface of the bundle (Figure 3.2b). This arrangement creates a hydrophobic core in the middle of the bundle along its length where the side chains are so closely packed that water is excluded.

(a)

(b)

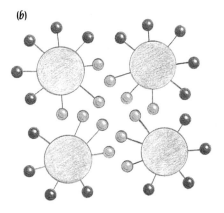

Figure 3.2 Four-helix bundles frequently occur as domains in α proteins. The α helices are arranged such that helices that are adjacent in the amino acid sequence are also adjacent in the three-dimensional structure. Side chains from all four helices are buried in the middle of the bundle where they form a hydrophobic core. (a) Schematic representation of the path of the polypeptide chain in a four-helix bundle domain. Red cylinders are α helices. (Adapted from P.C. Weber and F. R. Salemme, *Nature* 287: 83, 1980.) (b) Schematic view of a projection down the bundle axis. Large circles represent the main chain of the α helices; small circles are side chains. Green circles are buried side chains that are hydrophobic; red circles are side chains that are exposed on the surface of the bundle and that are mainly hydrophilic.

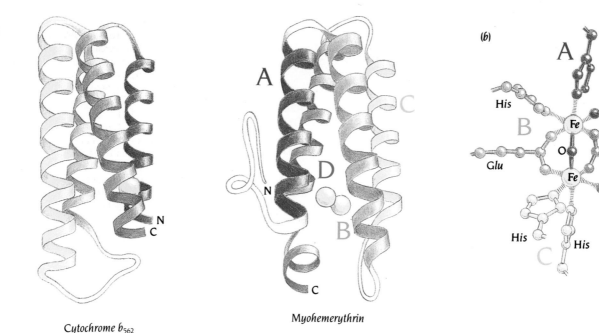

(a)

Cytochrome b_{562}

Myohemerythrin

The active site is between the α helices in four-helix bundle structures

The active sites of the myohemerythrin and the cytochromes are located in a pocket between the α helices at one end of the molecule as shown in Figure 3.3. To a first approximation these molecules thus have their active sites in the same region with respect to the common domain structure. The details are, however, different. Different sets of α helices participate in the formation of the active sites. Thus the haem group, which is the electron carrier in the cytochromes, is wedged between the first and the fourth α helix in the structure, whereas the Fe atoms in myohemerythrin are ligated by residues in all four α helices (see Figure 3.3).

The globin fold is present in myoglobin and hemoglobin

One of the most important α structures is the globin fold. This fold has been found in a large group of related proteins, including myoglobin, hemoglobins, and the light-capturing assemblies in algae, the phycocyanins. Functional and evolutionary aspects of these structures will not be discussed in this book; instead, we will examine some features that are of general structural interest.

The pairwise arrangements of the sequential α helices in the globin fold are quite different from the antiparallel organization found in the four-helix bundle α structures. The globin structure is a bundle of eight α helices, A–H, connected by rather short loop regions and arranged so that the helices form a pocket for the active site, which in myoglobin and the hemoglobins binds a heme group (see Figure 3.4). The lengths of the α helices vary considerably from 7 residues for the shortest helix (C) to 28 for the longest helix (H) in myoglobin. In the globin fold the α helices wrap around the core in different directions so that sequentially adjacent α helices are usually not adjacent to each other in the structure. This fold therefore is not built up from an assembly of smaller motifs. Thus it is quite difficult to visualize conceptually in spite of the fold's relatively small size and simplicity. The only exceptions are the last two α helices G and H, which form an antiparallel pair with extensive packing interactions between them. All other packing interactions are formed between pairs of α helices that are not sequentially adjacent.

Figure 3.3 (a) The polypeptide chains of myohemerythrin and cytochrome b_{562} both form four-helix bundle structures. Their active sites are located in similar regions of the structure. Cytochrome b_{562} functions as an electron carrier by having an Fe atom (yellow circle) in a heme group at the active site that can be oxidized and reduced. (b) The active site in myohemerythrin is formed by two Fe atoms (yellow circles) that are joined by one oxygen atom. This binuclear Fe complex is bound to the protein in such a way that one of the Fe atoms can reversibly bind an oxygen molecule. Side chains from all four helices [corresponding color and labels in (a) and (b)] bind to the two Fe atoms. Marine worms use this molecule for oxygen storage in a similar way that mammals use myoglobin. (Adapted from J. Richardson.) The structure of myohemerythrin was determined to 1.7 Å resolution in the laboratory of Wayne Hendrickson, Columbia University, New York, and that of cytochrome b_{562} to 2.5 Å resolution in the laboratory of Scott Mathews, Washington University, St. Louis.

Geometric aspects determine α-helix packing

When we compare the arrangements of the α helices for the four-helix bundle structure (Figure 3.3) and for the globin fold (Figure 3.4), it is obvious that the geometry of α-helix packing is quite different. In the four-helix bundle the α helices pack almost antiparallel to each other with an angle around 20° between the helical axes. In the globin fold these angles are usually larger, in most cases around 50°. These two structures illustrate the two main ways that α helices pack against each other, which are reflections of the geometry of the surfaces of α helices.

Ridges of one α helix fit into grooves of an adjacent helix

Since the side chains of an α helix are arranged in a helical row along the surface of the helix they form ridges separated by shallow furrows or grooves on the surface. α helices pack with the ridges on one helix packing into the grooves of the other and vice versa. The ridges and grooves that are used for these main packing arrangements are formed by the residues whose separation in the amino acid sequence is usually either four or three. This is illustrated in Figure 3.5, which shows slices through the surface of a polyalanine α helix where the directions of these ridges are marked out. In contrast to the ridges and grooves of the DNA double helix described in Chapter 6, which are formed by the sugar phosphate main chain atoms, those of an α helix are formed by the amino acid side chains. The detailed geometry of the ridges and grooves of an α helix is thus dependent not only on the geometry of the helix but also on the actual amino acid sequence.

The most common way of packing α helices is by fitting the ridges formed by a row of residues separated in sequence by four in one helix into the same type of grooves in the other helix. In this case the ridges and grooves form an angle of about 25° to the helical axis. In order to pack the two helices shown in Figure 3.6a (red and blue) against each other, one of these (the blue in Figure 3.6a) must be turned around 180° out of the plane of the paper and placed on top of the other (red). In the interface between the two α helices the directions of the ridges and grooves are then on opposite sides of the vertical axis, illustrated in Figure

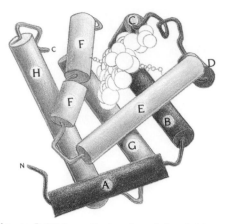

Figure 3.4 Schematic drawing of the globin domain. The eight α helices are labeled A to H. A–D are red, E and F green, and G and H blue. (Adapted from originals provided by A. Lesk.)

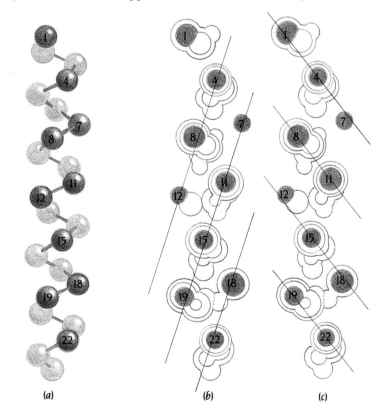

(a) (b) (c)

Figure 3.5 The side chains on the surface of an α helix form ridges separated by grooves in different directions along the surface, as schematically illustrated here. (a) An α helix with each residue represented by the first atom in the side chain, C_β. (b) The surface relief of a polyalanine α helix in the orientation shown in (a). Sections are cut through a space-filling model and superimposed. The residue numbers are placed on the side-chain atom. The ridges caused by the side chains separated by four residues are shown as lines. (c) The same as (b), but here the ridges are caused by side chains separated by three residues.

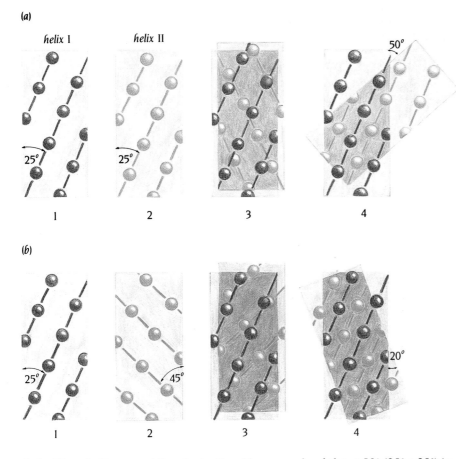

(a)

helix I   helix II

1   2   3   4

(b)

1   2   3   4

Figure 3.6 By fitting the ridges of side chains from one helix into the grooves between side chains of the other helix and vice versa, α helices pack against each other. (a) Two α helices, I and II, where ridges from side chains separated by four residues are marked out in red and blue, respectively. 1 and 2 are the same view of the two α helices. In 3 the blue α helix is turned over through 180° in order to form an interface with the red α helix. In 4 the orientation of the helices has been shifted 50° in order to pack the ridges of one α helix into the grooves of the other. (b) In the red α helix the ridges are formed by side chains separated by four residues and in the blue α helix by three residues. The α helices are here shifted by 20° in order to pack ridges into grooves, in a direction opposite that in (a). (Adapted from C. Chothia et al., *Proc. Natl. Acad. Sci. USA* 74: 4130, 1977.)

3.6a. The α helices must thus be inclined by an angle of about 50° (25° + 25°) in order for the ridges of one helix to fit into the grooves of the other and vice versa. This is the type of packing of several of the helix-helix interactions in the globin fold, and in many other helical structures.

The second frequently occurring packing mode is where the ridges formed by residues separated in sequence by three fit into the grooves of residues separated by four and vice versa. The direction of the first type of ridges forms an angle of about 45° to the helical axis, whereas the other type makes an angle of about 25° to the axis in the opposite direction (Figure 3.6b). In the interface, however, after one helix has been rotated 180°, these directions are on the same side of the helical axis. Thus an inclination of about 20° (45° – 25°) between the two α helices will fit these ridges and grooves into each other. This packing mode is observed in, for instance, the four-helix bundle structures (see Figure 3.3).

These two rules for fitting ridges into grooves are quite general and apply to almost all packing interactions between α helices. They explain the geometrical arrangements of adjacent α helices observed in protein structures.

The globin fold has been preserved during evolution

The three-dimensional structures of globin domains from many diverse organisms, including mammals, insects, and plant root nodules, have been determined independently of each other. They all show amino acid sequence homology in pairwise comparisons that range from 99% to 16%. The essential features of the globin fold are preserved even when the sequence homology is very low.

This family of structures is thus a good example of a situation where natural selection has produced proteins in which the amino acid sequences have diverged widely, although some homology is usually still recognizable, but the three-dimensional structure has been essentially preserved.

To study this situation, Arthur Lesk and Cyrus Chothia compared the family of globin structures. In particular, they tried to answer the following general questions: How can amino acid sequences that are very different form proteins that are very similar in their three-dimensional structure? What is the mechanism by which proteins adapt to mutations in the course of their evolution?

The hydrophobic interior is preserved

To answer the first question, they examined in detail the residues that are at structurally equivalent positions in all the nine then known globin structures and that are involved in helix-heme contacts and in packing the α helices against each other. The 59 positions they found were divided into 31 positions that were buried in the interior of the protein and 28 that were in contact with the heme group. These positions are the principal determinants of both the function and the three-dimensional structure of the globin family.

In a naive approach to sequence-structure relationships one would expect these positions to exhibit a higher degree of amino acid sequence homology than the rest of the molecule. This is, however, not the case for distantly related molecules that have low sequence homology and derive from distantly related species. The sequence homology of these residues is no greater than in the rest of the molecule, and for distantly related members it is even lower. Since the important residues involved in packing the α helices are not conserved, one used to assume that the changes that have occurred compensate each other in size. This is not the case either. The volumes occupied by the 31 buried residues vary considerably between individual members. Thus neither conserved sequence nor size compensatory mutations in the hydrophobic core are factors of importance to preserve three-dimensional structure during evolution. We now know that this is also true for other systems, for example, the immunoglobulins.

Lesk and Chothia did find, however, that there is a striking preferential conservation of the hydrophobic character of the amino acids at the buried positions, but that no such conservation occurs at the surface exposed positions. With a few exceptions hydrophobic residues have replaced hydrophilic on the surface and vice versa. However, the case of sickle-cell hemoglobin, which is described below, shows that a charge balance must be preserved to avoid hydrophobic patches on the surface. In summary, the evolutionary divergence of these nine globins has been constrained primarily by an almost absolute conservation of the hydrophobicity of the residues buried in the helix-to-helix and helix-to-heme contacts.

Helix movements accommodate interior side-chain mutations

Lesk and Chothia also found a simple answer to the question of how proteins adapt to changes in size of buried residues. The mode of packing the α helices are the same in all the globin structures: the same types of packing ridges into grooves occur in corresponding α helices in all these structures. However, the relative positions and orientations of the α helices change to accommodate changes in the volume of side chains involved in the packing.

The proteins thus adapt to mutations in buried residues by changes of overall structure, which in the globins involve movements of entire α helices relative to each other. The structure of loop regions changes so that the movement of one α helix is not transmitted to the rest of the structure. Only such movements that preserve the geometry of the heme pocket are accepted. Mutations that cause such major structural shifts are tolerated because many different combinations of side chains can produce well-packed helix-helix interfaces of similar but not identical geometry and because the shifts are coupled so that the geometry of the active site is retained.

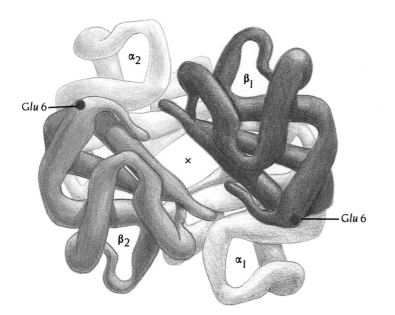

Figure 3.7 The hemoglobin molecule is built up of four polypeptide chains: two α chains and two β chains. Compare this with Figure 1.1 and note that for purposes of clarity parts of the α chains are not shown here. Each chain has a three-dimensional structure similar to that of myoglobin: the globin fold. In sickle-cell hemoglobin Glu 6 in the β chain is mutated to Val, thereby creating a hydrophobic patch on the surface of the molecule. The structure of hemoglobin was determined in 1968 to 2.8 Å resolution in the laboratory of Max Perutz at the M.R.C. Laboratory of Molecular Biology, Cambridge.

Sickle-cell hemoglobin confers resistance to malaria

Sickle-cell anemia is the classical example of an inherited disease that is caused by a change in protein structure. Linus Pauling proposed in 1949 that it was caused by a defect in the hemoglobin molecule, and he coined the term **molecular disease**. Seven years later Vernon Ingram showed that the disease was caused by a single mutation, a change in residue 6 of the β chain of hemoglobin from Glu to Val.

Hemoglobin is a tetramer, built up of two copies each of two different polypeptide chains, α- and β-globin chains in normal adults. Each of the four chains has the globin fold with a heme pocket. Residue 6 in the β chain is on the surface of α helix A of the β subunit, and it is also on the surface of the tetrameric molecule (Figure 3.7).

The hemoglobin concentration in red blood cells, erythrocytes, is extremely high, 340 mg/ml. This is almost as high as in the crystalline state: thus the hemoglobin molecules, which are spheroids of dimension 50 x 55 x 65 Å, are only 10 Å apart on the average in the cells. It is thus surprising that they can rotate and flow past one another without hindrance. The mutation in sickle-cell hemoglobin converts a charged to a hydrophobic residue, and as a result, it produces a hydrophobic patch on the surface. This patch happens to fit and can bind into a hydrophobic pocket in the deoxygenated form of another hemoglobin molecule (Figure 3.8a). In the oxygenated form of hemoglobin the shape of this pocket is slightly different. Thus no interaction occurs between hemoglobin molecules in the lungs. However, when hemoglobin in erythrocytes in the blood capillaries has delivered its oxygen, these hydrophobic interactions can occur and the highly concentrated hemoglobin in the cells polymerizes into fibers (Figure 3.8b and c). These fibers stiffen the erythrocytes and deform them into a sickle shape: hence the name sickle-cell anemia. In heterozygotes, where only one β-globin allele is mutated, this occurs only to a minor extent, whereas it is lethal for homozygotes, all of whose hemoglobin molecules carry the mutation.

We thus have here a case where a mutation on the surface of the globin fold, replacing a hydrophilic residue with a hydrophobic one, changes important properties of the molecule and produces a lethal disease. Why has the mutation survived during evolution? It turns out that the disease gives increased resistance to malaria by an intriguing molecular mechanism. The malarial parasite spends part of its life cycle within the red blood cell of its host. The parasite lowers the

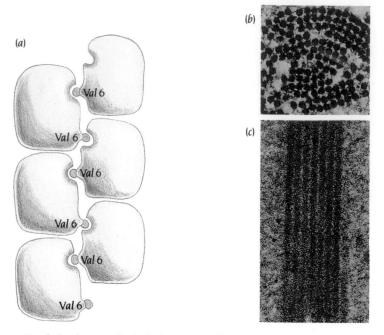

(a)

(b)

(c)

Figure 3.8 Sickle-cell hemoglobin molecules polymerize due to the hydrophobic patch introduced by the mutation Glu 6 to Val in the β chain. The diagram (a) illustrates how this hydrophobic patch (green) interacts with a hydrophobic pocket (green) in a second hemoglobin molecule, whose hydrophobic patch interacts with the pocket in a third molecule, and so on. Electron micrographs of sickle-cell hemoglobin fibers are shown in cross-section in (b) and along the fibers in (c). [(b) and (c) from J. T. Finch et al., *Proc. Natl. Acad. Sci. USA* 70: 718, 1973.]

pH of the host cell slightly, but sufficiently to make the cell more prone to sickling. When sickling occurs, the cell membrane becomes more permeable to potassium ions, which leak out into the surroundings. This drop in potassium ion concentration kills the malarial parasite. The resistance to malaria due to this mechanism has had a high survival value for heterozygotes, especially in Africa. In evolutionary terms, the death of homozygotes has been an acceptable price to pay for increased survival of heterozygotes in a malarial environment.

Conclusion

Alpha- (α-) domain structures consist of a bundle of α helices that are packed together to form a hydrophobic core. A common motif is the four-helix bundle structure, where four helices are pairwise arranged in an antiparallel fashion and packed against each other. The most intensively studied α structure is the globin fold, which has been found in a large group of related proteins, including myoglobin and hemoglobin. This structure comprises eight α helices that wrap around the core in different directions and form a pocket where the heme group is bound.

The mode of helix packing in these structures is determined by fitting ridges of side chains along one α helix into grooves between side chains of another helix. Rules have been derived for this packing that explain the different geometrical arrangements of α helices observed in α-domain structures. The globin fold has been used to study evolutionary constraints for maintaining structure and function. Evolutionary divergence is primarily constrained by conservation of the hydrophobicity of buried residues. In contrast, neither conserved sequence nor size compensatory mutations in the hydrophobic core are important. Proteins adapt to mutations in buried residues by small changes of overall structure that in the globins involve movements of entire helices relative to each other.

Selected readings

General

Chothia, C.; Lesk, A.M. Helix movements in proteins. *Trends Biochem. Sci.* 13: 116–118, 1985.

Chothia, C.; Levitt, M.; Richardson, D. Helix-to-helix packing in proteins. *J. Mol. Biol.* 145: 215–250, 1981.

Dickerson, R.E.; Geis, I. Hemoglobin: Structure, Function, Evolution and Pathology. Menlo Park, CA: Benjamin/Cummings, 1983.

Lesk, A.M.; Chothia, C. How different amino acid sequences determine similar protein structures: the structure and evolutionary dynamics of the globins. *J. Mol. Biol.* 136: 225–270, 1980.

Murzin, A.G.; Finkelstein, A.V. General architecture of the α-helical globule. *J. Mol. Biol.* 204: 749–769, 1988.

Pauling, L., et al. Sickle cell anemia: a molecular disease. *Science* 110: 543–548, 1949.

Perutz, M.F. Hemoglobin structure and respiratory transport. *Sci. Am.* 239(6): 92–125, 1978.

Specific structures

Argos, P.; Rossmann, M.G.; Johnsson, J.E. A four-helical super-secondary structure. *Biochem. Biophys. Res. Comm.* 75: 83–86, 1977.

Banner, D.W.; Kokkinidis, M.; Tsernoglou, D. Structure of the col1 rop protein at 1.7 Å resolution. *J. Mol. Biol.* 196: 657–675, 1987.

Bashford, D.; Chothia, C.; Lesk, A.M. Determinants of a protein fold. Unique features of the globin amino acid sequences. *J. Mol. Biol.* 196: 199–216, 1987.

Bloomer, A.C., et al. Protein disk of tobacco mosaic virus at 2.8 Å resolution showing the interactions within and between subunits. *Nature* 276: 362–368, 1978.

Blundell, T., et al. Solvent-induced distortions and the curvature of α-helices. *Nature* 306: 281–283, 1983.

Clegg, G.A., et al. Helix packing and subunit conformation in horse spleen apoferritin. *Nature* 288: 298–300, 1980.

Embury, S.H. The clinical pathophysiology of sickle-cell disease. *Annu. Rev. Med.* 37: 361–376, 1986.

Fermi, G., et al. The crystal structure of human deoxyhaemoglobin at 1.74 Å resolution. *J. Mol. Biol.* 175: 159–174, 1984.

Fermi, G.; Perutz, M.F. Atlas of Molecular Structures in Biology. 2. Haemoglobin and Myoglobin. Oxford, UK: Clarendon Press, 1981.

Finch, J.T., et al. Structure of sickled erythrocytes and of sickle-cell hemoglobin fibers. *Proc. Natl. Acad. Sci. USA* 70: 718–722, 1973.

Finzel, B.C., et al. Structure of ferricytochrome c′ from *Rhodospirillum molischianum* at 1.67 Å resolution. *J. Mol. Biol.* 186: 627–643, 1985.

Ingram, V.M. Gene mutation in human haemoglobin: the chemical difference between normal and sickle cell haemoglobin. *Nature* 180: 326–328, 1957.

Lederer, F., et al. Improvement of the 2.5 Å resolution model of cytochrome b_{562} by redetermining the primary structure and using molecular graphics. *J. Mol. Biol.* 148: 427–448, 1981.

Nordlund, P.; Sjöberg, B.-M.; Eklund, H. Three-dimensional structure of the free radical protein of ribonucleotide reductase. *Nature* 345: 593–598, 1990.

Pastore, A., et al. Structural alignment and analysis of two distantly related proteins: *Aplysia limacina* myoglobin and sea lamprey globin. *Proteins* 4: 240–250, 1988.

Pastore, A.; Lesk, A.M. Comparison of the structures of globins and phycocyanins: evidence for evolutionary relationship. *Proteins* 8: 133–155, 1990.

Phillips, S.E.V. Structure and refinement of oxymyoglobin at 1.6 Å resolution. *J. Mol. Biol.* 142: 531–554, 1980.

Presnell, S.R.; Cohen, F.E. Topological distribution of four-α-helix bundles. *Proc. Natl. Acad. Sci. USA* 86: 6592–6596, 1989.

Richmond, T.J.; Richards, F.M. Packing of α-helices: geometrical constraints and contact areas. *J. Mol. Biol.* 119: 537–555, 1978.

Sheriff, S.; Hendrickson, W.A.; Smith, J.L. Structure of myohemerythrin in the azidomet state at 1.7/1.3 Å resolution. *J. Mol. Biol.* 197: 273–296, 1987.

Watson, H.C. The stereochemistry of the protein myoglobin. *Progr. Stereochem.* 4: 299–333, 1969.

Weber, P.C.; Salemme, F.R. Structural and functional diversity in 4-α-helical proteins. *Nature* 287: 82–84, 1980.

Alpha/Beta Structures

4

The most frequent and most regular of the domain structures are the alpha/beta (α/β) domains, which consist of a central parallel or mixed β sheet surrounded by α helices. All the glycolytic enzymes are α/β structures as are many other enzymes as well as proteins that bind and transport metabolites. In α/β domains binding crevices are formed by loop regions. These regions do not contribute to structural stability but participate in binding and catalytic action.

Parallel β strands are arranged in barrels or sheets

There are two main classes of α/β proteins. In the first class there is a core of eight twisted parallel β strands arranged close together, like staves, into a barrel. The

Figure 4.1 α/β domains are frequently found in many proteins. They occur in two different classes: a closed barrel exemplified by schematic and topological diagrams of the enzyme triosephosphate isomerase (a) and an open twisted sheet with helices on both sides, such as in the coenzyme-binding domain of some dehydrogenases. (b) Both classes are built up from β-α-β motifs that are linked such that the β strands are parallel. Rectangles represent α helices and arrows β strands in the topological diagrams. [(a) Adapted from J. Richardson. (b) Adapted from B. Furugren.]

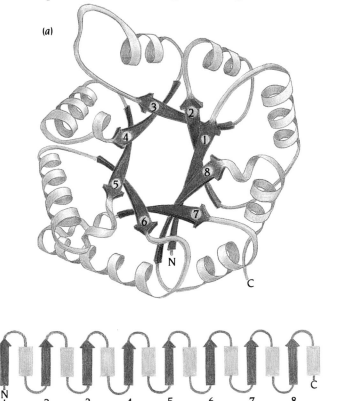

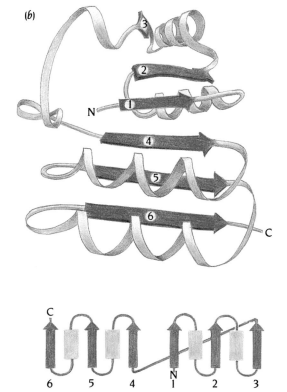

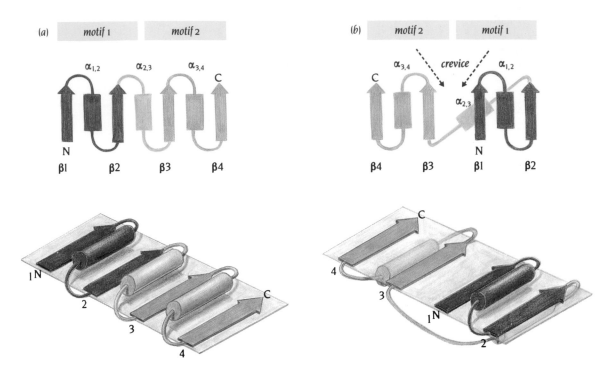

α helices that connect the parallel β strands are all on the outside of this barrel (see Figure 4.1a). This domain structure is often called the TIM barrel from the structure of the enzyme triosephosphate isomerase, where it was first observed. The second class contains an open twisted β sheet surrounded by α helices on both sides of the β sheet. A typical example, a nucleotide-binding domain found in a number of dehydrogenases and kinases, is shown in Figure 4.1b.

Both barrels and open sheets are built up from β-α-β motifs. To illustrate how they differ, let us consider two β-α-β motifs: β_1-$\alpha_{1,2}$-β_2 and β_3-$\alpha_{3,4}$-β_4 linked together by helix $\alpha_{2,3}$. There are two fundamentally different ways these two motifs can be connected into a β sheet of four parallel strands, as shown in Figure 4.2. Strand β_3 either can be aligned adjacent to strand β_2 giving the strand order 1 2 3 4 or can be adjacent to strand β_1 giving the strand order 4 3 1 2. In the first case the two motifs are joined with the same orientation. The resulting structure consists of consecutive β-α-β units, all in the same orientation. Since the β-α-β unit is a fixed right-handed structure, as described in Chapter 2, all three α helices (the two from the two motifs and the joining helix) are on the same side, above the β sheet in Figure 4.2a. In barrel structures all β-α-β motifs are linked in this way.

In the second case we must turn the second motif around in order to align β strands 1 and 3. As a result of the fixed, right-handed structure of the β-α-β motif, its α helix is on the other side of the β sheet (Figure 4.2b). In open-sheet structures there is always one or more such alignments and therefore there are α helices on both sides of the β sheet. These geometric rules apply because virtually all β-α-β motifs are right-handed. As pointed out in Chapter 2, this is an empirical rule, which almost always applies, even though no convincing explanation has been found why.

α/β barrels occur in many different enzymes

In α/β structures where the strand order is 1 2 3 4, all connections are on the same side of the β sheet. An open twisted β sheet of this sort with four or more parallel β strands would leave one side of the parallel β sheet exposed to the solvent and the other side shielded by the α helices. Such a domain structure is rarely observed, except as part of much more complex structures where loop regions and extra α helices cover the exposed side of the β sheet. Instead, a closed barrel

Figure 4.2 β-α-β motifs are rigid right-handed structures. Two such motifs can be joined into a four-stranded parallel β sheet in two different ways. They can either be aligned with the α helices on the same side of the β sheet (a) or on opposite sides (b). In case (a) the last β strand of motif 1 (red) is adjacent to the first β strand of motif 2 (blue) giving strand order 1 2 3 4. The motifs are aligned in this way in barrel structures (Figure 4.1a). In case (b) the first β strands of both motifs are adjacent giving strand order 4 3 1 2. Open twisted sheets (Figure 4.1b) contain at least one motif alignment of this kind. In both cases the motifs are joined by an α helix (green).

of twisted β strands is formed with all the connecting α helices on the outside of the barrel as shown in Figure 4.1a. More than four β strands, however, are needed to provide enough staves to form a closed barrel, and almost all the closed α/β barrels observed to date have eight parallel β strands. These are arranged such that β strand 8 is adjacent and hydrogen bonded to β strand 1. The seven cross-connections between these β strands are α helices in almost all cases; in addition, there is an eighth α helix formed by the chain after β strand 8.

This eight-stranded α/β-barrel structure is one of the largest and most regular of all domain structures. A minimum of about 200 residues are required to form this structure. It has so far been found in about 15 different proteins with completely different amino acid sequences and different functions (Table 4.1). Superimposing the structures of these proteins shows that around 160 residues are structurally equivalent. These residues form all eight of the β strands and α helices. The remaining residues form the loop regions that connect the β strands with the α helices. These loops have quite different lengths and conformations in the different proteins. In some cases they are very long and form independent domains in the overall subunit structure.

Table 4.1 Proteins whose structures contain at least one domain that is folded into a closed α/β barrel

Protein	Reference
Triosephosphate isomerase	Banner et al., *Nature* 255: 609, 1975
Pyruvate kinase	Stuart et al., *J. Mol. Biol.* 134: 109, 1979
Taka-amylase	Matsuura et al, *J. Biochem.* (Japan) 87: 1555, 1980
Phosphogluconate aldolase	Mavridis et al., *J. Mol. Biol.* 162: 419, 1982
Xylose (glucose) isomerase	Carrel et al., *J. Biol. Chem.* 259: 3230, 1984
Glycolate oxidase	Lindqvist & Branden, *Proc. Natl. Acad. Sci. USA* 82: 6855, 1985
Ribulose bisphosphate carboxylase/oxygenase	Schneider et al., *EMBO J.* 5: 3409, 1986
Trimethylamine dehydrogenase	Lim et al., *J. Biol. Chem.* 261: 15140, 1986
Muconate lactonizing enzyme	Goldman et al., *J. Mol. Biol.* 194: 143, 1987
Phosphoribosylanthranilate isomerase: Indoleglycerol phosphate synthase	Priestle et al., *Proc. Natl. Acad. Sci. USA* 84: 5690, 1987
Aldolase	Sygusch et al., *Proc. Natl. Acad. Sci. USA* 84: 7846, 1987
Tryptophane synthase	Hyde et al., *J. Biol. Chem.* 263: 17857, 1988
Enolase	Lebioda and Stec, *Nature* 333: 683, 1988, but see also *J. Biol. Chem.* 264: 3685, 1989

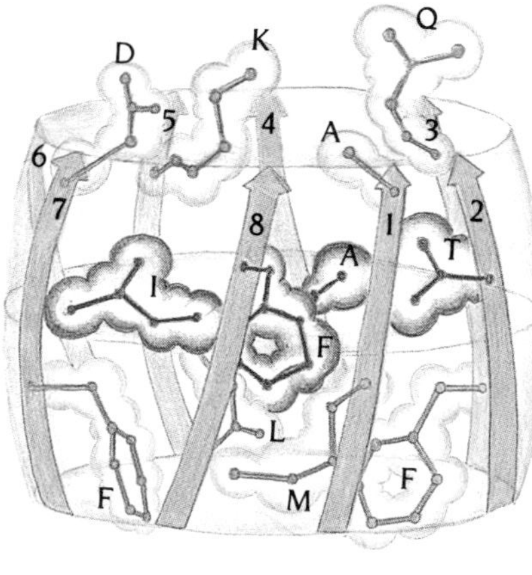

Figure 4.3 The eight β strands of the α/β barrel enclose a tightly packed hydrophobic core formed entirely by side chains from the β strands. The core is arranged in three layers, where each layer contains four side chains from alternate β strands. The schematic diagram shows this packing arrangement in the α/β barrel of the enzyme glycolate oxidase, the structure of which is described in Chapter 10.

Branched hydrophobic side chains dominate the core of α/β barrels

In barrels the hydrophobic side chains of the α helices are packed against hydrophobic side chains of the strands of the β sheet. The α helices are arranged such that they are antiparallel and adjacent to the β strands that they connect. Thus the barrel is provided with a shell of hydrophobic residues from the helices and the β strands.

Since the side chains of consecutive amino acids of a β strand are on opposite sides of the β sheet, every second residue of the β strands contributes to this hydrophobic shell. The other side chains of the β strands point inside the barrel to form a hydrophobic core; this core is therefore comprised exclusively of side chains of β-strand residues (Figure 4.3).

The packing interactions between α helices and β strands are dominated by the residues Val (V), Ile (I), and Leu (L), which have branched hydrophobic side chains. This is reflected in the amino acid composition. These three amino acids comprise approximately 40% of the residues of the β strands in parallel β sheets. The important role that these residues play in packing α helices against β sheets is particularly obvious in α/β-barrel structures, as Table 4.2 shows.

Table 4.2 The amino acid residues of the eight parallel β strands in the barrel structure of the enzyme triosephosphate isomerase from chicken muscle

Strand no.	Residue no.	Positions				
		1	2	3	4	5
1	6–10	Phe	Val	Gly	Gly	Asn
2	37–41	Glu	Val	Val	Cys	Gly
3	59–63	Gly	Val	Ala	Ala	Gln
4	89–93	Trp	Val	Ile	Leu	Gly
5	121–125	Gly	Val	Ile	Ala	Cys
6	158–162	Lys	Val	Val	Leu	Ala
7	204–208	Arg	Ile	Ile	Tyr	Gly
8	227–231	Gly	Phe	Leu	Val	Gly

The sequences are aligned such that residues in positions 1, 3, and 5 point into the barrel and residues in position 2 and 4 point toward the α helices on the outside and are involved in the hydrophobic interactions between the β strands and the α helices. The branched hydrophobic residues Val, Ile, and Leu (yellow) frequently occur in these interactions. They are important for packing β strands against α helices.

Pyruvate kinase contains several domains, one of which is an α/β barrel

All known eight-stranded α/β-barrel domains have enzymatic functions including isomerization of small sugar molecules, oxidation by flavin coenzymes,

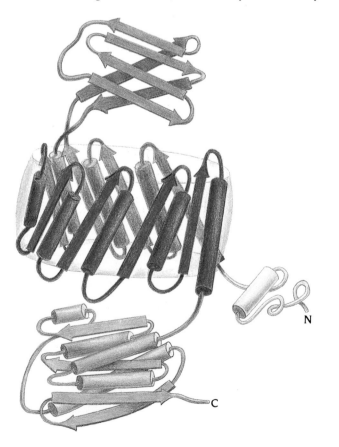

Figure 4.4 The polypeptide chain of the enzyme pyruvate kinase folds into several domains, one of which is an α/β barrel (red). One of the loop regions in this barrel domain is extended and comprises about 100 amino acid residues that fold into a separate domain (blue) built up from antiparallel β strands. The C-terminal region of about 140 residues forms a third domain (green), which is an open twisted α/β structure.

phosphate transfer, and degradation of sugar polymers. In some of these enzymes the barrel domain comprises the whole subunit of the protein; in others the polypeptide chain is longer and forms several additional domains. An enzymatic function in these multidomain subunits is, however, always associated with the barrel domain.

For example, each subunit of the dimeric glycolytic enzyme triosephosphate isomerase (Figure 4.1a) consists of one such barrel domain. The polypeptide chain has 248 residues in which the first β strand of the barrel starts at residue 6 and the last α helix of the barrel ends at residue 246. In contrast, the subunit of another glycolytic enzyme, pyruvate kinase (Figure 4.4), which was solved at 2.6 Å resolution in the laboratory of Hilary Muirhead, Bristol University, UK, is folded into four different domains. The polypeptide chain of this cat muscle enzyme has 530 residues. Residues 1–42 (yellow) form a small domain involved in subunit contacts in the tetrameric molecule. Residues 43–115 and 224–387 form an α/β-barrel domain (red) that binds substrate and provides the catalytic groups. Residues 116–223 loop out from the end of β-strand number 3 in the barrel domain and are folded into a separate domain consisting of an antiparallel β sheet (blue). Finally, residues 388–530 form an open twisted α/β domain (green). This structure illustrates perfectly how a long polypeptide chain can be arranged in domains of different structural types.

Double barrels have occurred by gene fusion

PRA-isomerase: IGP-synthase, a bifunctional enzyme from *E. coli*, which catalyzes two reactions in the synthesis of tryptophane, has a polypeptide chain that forms two α/β barrels (Figure 4.5). The structure of this enzyme, solved at 2.8 Å in the laboratory of Hans Jansonius in Basel, showed that residues 48–254 form one barrel with IGP-synthase activity, while residues 255–450 form the second barrel with PRA-isomerase activity.

In *Bacillus subtilis*, these two reactions are catalyzed by two separate enzymes that have amino acid sequences homologous to the corresponding regions of the bifunctional enzyme from *E. coli* and thus each form barrel structures. There is apparently no functional advantage at the enzyme level for *E. coli* to have these two enzymatic activities in one polypeptide chain, since the active sites of these two barrels are on opposite sides of the molecule facing away from each other. With this arrangement the two reactions are independent of each other. A third organism, *Neurospora crassa*, has a trifunctional enzyme within the same polypeptide chain; here two domains similar to those of the *E. coli* enzyme are linked to a third domain that has yet another enzymatic function in the same biosynthetic pathway. These differences between species reflect different ways to organize the genome. Pieces of DNA that code for domains with different functions are organized into separate genes in one organism and fused into a single gene in another. Although the three-dimensional structures of these enzymes in *B. subtilis* and *N. crassa* have not been solved by crystallography, we can be certain that they are α/β-barrel domains because of their sequence homologies to the *E. coli* proteins.

The active site is formed by loops at one end of the α/β barrel

In all these α/β-barrel domains the active site is in very similar positions. It is situated in the bottom of a funnel-shaped pocket created by the eight loops that connect the carboxy end of the β strands with the amino end of the α helices (see Figure 4.6). Residues that participate in binding and catalytic activity are in these loop regions. In other words, these enzymes are modeled on a common stable scaffold of eight parallel β strands surrounded by eight α helices. The specific enzymatic activity is, in each case, determined by the eight loop regions at the carboxy end of the the β strands, which do not contribute to the structural stability of the scaffold. In some cases an additional loop region from a second domain or a different subunit comes close to this active site and also participates in binding.

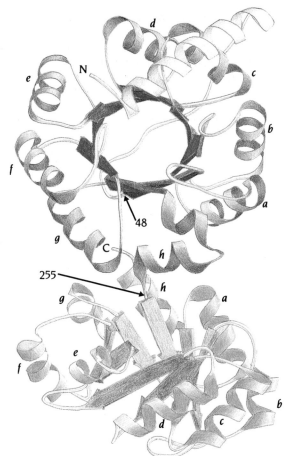

Figure 4.5 Two of the enzymatic activities involved in the biosynthesis of tryptophane in *E. coli*, PhosphoRibosyl Anthranilate (PRA) isomerase and IndoleGlycerol Phosphate (IGP) synthase, are performed by two separate domains in the polypeptide chain of a bifunctional enzyme. Both these domains are α/β-barrel structures, oriented such that their active sites are on opposite sides of the molecule. The two catalytic reactions are therefore independent of each other. The diagram shows the IGP-synthase domain (residues 48–254) with dark colors and the PRA-isomerase domain with light colors. The α helices are sequentially labeled a to h in both barrel domains. Residue 255 (arrowed) is the first residue of the second domain. (Adapted from J. P. Priestle et al., *Proc. Natl. Acad. Sci. USA* 84: 5692, 1987.)

(a)

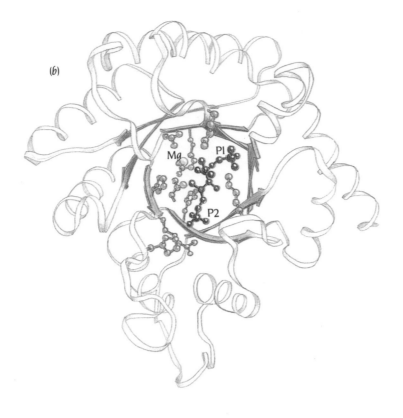

(b)

Figure 4.6 The active site in all α/β barrels is in a pocket formed by the loop regions that connect the carboxy ends of the β strands with the adjacent α helices, as shown schematically in (a), where only two such loops are shown. Figure (b) shows a view from the top of the barrel of the active site of the enzyme RuBisCo (ribulose bisphosphate carboxylase), which is involved in CO_2 fixation in plants. A substrate analogue (red) binds across the barrel with the two phosphate groups, P1 and P2, on opposite sides of the pocket. A number of charged side chains (blue) from different loops as well as a Mg^{++} ion (yellow) form the substrate binding site and provide catalytic groups. The structure of this 500kD enzyme was determined to 2.4 Å resolution in the laboratory of Carl Branden, Uppsala, Sweden. (Adapted from original provided by Bo Furugren.)

Stability and function are separated

These structures provide very clear examples of the physical separation of those residues that contribute to the stability of the domain from those that are responsible for the specific function. Such an arrangement has obvious evolutionary advantages for changing function without affecting the structural stability of the scaffold. As a result, these enzymes are excellent targets for genetic redesign *in vitro*. By changing the lengths and the residues of the eight active site loop regions it might be possible to produce novel enzymatic activities without affecting the stability of the structural framework and therefore the enzyme.

Was there an ancestral barrel?

Since all α/β-barrel structures are so similar and the active sites are always at structurally similar positions, it is reasonable to ask whether or not these structures have evolved from a common ancestral barrel domain. Evidence relevant to that question is scanty but negative. There is no sequence homology at either the gene or protein level amongst the α/β-barrel proteins, except between homologues that perform the same function in different species. In other words, the barrels seem not to be evolutionarily related. What about the active sites? Are there traces of evolutionary relationships in the details of the active-site regions? In particular, are similar residues present at the same sites in the polypeptide chains of different proteins. The answer is again negative, since different loops and therefore different regions of the polypeptide chains provide the catalytic groups in different enzymes.

This implies that these barrel structures have evolved not from a common barrel ancestor but independently of each other. The α/β barrel, therefore, provides a clear example of convergent evolution at the level of protein structure. The common structure, eight β strands and eight α helices, must, as far as we can tell, have arisen independently several times during evolution. The structural similarity is deceptive. The moral is obvious: at the molecular level, just as at the level of gross anatomy, similarity of structure is not necessarily evidence of common evolutionary ancestry.

α/β twisted open-sheet structures contain α helices on both sides of the β sheet

In the second class of α/β structures there are α helices on both sides of the β sheet. This has at least three important consequences. First, a closed barrel cannot be formed unless the β strands completely enclose the α helices on one side of the β sheet. Such structures have never been found and are very unlikely to occur since a large number of β strands would be required to enclose even a single α helix. Instead, the β strands are arranged into an open twisted β sheet such as shown in Figure 4.1b.

Second, there are always two adjacent β strands (numbers 1 and 3 in Figure 4.2b) in the interior of the β sheet where the connections to the next β strand are on opposite sides of the β sheet. One of the loops from one of these two β strands goes above the β sheet, whereas the other loop goes below. This creates a crevice outside the edge of the β sheet between these two loops (see Figure 4.7). Almost all binding sites in this class of α/β proteins are located in crevices of this type at the carboxy edge of the β sheet. (We define the carboxy edge of the sheet as the edge that is formed by the carboxy ends of the parallel β strands in the sheet.)

Third, in open-sheet structures the α helices are packed against both sides of the β sheet. Each β strand thus contributes hydrophobic side chains to pack against α helices in two similar hydrophobic core regions, one on each side of the β sheet.

Open β-sheet structures have different topologies

We have seen that all members of the α/β-barrel domain structures are very similar with the same basic arrangement of eight α helices and eight β strands. Within the open α/β sheets there is much more variation in structure. It is obvious from pure geometric considerations that there is scope for a greater variability. Since they form an open β sheet, there are no geometric restrictions on the number of β strands involved. In fact, the number varies from four up to ten. Furthermore, the two β strands that are joined by a crossover connection need not be adjacent in the β sheet, even though the β-α-β motif where the two β strands are adjacent is a preferred structural building block. In addition, there can be mixed β sheets where hairpin connections give rise to some antiparallel β strands mixed with the parallel β strands. All these variations occur in actual structures, some of which are illustrated in Figures 4.8a–d. There are thus many variations on the regular theme containing six parallel β strands that is found in the structure of the nucleotide-binding fold illustrated in Figure 4.1b.

(a)   (b)

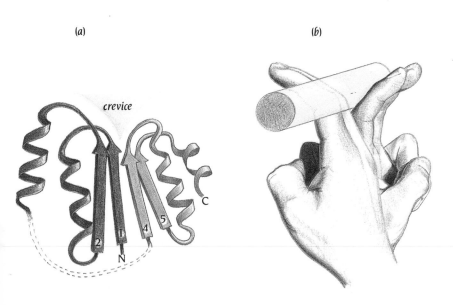

Figure 4.7 (a) The active site in open twisted α/β domains is in a crevice outside the carboxy ends of the β strands. This crevice is formed by two adjacent loop regions that connect the two strands with α helices on opposite sides of the β sheet. This is illustrated by the curled fingers of two hands (b) where the top halves of the fingers represent loop regions and the bottom halves represent the β strands. The rod represents a bound molecule lying in the binding crevice.

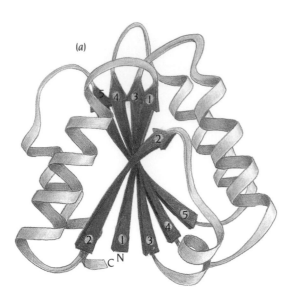

(a)

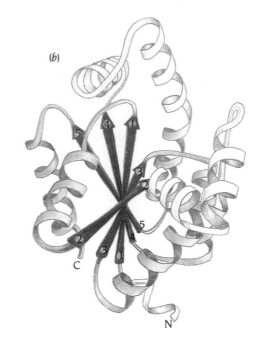

(b)

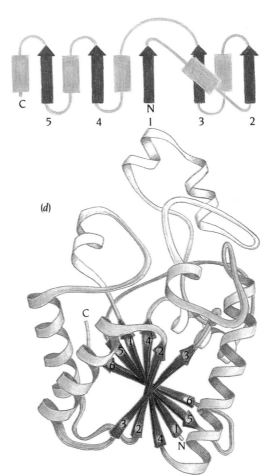

(c)

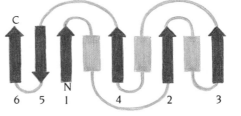

(d)

Figure 4.8
Examples of different types of open twisted α/β structures. Both schematic and topological diagrams are given. Arrows denote strands of β sheet and rectangles denote α helices. (a) The FMN-binding redox protein flavodoxin. (b) The enzyme adenylate kinase, which catalyzes the reaction AMP + ATP = 2ADP. The structure was determined to 3.0 Å resolution in the laboratory of Georg Schulz in Heidelberg, Germany. (c) The ATP-binding domain of the glycolytic enzyme hexokinase, which catalyzes the phosphorylation of glucose. The structure was determined to 2.8 Å resolution in the laboratory of Tom Steitz, Yale University. (d) The glycolytic enzyme phosphoglycerate mutase, which catalyzes transfer of a phosphoryl group from carbon 3 to carbon 2 in glycerate. The structure was determined to 2.5 Å resolution in the laboratory of Herman Watson, Bristol University, UK. (Adapted from J. Richardson.)

The position of the active sites can be predicted in α/β structures

We have described a general relationship between structure and function for the α/β-barrel structures. They all have the active site at the same position with respect to their common structure in spite of having different functions as well as different amino acid sequences. We can now ask if similar relationships occur also for the open α/β-sheet structures in spite of the much greater variation in structure. We will see that the answer is positive and, therefore, the position of the active site can be predicted from the structure for all α/β proteins.

A survey of the 20 different known α/β structures of this class was made in 1980 by Carl Branden at Uppsala, Sweden. He found that the active site is at the carboxy edge of the β sheet in all these structures. Functional residues are provided by the loop regions that connect the carboxy end of the β strands with the amino end of the α helices. In this one respect there is therefore a fundamental similarity between the α/β-barrel structures and the open α/β-sheet structures.

The general shape of the active sites are quite different, however. Open α/β structures cannot form funnel-shaped active sites like the barrel structures. Instead, they form crevices at the edge of the β sheet. Such crevices occur when there are two adjacent connections that are on opposite sides of the β sheet. One of the loop regions in these two connections goes out from its β strand above the β sheet and the other below, creating a crevice between them (see Figure 4.7). In all these 20 structures the active site or part of it was found in such a crevice. The position of such crevices is determined by the topology of the β sheet and can be predicted from a topology diagram. The crevices occur when the strand order is reversed and can easily be identified by looking, in a topology diagram, for the place where connections from the carboxy ends of two adjacent β strands go in opposite directions, one to the left and one to the right. Let us examine the first two diagrams given in Figure 4.8. The first structure, flavodoxin (Figure 4.8a), has one such position, between strands 1 and 3. The connection from strand 1 goes to the right and that from strand 3 to the left. In the schematic diagram in Figure 4.8a we can see that the corresponding α helices are on opposite sides of the β sheet. The loops from these two β strands, 1 and 3, to their respective α helices form the major part of the binding cleft for the coenzyme FMN.

The second structure, adenylate kinase (Figure 4.8b), has two such positions, one on each side of β strand 1. The connection from strand 1 to strand 2 goes to the right, whereas the connection from the flanking strands 3 and 4 both go to the left. Crevices are formed both between β strands 1 and 3 and between strands 1 and 4. One of these crevices forms part of an AMP binding site, and the other crevice forms part of an ATP binding site for this enzyme that catalyzes the formation of ADP from AMP and ATP.

Such positions in a topology diagram are called **topological switch points**. It was postulated in 1980 that the position of the active sites could be predicted from such switch points. Since then at least one part of the active site has been found in crevices defined by such switch points in all new α/β structures that have been determined. Thus we can predict the approximate position of the active site and possible loop regions that form this site in all α/β proteins. This is in contrast to proteins of the other two main classes—α-helical proteins and antiparallel β proteins—where no such predictive rules have been found. We will now examine a few examples that illustrate the relationship between the topology diagrams of some α/β proteins, their switch points, and the active-site residues. These examples have been chosen because they represent different types of α/β open-sheet structures.

Tyrosyl-tRNA synthetase has two different domains (α/β + α)

One of the crucial steps in protein synthesis is performed by the group of enzymes called aminoacyl-tRNA synthetases. These enzymes connect each amino acid with its specific transfer RNA molecule in a two-step reaction. First the amino acid is activated by ATP to give enzyme-bound amino acid adenylate; then this complex is attacked by the tRNA to give the charged tRNA molecule.

The structure of the synthetase specific for the amino acid tyrosine was determined to 2.7 Å resolution in the laboratory of David Blow in London. A schematic diagram (Figure 4.9) shows the first 320 residues of a single subunit of this dimeric molecule. The last 100 residues are disordered in the crystal and not visible. There are essentially two different domains, one α/β domain (red and green in Figure 4.9) that binds ATP and tyrosine and one α-helical domain (blue), the function of which is not known. A topological diagram is also given in Figure 4.9.

Let us now apply the rules for predicting active sites of α/β structures to this topological diagram. The β sheet has six strands, one of which, number 1, is antiparallel to the others. The remaining five parallel β strands are arranged in a way rather similarly to the nucleotide-binding fold (Figure 4.1b) but here the strand order is 6 5 2 3 4. α helix 2,3 (which connects β strands 2 and 3) and α helix 3,4 are on one side of the β sheet (red helices in Figure 4.9), whereas α helices 4,5 and 5,6 are on the other side (green helices). The switch point is thus between β strands 2 and 5. We would predict that the active site is outside the carboxy end of β strands 2 and 5 and that the loop regions that connect these strands with their respective α helices participate in binding the substrates. These loop regions comprise residues 38–47 and 190–193, respectively. The active site has been identified in the crystal structure by diffusing tyrosine and ATP into the crystals. The enzyme molecules in the crystals are active so tyrosyl adenylate is formed and stays bound to the enzyme since no tRNA is present. The position of bound tyrosyl adenylate was determined from an electron density map of this complex.

The position of this ligand was determined just after the predictive rules were formulated. The region where it binds proved to be as predicted. Loop regions 38–47, after β strand 2, and 190–193, after β strand 5, line a cleft where the substrate binds (Figure 4.10). The phosphate and the sugar moieties are hydrogen bonded to the main chain nitrogen atoms of residues 38 and 192, respectively. This part of the substrate is thus very close to the switch point. The substrate straddles the edge of the β sheet so that the tyrosine and adenine ends

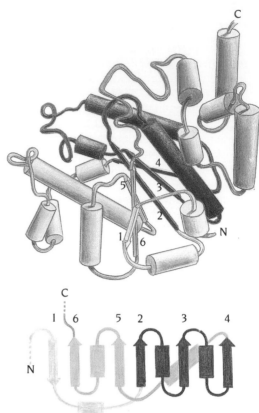

Figure 4.9 Schematic diagram of the enzyme tyrosyl-tRNA synthetase, which couples tyrosine to its cognate transfer RNA. The central region of the catalytic domain (red and green) is an open twisted α/β structure with five parallel β strands. The active site is formed by the loops from the carboxy ends of β strands 2 and 5. These two adjacent strands are connected to α helices on opposite sides of the β sheet. Where more than one α helix connects two β strands, for example, between strands 4 and 5, they are represented as one cylinder in the topology diagram. (Adapted from T. N. Bhat et al., *J. Mol. Biol.* 158:702, 1982.)

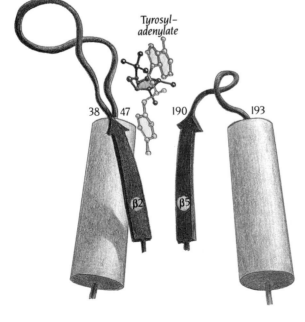

Figure 4.10 A schematic view of the active site of tyrosyl-tRNA synthetase. Tyrosyl-adenylate, the product of the first reaction catalyzed by the enzyme, is bound to two loop regions: residues 38–47, which form the loop after β strand 2, and residues 190–193, which form the loop after β strand 5. The tyrosine and adenylate moieties are bound on opposite sides of the β sheet outside the carboxy ends of β strands 2 and 5.

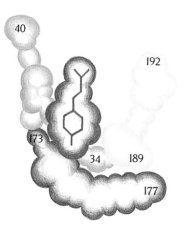

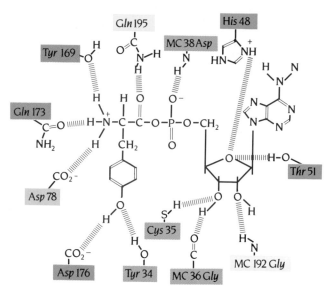

Figure 4.11 Schematic diagram of bound tyrosine to tyrosyl-tRNA synthetase. Colored regions correspond to van der Waals radii of atoms within a layer of the structure through the tyrosine ring. Red is bound tyrosine; green is the end of β strand 2 and the beginning of the following loop region; yellow is the loop region 189–192; and brown is part of the α helix in region 173–177.

Figure 4.12 Side chains of the tyrosyl-tRNA synthetase that form hydrogen bonds to tyrosyl adenylate. Green residues are from β strand 2 and the following loop regions, yellow from the loop after β strand 5, and brown from the α helix before β strand 5. (Adapted from T. Wells and A. Fersht, *Nature* 316: 656, 1985.)

are on opposite sides of the β sheet. The substrate also interacts with some of the other regions at this end of the β sheet, especially residues 173–177 of the α helix that connects β strands 4 and 5 and, in addition, some residues within β strand 2. Figure 4.11 shows a schematic diagram of the position of bound tyrosine (red) in relation to these regions of the protein, and Figure 4.12 gives the important hydrogen bonds to the substrate, knowledge of which formed the basis for the beautiful site-directed mutagenesis experiments on this system by Alan Fersht in London described in Chapter 16 of this book.

Carboxypeptidase is an α/β protein with mixed β sheet

Carboxypeptidases are zinc-containing enzymes that catalyze the hydrolysis of polypeptides at the C-terminal peptide bond. The bovine enzyme form A is a monomeric protein comprising 307 amino acid residues. The structure was determined in the laboratory of William Lipscomb, Harvard University, in 1970 and later refined to 1.5 Å resolution. Biochemical and x-ray studies have shown that the zinc atom is essential for catalysis by binding to the carbonyl oxygen of the substrate. This binding weakens the C´=O bond by abstracting electrons from the carbon atom and thus facilitates cleavage of the adjacent peptide bond. Carboxypeptidase is an unusually large single domain structure comprising a mixed β sheet of eight β strands (see Figure 4.13) with α helices on both sides. Some of the loop regions are very long and curl around the central theme of the structure.

The four central strands of the β sheet are parallel with strand order 8 5 3 4 (see Figure 4.13). If all other β strands are ignored, we can consider the strand order of these four parallel β strands to be 4 3 1 2. The strand order is thus reversed once, and there is a switch point in the middle of this β sheet between β strands 5 and 3 where we would expect the active site to be located.

This is precisely where the catalytically essential zinc atom is found. This zinc is firmly anchored to the protein by three side-chain ligands, His 69, Glu 72 and His 196 (Figure 4.14). The last residue of β strand 3 is residue 66. The two zinc ligands His 69 and Glu 72 are thus at the beginning of the loop region that connects this β strand with its corresponding α helix. The last residue of β strand 5 is the third zinc ligand, His 196.

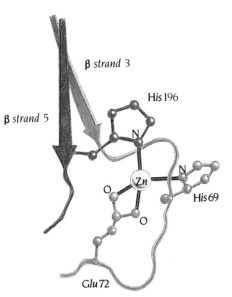

Figure 4.13 Schematic and topological diagrams for the structure of the enzyme carboxypeptidase. The central region of the mixed β sheet contains four adjacent parallel β strands, numbers 8, 5, 3, and 4, where the strand order is reversed between strands 5 and 3. The active-site zinc atom (yellow circle) is bound to side chains in the loop regions outside the carboxy ends of these two β strands. The first part of the polypeptide chain is red followed by green, blue, and brown. (Adapted from J. Richardson.)

In this structure the loop regions adjacent to the switch point do not provide a binding crevice for the substrate but instead accommodate the active-site zinc atom. The essential point here is that this zinc atom and the active site are in the predicted position outside the switch point for the four central parallel β strands, even though these β strands are only a small part of the total structure. This sort of arrangement, in which an active site formed from parallel β strands is flanked by antiparallel β strands, has been found in a number of other α/β proteins with mixed β sheets.

Arabinose-binding protein has two similar α/β domains

The arabinose-binding protein is one of the group of proteins that occur in the periplasmic space between the inner and outer cell membranes of Gram negative bacteria, such as *E. coli*. These proteins are components of active transport systems for various sugars, amino acids, and ions. Arabinose-binding protein is involved in arabinose transport. It is a single polypeptide chain of 306 amino acids folded into two domains of similar structure and topology (Figure 4.15) as was shown in the laboratory of Florante Quiocho in Houston, Texas, by a structure determination of the protein with bound arabinose at 1.7 Å resolution. The first five β strands in both domains are parallel; the strand order is reversed once, and the switch points are between β strands 1 and 3 in both domains.

There are a total of 12 amino acid residues from both domains involved in forming a complicated network of hydrogen bonds to the oxygen atoms of bound arabinose (Figure 4.16). We predict from the topology of these domains that loop residues from β strands 1 and 3 of both domains should participate in

Figure 4.14 Detailed view of the zinc environment in carboxypeptidase. The active-site zinc atom is bound to His 69 and Glu 72, which are part of the loop region outside β strand 2. In addition, His 196, which is the last residue of β strand 5, is also bound to Zn.

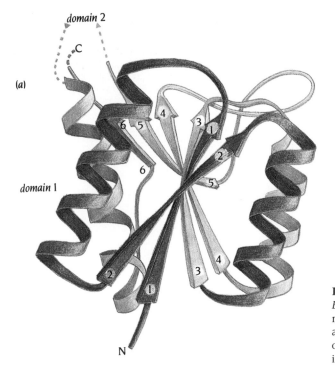

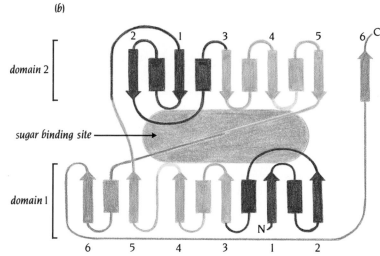

Figure 4.15 The polypeptide chain of the arabinose-binding protein in *E. coli* contains two open twisted α/β domains of similar structure. A schematic diagram of one of these domains is shown in (a). The two domains are oriented such that the carboxy ends of the parallel β strands face each other on opposite sides of a crevice in which the sugar molecule binds, as is illustrated in the topology diagram (b). [(a)Adapted from J. Richardson.]

these interactions. Five residues in loop regions after β strands 1 in both domains (pink color in Figure 4.16) and three residues after β strand 3 in the second domain (green color) participate in binding. The remaining four residues (yellow color) are all from loop regions after β strands 4 in both domains. These are one β strand removed from the switch point. Both domains thus participate to approximately equal extents in binding the sugar at the carboxy edge of the β sheets, outside the switch point where the strand order is reversed. A schematic diagram relating binding to the topology of the domains is shown in Figure 4.15b.

A number of proteins consist, like the arabinose-binding protein, of two α/β open-sheet domains formed from a single polypeptide chain. In almost all these cases the active sites are found in cleft regions between these two domains. The domains are oriented in such a way that the carboxy edge of both β sheets points toward the active site. Loop regions adjacent to the switch points of both domains participate in forming the active site. In enzymatic reactions where

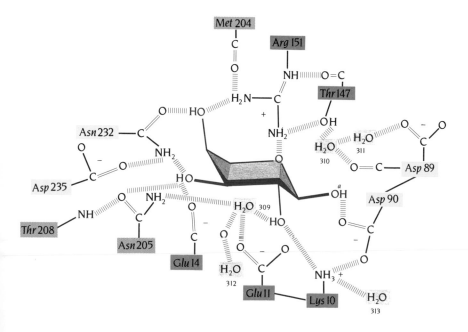

Figure 4.16 Schematic diagram of the complex networks of hydrogen bonds formed by polar side chains from the arabinose-binding protein and L-arabinose. The residues that interact with the sugar are in turn hydrogen bonded to each other or to other residues or isolated water molecules. The pink and green residues are in loop regions that from the topology diagram are predicted to form the binding site. The yellow residues are from adjacent loop regions. (Adapted from C. F. Sams et al., *Nature* 310:429, 1984.)

two different substrates participate, they are bound to different domains and brought together for catalytic reactions by the orientation of these domains. In other proteins the two domains bind different regions of the same ligand. Sugar-binding proteins from bacteria are examples of this second case.

Conclusion

Alpha/beta (α/β) structures are the most frequent and most regular of the protein structures. They fall into two classes: one with a central core of eight parallel β strands arranged close together, like staves, into a barrel that is surrounded by eight α helices and a second class that comprises an open twisted parallel or mixed β sheet with α helices on both sides of the β sheet.

The α/β-barrel structure is one of the largest and most regular of all domain structures, comprising about 250 amino acids. It has so far been found in more than 15 different proteins, with completely different amino acid sequences and different functions. They are all enzymes that are modeled on this common scaffold of eight parallel β strands surrounded by eight helices. They all have their active sites in very similar positions, at the bottom of a funnel-shaped pocket created by the loops that connect the carboxy end of the β strands with the amino end of the α helices. The specific enzymatic activity is, in each case, determined by the lengths and amino acid sequences of these loop regions, which do not contribute to the stability of the fold.

The open α/β-sheet structures vary considerably in size, number of β strands, and their strand order. Independent of these variations, they all have their active sites at the carboxy edge of the β strands and these active sites are lined by the loop regions that connect the β strands with the α helices. In this respect, they are similar to the α/β-barrel structures. The active-site region is, however, created differently in these structures. They are formed in those regions outside the carboxy edge of the β sheet where two adjacent loops are on opposite sides of the β sheet. The positions of these regions can be predicted from topology diagrams. The rules that relate the general position of functional binding sites to the overall structure of the protein are thus known for all α/β proteins.

Selected readings

General

Brändén, C.-I. Relation between structure and function of α/β proteins. *Q. Rev. Biophys.* 13: 317–338, 1980.

Cohen, F.E.; Sternberg, M.J.E.; Taylor, W.R. Analysis and prediction of the packing of α-helices against a β-sheet in the tertiary structure of globular proteins. *J. Mol. Biol.* 156: 821–862, 1982.

Farber, G.; Petsko, G.A. The evolution of α/β-barrel enzymes. *Trends Biochem. Sci.* 15: 228–234, 1990.

Fothergill-Gilmore, L.A. Domains of glycolytic enzymes. *In* Multidomain proteins—structure and evolution. D.G. Hardie; J.R. Coggins, eds. Pp. 85–174, Amsterdam: Elsevier, 1986.

Janin, J.; Chothia, C. Packing of α-helices onto β-pleated sheets and the anatomy of α/β-proteins. *J. Mol. Biol.* 143: 95–128, 1980.

Muirhead, H. Triose phosphate isomerase, pyruvate kinase and other α/β-barrel enzymes. *Trends Biochem. Sci.* 8: 326–330, 1983.

Richardson, J.S. β-sheet topology and the relatedness of proteins. *Nature* 268: 495–500, 1977.

Sternberg, M.J.E.; Thornton, J.M. On the conformation of proteins: the handedness of the connection between parallel β-strands. *J. Mol. Biol.* 110: 269–283, 1977.

Specific structures

Brick, P.; Bhat, T.N.; Blow, D.M. Structure of tyrosyl-tRNA synthetase refined at 2.3 Å resolution. Interaction of the enzyme with the tyrosyl adenylate intermediate. *J. Mol. Biol.* 208: 83–98, 1988.

Buisson, G., et al. Three-dimensional structure of porcine pancreatic α-amylase at 2.9 Å resolution. Role of calcium in structure and activity. *EMBO J.* 6: 3909–3916, 1987.

Campbell, J.W.; Watson, H.C.; Hodgson, G.I. Structure of yeast phosphoglycerate mutase. *Nature* 250: 301–303, 1974.

Carrel, H.L., et al. X-ray structure of D-xylose isomerase from *Streptomyces rubiginosus* at 4 Å resolution. *J. Biol. Chem.* 259: 3230–3236, 1984.

Dreusicke, D.; Karplus, P.A.; Schulz, G.E. Refined structure of porcine adenylate kinase at 2.1 Å resolution. *J. Mol. Biol.* 199: 359–371, 1988.

Farber, G.K.; Petsko, G.A.; Ringe, D. The 3.0 Å crystal structure of xylose isomerase from *Streptomyces olivochromogenes*. *Protein Eng.* 1: 459–466, 1987.

Gilliland, G.L.; Quiocho, F.A. Structure of the L-arabinose-binding protein from *Escherichia coli* at 2.4 Å resolution. *J. Mol. Biol.* 146: 341–362, 1981.

Goldman, A.; Ollis, D.L.; Steitz, T.A. Crystal structure of muconate lactonizing enzyme at 3 Å resolution. *J. Mol. Biol.* 194: 143–153, 1987.

Henrick, K.; Collyer, C.A.; Blow, D.M. Structures of D-xylose isomerase from *Arthrobacter* strain B3728 containing the inhibitors xylitol and D-sorbitol at 2.5 Å and 2.3 Å resolution, respectively. *J. Mol. Biol.* 208: 129–157, 1989.

Hofmann, B.E.; Bender, H.; Schulz, G.E. Three-dimensional structure of cyclodextrin glycosyltransferase from *Bacillus circulans* at 3.4 Å resolution. *J. Mol. Biol.* 209: 793–800, 1989.

Hyde, C.C., et al. Three-dimensional structure of the tryptophan synthase $\alpha 2 \beta 2$ multienzyme complex from *Salmonella typhimurium*. *J. Biol. Chem.* 263: 17857–17871, 1988.

Knight, S.; Andersson, I.; Brändén, C.-I. Crystallographic analysis of ribulose-1,5-bisphosphate carboxylase from spinach at 2.4 Å resolution. Subunit interactions and active site. *J. Mol. Biol.* 215: 113–160, 1990.

Lasters, I.; Wodak, S.J.; Pio, F. The design of idealized α/β-barrels: analysis of β-sheet closure requirements. Proteins 7: 249–256, 1990.

Lebioda, L.; Stec, B.; Brewer, J.M. The structure of yeast enolase at 2.5 Å resolution. An 8-fold $\beta + \alpha$ barrel with a novel $\beta\beta\alpha\alpha(\beta\alpha)_6$ topology. *J. Biol. Chem.* 264: 3685–3693, 1989.

Lesk, A. M.; Brändén, C.-I.; Chothia, C. Structural principles of α/β-barrel proteins: the packing of the interior of the sheet. *Proteins* 5: 139–148, 1989.

Lim, L.W., et al. Three-dimensional structure of the iron-sulfur flavoprotein trimethylamine dehydrogenase at 2.4 Å resolution. *J. Biol. Chem.* 261: 15140–15146, 1986.

Lindqvist, Y. Refined structure of spinach glycolate oxidase at 2 Å resolution. *J. Mol. Biol.* 209: 151–166, 1989.

Matsuura, Y., et al. Structure and possible catalytic residues of taka-amylase A. *J. Biochem* 95: 697–702, 1984.

Mavridis, I.M., et al. Structure of 2-keto-3-deoxy-6-phosphogluconate aldolase at 2.8 Å resolution. *J. Mol. Biol.* 162: 419–444, 1982.

Muirhead, H., et al. The structure of cat muscle pyruvate kinase. *EMBO J.* 5: 475–481, 1986.

Neidhart, D.J., et al. Mandelate racemase and muconate lactonizing enzyme are mechanistically distinct and structurally homologous. *Nature* 347: 692–694, 1990.

Ohlsson, I.; Nordström, B.; Brändén, C.-I. Structural and functional similarities within the coenzyme binding domains of dehydrogenases. *J. Mol. Biol.* 89: 339–354, 1974.

Priestle, J.P., et al. Three-dimensional structure of the bifunctional enzyme N-(5′-phosphoribosyl) anthranilate isomerase-indole-3-glycerol-phosphate synthase from *Escherichia coli*. *Proc. Natl. Acad. Sci. USA* 84: 5690–5694, 1987.

Rees, D.C.; Lewis, M.; Lipscomb, W.N. Refined crystal structure of carboxypeptidase A at 1.54 Å resolution. *J. Mol. Biol.* 168: 367–387, 1983.

Rey, F., et al. Structural analysis of the 2.8 Å model of xylose isomerase from *Actinoplanes missouriensis*. *Proteins* 4: 165–172, 1988.

Schneider, G.; Lindqvist, Y.; Lundqvist, T. Crystallographic refinement and structure of ribulose-1,5-bisphosphate carboxylase from *Rhodospirillum rubrum* at 1.7 Å resolution. *J. Mol. Biol.* 211: 989–1008, 1990.

Steitz, T.A., et al. High resolution x-ray structure of yeast hexokinase, an allosteric protein exhibiting a non-symmetric arrangement of subunits. *J. Mol. Biol.* 104: 197–222, 1976.

Sternberg, M.J.E., et al. Analysis and prediction of structural motifs in the glycolytic enzymes. *Phil. Trans. R. Soc. Lond.* B293: 177–189, 1981.

Sygusch, J.; Beaudry, D.; Allaire, M. Molecular architecture of rabbit skeletal muscle aldolase at 2.7 Å resolution. *Proc. Natl. Acad. Sci. USA* 84: 7846–7850, 1987.

Xia, Z.-X., et al. Three-dimensional structure of flavocytochrome b_2 from baker's yeast at 3.0 Å resolution. *Proc. Natl. Acad. Sci. USA* 84: 2629–2633, 1987.

Antiparallel Beta Structures

5

Antiparallel beta (β) structures comprise the second large group of protein domain structures. Functionally, this group is the most diverse; it includes enzymes, transport proteins, antibodies, and virus coat proteins. The cores of these domains are built up by a number of β strands that can vary from four or five to over ten. The β strands are arranged in a predominantly antiparallel fashion and usually in such a way that they form two β sheets that are joined together and packed against each other.

The β sheets have the usual twist, and when two such twisted β sheets are packed together, they form a barrel-like structure (Figure 5.1). Antiparallel β structures, therefore, in general have a core of hydrophobic side chains inside the barrel provided by residues in the β strands. The surface is formed by residues from the loop regions and from the strands. The aim of this chapter is to examine a number of antiparallel β structures and demonstrate how these rather complex structures can be separated into smaller comprehensible motifs.

Figure 5.1 The enzyme superoxide dismutase (SOD). SOD is a β structure comprising eight antiparallel β strands (a). In addition, SOD has two metal atoms, Cu and Zn (yellow circles), that participate in the catalytic action; conversion of the superoxide radical to hydrogen peroxide and oxygen. The eight β strands are arranged around the surface of a barrel, which is viewed along the barrel axis in (b) and perpendicular to this axis in (c). [(a) Adapted from J. S. Richardson. The structure of SOD was determined in the laboratory of J. S. and D. R. Richardson, Duke University.]

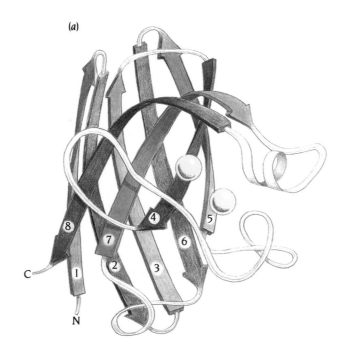

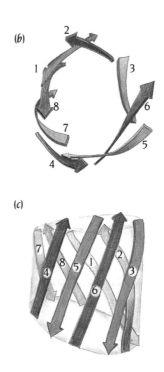

In Chapter 2 we described the twelve different ways that two β-loop-β units can form a four-stranded β sheet. The number of possible ways to form anti-parallel β-sheet structures rapidly increases as the number of strands increases. It is thus surprising, but reassuring, that the number of topologies actually observed is small and that most β structures fall into a few groups of common or similar topology. The three most frequently occurring groups—up-and-down barrels, Greek key, and jelly roll barrels—can all be related to simple ways of connecting antiparallel β strands arranged in a barrel structure.

Up-and-down barrels have a simple topology

The simplest topology is obtained if each successive β strand is added adjacent to the previous strand until the barrel is closed and the last strand is joined by hydrogen bonds to the first strand (Figure 5.2). These are called **up-and-down β sheets** or **barrels**. The arrangement of β strands is similar to that in the α/β-barrel structures we have just described in Chapter 4, except that here the strands are antiparallel and all the connections are hairpins. The structural and functional versatility of even this simple arrangement will be illustrated by two examples.

Retinol-binding protein folds into an up-and-down β barrel

The first example is the plasma-borne **retinol-binding protein**, RBP, which is a single polypeptide chain of 182 amino acid residues. This protein is responsible for transporting the lipid alcohol **vitamin A** (retinol) from its storage site in the liver to the various vitamin-A-dependent tissues. It is a disposable package in the sense that each RBP molecule transports only a single retinol molecule and is then degraded.

RBP is synthesized in the hepatocytes, where it picks up one molecule of retinol in the endoplasmic reticulum. Both its synthesis and its secretion from the hepatocytes to the plasma is regulated by retinol. In plasma, the RBP-retinol complex binds to a larger protein molecule, prealbumin, which further stabi-

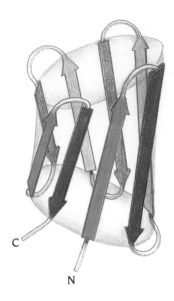

Figure 5.2 Schematic and topological diagrams of an up-and-down β barrel. The eight β strands are all antiparallel to each other and are connected by hairpin loops. β strands that are adjacent in the amino acid sequence are also adjacent in the three-dimensional structure of up-and-down barrels.

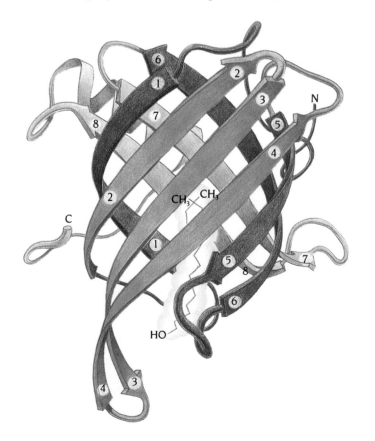

Figure 5.3 Schematic diagram of the structure of human plasma retinol-binding protein (RBP), which is an up-and-down β barrel. The eight antiparallel β strands twist and curl such that the structure can also be regarded as two β sheets (green and blue) packed against each other. Some of the twisted β strands (red) participate in both β sheets. A retinol molecule, vitamin A (yellow), is bound inside the barrel, between the two β sheets, such that its only hydrophilic part (an HO tail) is at the surface of the molecule. The topological diagram of this structure is the same as that in Figure 5.2. (Courtesy of Alwyn Jones, Uppsala, Sweden.)

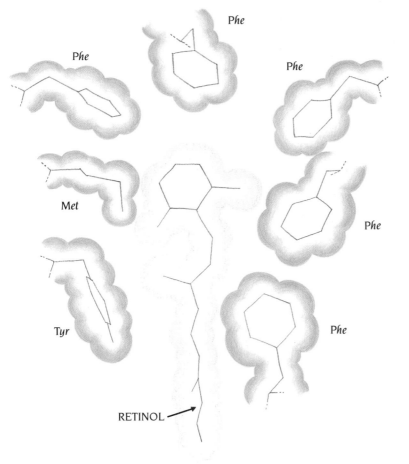

Figure 5.4 The binding site for retinol inside the RBP barrel is lined with hydrophobic residues. They provide a hydrophobic surrounding for the hydrophobic part of the retinol molecule.

lizes it and prevents its loss via the kidney. Recognition of this complex by a cell-surface receptor causes RBP to release the retinol and, as a result, to undergo a conformational change that drastically reduces its affinity for prealbumin. The free RBP molecule is then excreted through the kidney glomerus, reabsorbed in the proximal tubule cells, and degraded.

Retinol is bound inside the β barrel

The structure of RBP with bound retinol has been determined in the laboratory of Alwyn Jones in Uppsala, Sweden, to 2.0 Å resolution. Its most striking feature is a β-barrel core consisting of eight up-and-down antiparallel β strands as shown in Figure 5.3. In addition, there is a four-turn α helix at the carboxy end of the polypeptide chain that is packed against the outside of the β barrel. The barrel is wrapped around the retinol molecule. One end of the barrel is open to the solvent whereas the other end is closed by tight side-chain packing. The tail of the retinol molecule is at the open end of the barrel.

The β strands are curved and twisted, providing space inside the barrel to accommodate a retinol molecule. This hydrophobic molecule is packed against hydrophobic side chains from the β strands in the barrel's core (Figure 5.4). The removal of the retinol molecule from this structure would thus leave a large empty hole in the middle of the core. It is not known whether or not the barrel is stable without retinol or whether a structural change is required to remove this hole by bringing the sheets closer together. This will remain unknown until the structure of RBP minus retinol is determined.

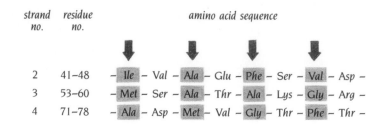

strand no.	residue no.	amino acid sequence							
2	41–48	– Ile –	Val –	Ala –	Glu –	Phe –	Ser –	Val –	Asp –
3	53–60	– Met –	Ser –	Ala –	Thr –	Ala –	Lys –	Gly –	Arg –
4	71–78	– Ala –	Asp –	Met –	Val –	Gly –	Thr –	Phe –	Thr –

Table 5.1 Amino acid sequence of β strands 2 3 4 in human plasma retinol-binding protein

The sequences are listed in such a way that residues which point into the barrel are aligned. These hydrophobic residues are arrowed and colored green. The remaining residues are exposed to the solvent.

Amino acid sequence reflects β structure

On a large part of the surface of RBP (the front face in Figure 5.3), side chains from residues in the β strands are exposed to the solvent. This is achieved by alternating hydrophobic with polar or charged hydrophilic residues in the amino acid sequences of the β strands; in other words, side chains of the β strands form the hydrophobic core of the barrel as well as part of the hydrophilic outer surface. Strands 2 3 4 of RBP clearly illustrate this arrangement (see Table 5.1 where the core residues are colored green). This structure also very clearly illustrates that an antiparallel β barrel is built up from two β sheets that are packed against each other (see Figure 5.3). β strands 1 2 3 4 5 6 (blue and red color) form one sheet, and strands 1 8 7 6 5 (green and red color) form the second sheet. Strands 1 5 6 thus contribute to both sheets by having sharp corners where they can turn over from one sheet to the other.

The retinol-binding protein belongs to a superfamily of protein structures

RBP is one member of a **superfamily** of proteins with different functions, marginally homologous amino acid sequences, but similar three-dimensional structures. The x-ray structures have been determined for other members of this superfamily; an insect protein that binds a blue pigment, biliverdin, and β-lactoglobulin, a protein of unknown function that is abundant in milk. All three proteins have polypeptide chains of approximately the same lengths that are wrapped into very similar up-and-down eight-stranded antiparallel β barrels. They all tightly bind hydrophobic ligands inside this barrel.

There is also a second superfamily of small lipid-binding proteins, the **P2 family**, which include among others cellular retinol and fatty acid binding proteins as well as a protein, P2, from myelin in the peripheral nervous system. Members of this family of proteins show no amino acid sequence homology to members of the RBP superfamily. Nevertheless, their three-dimensional structures have similar architecture and topology, up-and-down β barrels. This second family has, however, ten antiparallel β strands in their barrels compared to the eight strands found in the barrels of the RBP superfamily.

Retinol binding in humans and biliverdin binding in insects show evolutionary relationship

Several species of insects, for example, cabbage white butterflies and tobacco horn worms, use **biliverdin-binding proteins**, BBP, as one component in camouflage which is an essential ingredient in their survival strategy. The remarkable matching of the color of these insects to their background arises from a combination of two pigments: one yellow from carotenoids and one blue from biliverdins. The latter are open tetrapyrrole chains, derived from heme groups. Both types of pigments are present as complexes with specific pigment-binding proteins.

The x-ray structures of two different biliverdin-binding proteins have been determined, one from the butterfly *Pieris brassicae* to 2.35 Å resolution in the laboratory of Robert Huber in Munich, Germany, and a second from the tobacco hornworm *Manduca sexta* to 2.6 Å resolution in the laboratory of Ivan Rayment in Tucson, Arizona. These structures are very similar and contain the same structural motif as RBP: an eight-stranded up-and-down β barrel, flanked by an

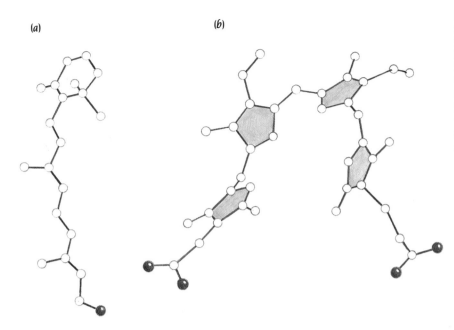

(a)   (b)

Figure 5.5 Comparison of the conformations of (a) retinol and (b) biliverdin bound inside the up-and-down β barrels of RBP and insect biliverdin-binding protein, respectively. Biliverdin is a degradation product of heme with a linear chain of four tetrapyrrole rings (green). The polar ends (red) of biliverdin are at the open end of the barrel. The structures of the two binding proteins are quite similar, but they can, nevertheless, accommodate two quite different ligands inside their barrels.

α helix. Even though the amino acid sequences of both RBP and the biliverdin-binding protein from *Manduca sexta* were known prior to the x-ray structure determination, the weak sequence homology of 17% identity between these two proteins was not recognized as significant until the similarity of their x-ray structures had been revealed.

The chromophore biliverdin binds inside the barrel in the same region that retinol is bound inside the barrel of RBP. Although it is a linear chain of four tetrapyrrole rings (Figure 5.5), biliverdin does not bind in an extended conformation, like retinol, but in a compact conformation, much like heme. By contrast, when it is acting as a photoreceptor in algal **phycobilisomes**, biliverdin does bind in an extended conformation. In these photosynthetic organelles, the protein that binds biliverdin has a completely different structure, being an α-helical protein with the globin fold. The different conformations of the biliverdin molecule when it is bound to the insect BBP and to algal phycobilisomes is an example of how a protein can influence the properties of a pigment by providing it with a specific chemical context. In the phycobilisomes the biliverdin molecules are arranged in arrays that capture photons and funnel them to the photosynthetic reaction center where the light energy is converted to chemical energy as described in Chapter 13. In insects the same pigment provides a protective color.

The structural similarity and the amino acid sequence homology between RBP and the biliverdin-binding protein strongly indicate an evolutionary relationship between them. The common function of these two proteins is to bind large hydrophobic molecules, for which the up-and-down β-barrel structural motif seems to be particularly suited. This may help assign functions to the other members of this superfamily; β-lactoglobulin from milk and protein HC, a glycoprotein present in high concentrations in both blood and urine of patients on renal dialysis.

Structure suggests function for β-lactoglobulin

The structure of β-**lactoglobulin** from bovine milk was determined to 2.8 Å resolution by Lindsay Sawyer in Edinburgh and Tony North in Leeds, UK, in 1985. The structural similarity to RBP was so unexpected that it was not immediately appreciated as soon as the structure was solved. In fact, it was only first recognized by David Phillips, Oxford, when the structure was presented in public.

Significant amino acid sequence homology, amounting to 24% identity, between human RBP and bovine β-lactoglobulin was independently reported at the same time, in 1985, some four years after the two protein sequences became available. It was then realized that protein HC, which had also been sequenced in 1981, exhibited a weak, 18%, sequence identity to both RBP and β-lacto-globulin. These accidental discoveries are excellent illustrations of the combined use of protein structure and DNA and protein sequence databases. Weak amino acid sequence similarities, such as that between HC and RBP, are by themselves not significant and quite meaningless. However, when combined with a three-dimensional structure, they can become highly significant and provide important biological information if the sequence similarity conserves crucial structural properties.

The observation that β-lactoglobulin is evolutionarily related to RBP prompted an investigation of the functional consequences of its ability to bind retinol. It was previously known that *in vitro* at least β-lactoglobulin binds retinol more tightly than RBP. The studies strongly indicated that specific receptors for the β-lactoglobulin-retinol complex exist in the intestine of young calves. These structural comparisons have thus provided important clues as to the function of milk β-lactoglobulin, which appears to participate in the transport and uptake of vitamin A in suckling animals.

Neuraminidase folds into up-and-down β sheets

The second example of up-and-down β sheets is the protein **neuraminidase** from **influenza virus**. Here the packing of the sheets is different from that in RBP. They do not form a simple barrel but instead six small sheets, each with four β strands, form a superbarrel, with the loops at one end forming an enzyme active site.

Influenza virus is an RNA virus with an outer lipid envelope. There are two viral proteins anchored in this membrane, a neuraminidase and a hemagglutinin. They are both transmembrane proteins with a few residues inside the membrane, a transmembrane region, and a stalk and a headpiece outside the membrane. The heads are exposed on the surface of the virion and thus provide the antigenic determinants of this epidemic virus. The function of hemagglutinin, which is glycosylated, is to mediate the binding of virus particles to host cells by recognizing and binding to sialic acid residues on cellular membrane glycoproteins, as we shall discuss in more detail at the end of this chapter.

The role of the viral neuraminidase, conversely, seems to be to facilitate the release of progeny virions from infected cells by cleaving sialic acid residues from the carbohydrate side chains both of the viral hemagglutinin and of the glycosylated cellular membrane proteins. This helps prevent progeny virions from binding to and reinfecting the cells from whose surface they have just budded. From the point of view of the viruses, reinfecting an already infected cell is, of course, a waste of time.

The neuraminidase molecule is a homotetramer made up of four identical polypeptide chains, each of around 470 amino acids; the exact number varies depending on the strain of the virus. If influenza virus is treated with the proteolytic enzyme pronase, the head of the neuraminidase, which is soluble, is cleaved off from the stalk projecting from the viral envelope. The soluble head comprising four subunits, each of about 400 amino acids, can be crystallized.

Folding motifs form a superbarrel in neuraminidase

The structure of these tetrameric neuraminidase heads was determined in the laboratory of Peter Colman in Parkville, Australia, to 2.9 Å resolution. Each of the four subunits of the tetramer is folded into a single domain built up from six closely packed, similarly folded motifs. The motif is a simple up-and-down antiparallel β sheet of four strands (Figure 5.6). The strands have a rather large twist such that the directions of the first and the fourth strands differ by 90°.

To a first very rough approximation the six motifs are arranged within each subunit with an approximate sixfold symmetry around an axis through the

Figure 5.6 Schematic and topological diagrams of the folding motif in neuraminidase from influenza virus. The motif is built up from four antiparallel β strands joined by hairpin loops, an up-and-down open β sheet.

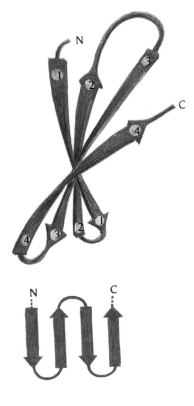

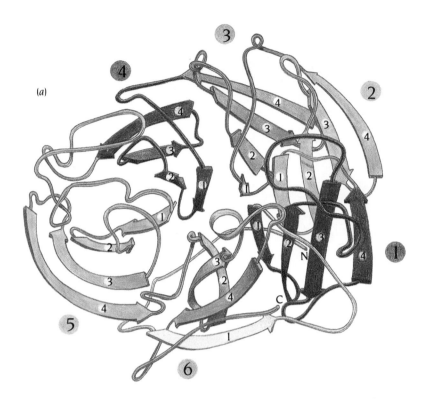

(a)

(b)

1 2 3 4   1 2 3 4   1 2 3 4   1 2 3 4   1 2 3 4   2 3 4 1

   C N

①   ②   ③   ④   ⑤   ⑥

Figure 5.7 The subunit structure of the neuraminidase headpiece (residues 84–469) from influenza virus is built up from six similar, consecutive motifs (Figure 5.6) of four up-and-down antiparallel β strands. The motifs are connected by loop regions from β strand 4 in one motif to β strand 1 in the next motif. The schematic diagram (a) is viewed down an approximate sixfold axis that relates the centers of the motifs. Four such superbarrel subunits are present in each complete neuraminidase molecule (see Figure 5.8). In the topological diagram (b) the yellow loop that connects the N-terminal β strand to the first β strand of motif 1 is not to scale. In the folded structure it is about the same length as the other loops that connect the motifs. (Adapted from J. Varghese et al., *Nature* 303: 35, 1983.)

center of the subunit (see Figure 5.7a). These six β sheets can be regarded as six motifs that form a **superbarrel**.

This is much the biggest structure we have discussed thus far. The whole molecule has almost 1600 amino acid residues. It is composed of four identical polypeptide chains, each of which is folded into a superbarrel with 24 β strands (see Figure 5.8). These 24 β strands are arranged in six similar motifs, each of which contains 4 β strands that form the staves of the superbarrel.

The active site is at one end of the superbarrel

Not only are the topologies of the six β sheets in each subunit identical, but so are their connections to each other, with the exception of the last β sheet (see Figure 5.7b). The fourth strand of each β sheet is connected across the top of the subunit (as seen in Figure 5.7a) to the first strand of the next sheet. This loop enters the β sheet at the same edge of the sheet as the loop that connects strands 2 and 3 within the sheet.

Furthermore, because of the approximate sixfold symmetry of the β-sheet motifs, these 12 loop regions, derived from the six β sheets, are at the same end of the molecule, as can be seen in Figure 5.9a, where we see a single polypeptide chain (one of the four subunits) from the side of the superbarrel. Together they form a wide funnel-shaped pocket (Figure 5.9b) where the active site is located. The β sheets are arranged cyclically around an axis through the center of the molecule. The loop regions at the top of this barrel are extensive (Figure 5.9a)

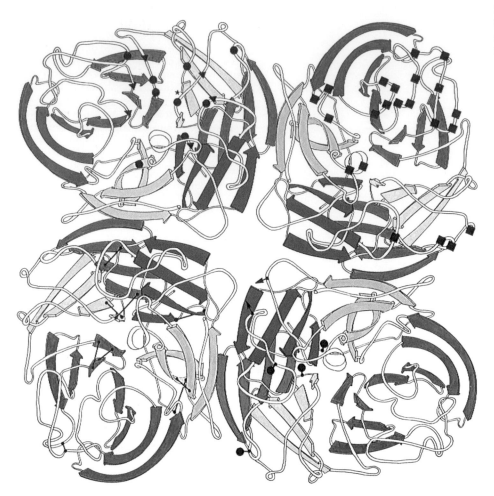

Figure 5.8 Schematic view down the fourfold axis of the tetrameric molecule of neuraminidase as it appeared on the cover of *Nature*, May 5, 1983.

and form a wide funnel-shaped active site analogous to the active site formed by the loop regions at the top of the α/β-barrel structures.

Greek key motifs occur frequently in antiparallel β structures

We saw in Chapter 2 that the **Greek key motif** provides a simple way to connect antiparallel β strands that are on opposite sides of a barrel structure. We will now look at how this motif is incorporated into some of the simple antiparallel β-barrel structures and show that an antiparallel β sheet of eight strands can be built up only by hairpin and/or Greek key motifs, if the connections do not cross between the two ends of the β sheet.

Assume that we have eight antiparallel β strands arranged in a barrel structure. We decide that we want to connect strand number n to an antiparallel strand at the same end of the barrel. We do not want to connect it to strand number n + 1 as in the up-and-down barrels just described, nor do we want to connect it to strand number n – 1 which is equivalent to turning the up-and-down barrel in Figure 5.2 upside down. What alternatives remain?

It is easy to see from Figure 5.10 that there are only two alternatives. We can connect it either to strand number n + 3 or to n – 3. Both cases require only short loop regions that traverse the end of the barrel. How do we now continue the connections? The simplest way to connect the strands that were skipped over is to join them by up-and-down connections, as illustrated in Figure 5.10. We have now connected four adjacent strands of the barrel in a simple and logical fashion requiring only short loop regions. The result is the Greek key motif described in Chapter 2, which is found in a large majority of antiparallel β structures. The two cases represent the two possible different hands, but in all known structures one observes the hand that corresponds to the case where β strand n is linked to β strand n + 3 as in Figure 5.10a.

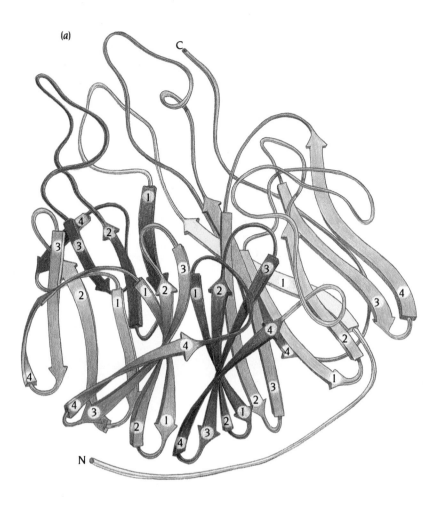

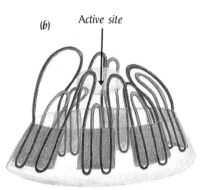

Figure 5.9 The 24 β strands of the six motifs in a single subunit of neuraminidase form the wall of a superbarrel structure. A schematic diagram of the subunit structure shows the superbarrel viewed from its side (a). An idealized superbarrel structure is shown in (b). The loop regions that connect the motifs (red in b) in combination with the loops that connect strands 2 and 3 within the motifs (green in b) form a wide funnel-shaped active site pocket. [(a) Adapted from P. Colman et al., *Nature* 326: 358, 1987.]

The remaining four strands of the barrel can be joined either by up-and-down connections before and after the motif or by another Greek key motif. We will examine examples of both cases.

The γ-crystallin molecule has two domains

The transparency and refractive power of the lenses of our eyes depend on a smooth gradient of refractive index for visible light. This is achieved partly by

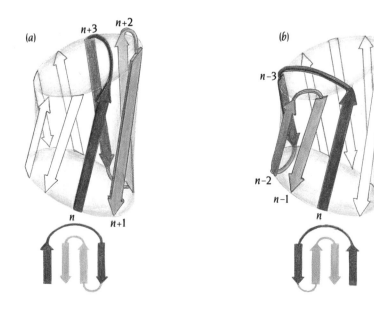

Figure 5.10 Idealized diagrams of the Greek key motif. This motif is formed when one of the connections of four antiparallel β strands is not a hairpin connection. The motif occurs when strand number n is connected to strand n + 3 (a) or n − 3 (b) instead of n + 1 or n − 1 in an eight-stranded antiparallel β sheet or barrel. The two different possible connections give two different hands of the Greek key motif. In all protein structures known so far only the hand shown in (a) has been observed.

a regular packing arrangement of the cells in the lens and partly by a smoothly changing concentration gradient of lens-specific proteins, the crystallins.

There are at least three different classes of crystallins. The α and β crystallins are heterogeneous assemblies of different subunits, whereas the **gamma (γ) crystallins** are monomeric proteins with a polypeptide chain of around 170 amino acid residues. The structure of one such γ crystallin has been determined in the laboratory of Tom Blundell in London to 1.9 Å resolution. A picture of this molecule generated from a graphics display is shown in Figure 5.11.

Let us now examine this molecule and dissect it into its structural components to see if we can understand how these are put together. We will reduce this rather complex, and at first sight bewildering, structure to its simplest representation as a series of motifs. This will help us to understand the structure and see its relationships to other structures.

We can immediately discern from Figure 5.11 that the molecule is divided into two clearly separated domains that seem to be of similar size. For the next step we would need a stereopicture of the model or, much better, a graphics display where we could manipulate the model and look at it from different views. Here instead we have made a schematic diagram of one domain (Figure 5.12), which is normally not done until the analysis is completed and the structural principles are clear.

The domain structure has a simple topology

We will now follow the main polypeptide chain and trace out a **topological diagram** for this domain. We can immediately see from Figure 5.12 that the only secondary structure in the molecule is the β strands, which are arranged in an antiparallel fashion into two separate β sheets. β strands 1 2 4 7 form one antiparallel β sheet with the strand order 2 1 4 7. We thus draw the left four arrows in Figure 5.13 and connect strands 1 and 2. Similarly, we see that β strands 3 5 6 8 form another antiparallel β sheet with the strand order 6 5 8 3. We notice that strands 7 and 6 are adjacent although not hydrogen bonded to each other on the back side of the domain. We thus position strand 6 adjacent to strand 7 in the topology diagram but make a space between them to indicate that they belong to different β sheets. Alternatively, we could have positioned strands 2 and 3 adjacent to each other, which would have given a topologically identical diagram. We then connect the strands in consecutive order along the polypeptide chain.

Two Greek key motifs form the domain

The topological diagram of Figure 5.13 has been drawn to reflect the observation that the two β sheets are separate: β strands 2 and 3 are not hydrogen bonded to each other, nor are strands 6 and 7. However, the connections look unnecessarily complicated, almost like the route of a jet set scientist on a lecture tour. But notice from the schematic diagram of the domain in Figure 5.12 that the two β sheets are packed against each other such that they form a distorted barrel. To see if the diagram can be simplified, we idealize the barrel and plot the strands along the surface of the barrel as shown in Figure 5.14. It is then immediately obvious that strands number 1 2 3 4 form a Greek key motif, as do strands number 5 6 7 8. These two motifs are joined by a loop across the bottom of the barrel, between strands number 4 and 5.

On the basis of this new insight we can draw the topology diagram shown in the left half of Figure 5.15b. What is the difference between this and the previous topological diagram we made? The only change we have made is to move β strand 3 from the right edge to the left edge of the domain topology and to close the gap between strands 7 and 6. We have changed neither the strand order nor the connections between the strands; thus the two diagrams are topologically identical.

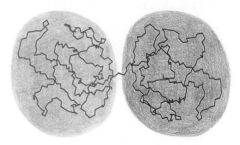

Figure 5.11 A computer-generated diagram of the structure of γ crystallin comprising one polypeptide chain of 170 amino acid residues. The diagram illustrates that the polypeptide chain is arranged in two domains (blue and red). Only main chain (N, C´, Cα) atoms and no side chains are shown.

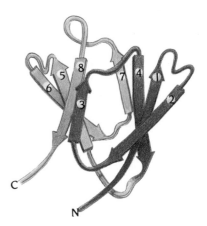

Figure 5.12 Schematic diagram of the path of the polypeptide chain in one domain (the blue region in Figure 5.11) of the γ-crystallin molecule. The domain structure is built up from two β sheets of four antiparallel β strands, sheet 1 from β strands 1 2 4 7 and sheet 2 from strands 3 5 6 8.

Figure 5.13 A preliminary topological diagram of the structure of one domain of γ crystallin shown in Figure 5.12, illustrating that the two β sheets are separate within the domain.

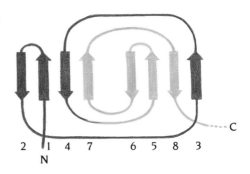

The two domains have identical topology

Using a graphics display, we could do the same thing for the second domain and arrive at the full topology diagram in Figure 5.15. From this diagram it is obvious that the two domains have identical topology and thus in all probability similar structures. This realization is not at all trivial. To be able to see it when one looks at the structure on the display requires considerable experience because the two domains are in different orientations in the molecule. The brain therefore has to store the image of one domain while examining different orientations of the second domain. A topology diagram, on the other hand, immediately reveals similarities in domain structures. This illustrates one very important use of topology diagrams—namely, to reduce a complicated pattern to a simpler one, one from which conclusions can be drawn that are valid also for the complicated pattern.

The two domains have similar structures

A relevant question to ask at this stage is, do the topological identities displayed in the diagram reflect structural similarity? We can now see that topologically the polypeptide chain is divided into four consecutive Greek key motifs arranged in two domains. How similar are the domain structures to each other, and how similar are the two motifs within each domain?

Tom Blundell has answered these questions by superposing the C_α atoms for both the domains and the motifs with each other. For each pair of motifs he found that 40 C_α atoms from each motif superpose with a mean distance of 1.4 Å. These 40 C_α atoms within each motif are therefore structurally equivalent.

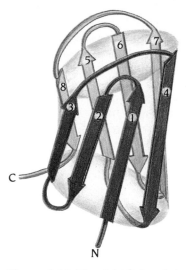

Figure 5.14 The eight β strands in one domain of the crystallin structure are in this idealized diagram drawn along the surface of a barrel. From this diagram it is obvious that the β strands are arranged in two Greek key motifs, one (red) formed by strands 1–4 and the other (green) by strands 5–8. Notice that the β strands that form one motif contribute to both β sheets as shown in Figure 5.12.

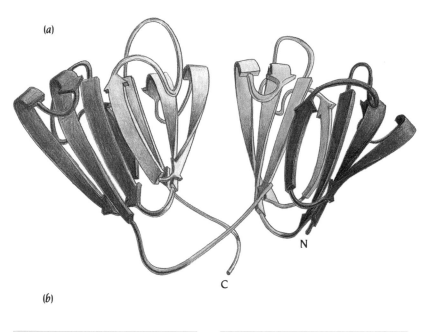

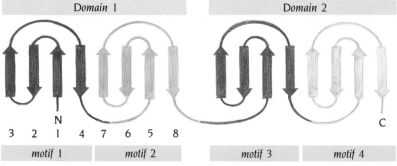

Figure 5.15 Schematic diagram (a) and topology diagram (b) for the γ-crystallin molecule. The two domains of the complete molecule have the same topology; each is composed of two Greek key motifs that are joined by a short loop region. [(a) Adapted from T. Blundell, et al. *Nature* 289: 771–777, 1981.]

These comparisons show that the structures of the complete motifs are very similar since each motif comprises only 43 or 44 amino acid residues in total. Not only are the individual motifs similar in structure, but they are also pairwise arranged into the two domains in a similar way since superposition of the domains showed that about 80 C_α atoms of each domain were structurally equivalent.

This structural similarity is also reflected in the amino acid sequences of the domains that show 40% identity. They are thus clearly homologous to each other. The motif structures within the domains superpose equally well but here the sequence homology is less, around 30% between motifs 1 and 2 and 20% between 3 and 4. This study, however, clearly shows that the topological description in terms of four Greek key motifs is also valid at the structural and amino acid sequence levels.

The Greek key motifs are evolutionarily related in γ crystallin

These comparisons strongly suggest that the four Greek key motifs are evolutionarily related. We can guess from the amino acid sequence comparison that this protein evolved in two stages, beginning with the duplication of a primordial gene coding for one motif of about 40 amino acid residues, followed by fusion of the duplicated genes to give a single gene encoding one domain. The gene for this domain we may imagine, later duplicated in turn and fused to give the full gene for the present γ-crystallin polypeptide. The evidence that this was the second step lies in the fact that the amino acid sequence homology is greater between the domains than between the motifs within each domain.

Intron positions separate the four Greek key motifs

There is some circumstantial evidence in the organization of the crystallin gene for the evolutionary history that we have reconstructed. The crystallin gene, like many eucaryotic genes, has its coding sequence in separate pieces (**exons**) interrupted by noncoding DNA (**introns**). When this feature of eucaryotic genes was first discovered, Walter Gilbert at Harvard University suggested in 1978 that genes for large proteins might have evolved by the accidental juxtaposition of exons coding for specific functions or structural motifs or even domains.

The gene of the first crystallin that was examined, that of a mouse β crystallin, has four exons. The amino acid sequence of the protein is homologous to that of γ crystallin and shows the same four homologous motifs. The three intron positions are at the junctions between the motifs. Here we have a case where all four motifs are coded in separate exons, supporting Gilbert's ideas about a correlation between exons and structural motifs. The introns at these positions could, therefore, be evolutionary remnants of the gene duplication and fusion events.

The Greek key motifs can form jelly roll barrels

In antiparallel barrel structures with the Greek key motif one of the connections in the motif is made across one end of the barrel. Such connections can be made several times in a β barrel giving different variations and combinations of the Greek key motif. In the structure of crystallin there are two consecutive Greek key motifs that form a barrel with two such connections. There is a different but frequently occurring motif, the **jelly roll motif**, in which there are four connections of this type. It is called jelly roll, or Swiss roll, because the polypeptide chain is wrapped around a barrel core like a jelly roll. This motif has been found in a variety of different structures including the coat proteins of most of the spherical viruses examined thus far by x-ray crystallography, the plant lectin concanavalin A and the hemagglutinin protein from the influenza virus.

The jelly roll motif is wrapped around a barrel

To illustrate how this rather complicated structure is built up, we will start by wrapping a piece of string around a barrel as shown in Figure 5.16. The string goes up and down the barrel four times, crosses over once at the bottom and twice at the top of the barrel. This configuration is the basic pattern for the jelly roll motif.

Let us do the same thing with a strip of paper the width of which is approximately one-eighth of the circumference of the top of the barrel. We assume that a polypeptide chain follows the edges of this strip, starting at the bottom right corner of the strip and ending at the bottom left corner (Figure 5.17a). The polypeptide chain has eight straight sections, β strands, interrupted by loop regions. The β strands are arranged in a long anti-parallel hairpin such that strand 1 is hydrogen bonded to strand 8, 2–7 and so on.

We now wrap the strip around the barrel following the path of the string in Figure 5.16 and in such a way that the β strands go along the sides of the barrel and the loop regions form the connections at the top and bottom of the barrel (Figure 5.17b).

The hydrogen-bonded antiparallel β strand pairs 1:8, 2:7, 3:6, and 4:5 are now arranged such that β strand 1 is adjacent to strand 2; 7 is adjacent to 4; 5 to 6; and 3 to 8. These can also form hydrogen bonds to each other. All adjacent β strands are antiparallel. This is the basic jelly roll β-barrel structure for eight β strands (Figure 5.18a). Most such barrels have eight strands, but any even number of strands greater than four can form a jelly roll barrel. In eight-stranded barrels there are two connections across the top of the barrel and two across the bottom. In addition, there are two connections between adjacent β strands at the top and one at the bottom. A topological diagram of this fold is given in Figure 5.18b.

The jelly roll barrel is thus conceptually simple, but it can be quite puzzling if it is not considered in this way. These structures will be extensively discussed and exemplified in this chapter by hemagglutinin, in Chapter 7 by the DNA-binding protein CAP, and in Chapter 11 by viral coat proteins.

The jelly roll barrel is usually divided into two sheets

The barrels we have used to illustrate both the Greek key and the jelly roll structures provide a topological description. As such, they are idealizations of real structures. A topological description accurately represents the connectivity and the strand order around the barrel and is thus very useful. However, as we

Figure 5.16 A schematic diagram of a piece of string wrapped around a barrel to illustrate the basic pattern of a jelly roll motif.

(a)

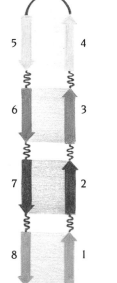

(b)

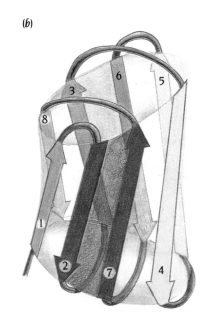

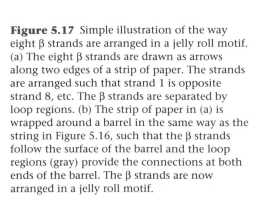

Figure 5.17 Simple illustration of the way eight β strands are arranged in a jelly roll motif. (a) The eight β strands are drawn as arrows along two edges of a strip of paper. The strands are arranged such that strand 1 is opposite strand 8, etc. The β strands are separated by loop regions. (b) The strip of paper in (a) is wrapped around a barrel in the same way as the string in Figure 5.16, such that the β strands follow the surface of the barrel and the loop regions (gray) provide the connections at both ends of the barrel. The β strands are now arranged in a jelly roll motif.

(a)

(b)

C N

8 1 2 7 4 5 6 3

Figure 5.18 Topological diagrams of the jelly roll structure. The same color scheme is used as in Figure 5.17.

saw in the crystallin structure, when one analyzes the pattern of hydrogen bonds between the β strands of such barrels, one finds that they usually form two sheets with few if any hydrogen bonds between strands that belong to the different β sheets. The barrel is distorted and adjacent β strands are separated from each other in two places across the barrel. The division of β strands into these two sheets does not necessarily follow the division into topological motifs. The β strands in jelly roll barrels are also usually arranged in two sheets that are packed against each other. This does not, however, change either the topology or the usefulness of the description of these structures as barrels as long as one keeps in mind that these barrels are distorted, flattened.

A folding scheme has been suggested for the jelly roll barrel structure

It seems probable that the jelly roll structure is actually formed from one long antiparallel hairpin in much the same way that we derived the jelly roll structure from a set of paired β strands along a strip of paper. In that case, folding after synthesis of the polypeptide chain would begin with the pairing of the β strands from a point about halfway down the polypeptide to form a long antiparallel hairpin like the one illustrated in Figure 5.17b. This long antiparallel hairpin could then curl up to produce the additional β sheet interactions as schematically illustrated in Figure 5.19.

This process is similar to that discussed for the Greek key motifs in Chapter 2, and the availability of a relatively stable intermediate folded state provides a simple explanation for the fact that Greek keys and jelly rolls occur in proteins much more frequently than other possible motifs of antiparallel β strands.

The hemagglutinin polypeptide chain folds into a complex structure

Hemagglutinin and neuraminidase are the two envelope proteins of influenza virus. The structure of neuraminidase has already been described as an example of an up-and-down antiparallel β motif. In hemagglutinin one domain of the polypeptide chain is folded into a jelly roll motif. It is appropriate to end these first chapters, which deal with structural principles, with a description of the structure of influenza hemagglutinin since it is one of the most complex subunit structures known and displays some rather unusual features.

The hemagglutinin polypeptide chain of human influenza virus is synthesized on membrane-bound polysomes of the rough endoplasmic reticulum and then cotranslationally inserted into the membrane. During translocation of the polypeptide chain across the membrane an aminoterminal signal sequence is

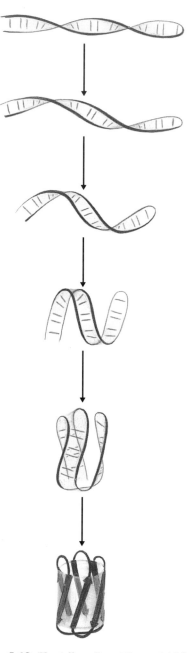

Figure 5.19 The jelly roll motif may fold from a long antiparallel hairpin structure that serves as a relatively stable folding intermediate. This hairpin structure can then twist and curl as shown in the diagram to form the jelly roll structure of six strands in a way similar to the illustration in Figure 5.17. (Adapted from J. Richardson.)

cleaved off. The remaining polypeptide chain of 550 amino acids is anchored to the membrane with the bulk of the sequence within the lumen of the endoplasmic reticulum. This polypeptide chain is cleaved again at position 329 to yield two chains called HA_1 and HA_2, which are held together by disulfide bonds (Figure 5.20). Hemagglutinin is a glycoprotein with *N*-linked sugar chains that are added in the endoplasmic reticulum and modified in the Golgi complex. Three monomers, each with one HA_1 and one HA_2 chain, trimerize within the rough endoplasmic reticulum and are transported from there through the Golgi apparatus to the plasma membrane, where the functional part of the molecule is now outside the cell.

Progeny virus particles bud from patches of the infected cell's plasma membrane that contain both hemagglutinin and the viral neuraminidase. The viral envelopes therefore contain both viral membrane proteins but no cellular membrane proteins.

Protease treatment of influenza virus particles cleaves the three HA_2 chains of the trimeric hemagglutinin molecules. Such cleavage leaves three C-terminal chains, 47 amino acids long, still inserted into the viral envelope and a second much larger soluble fragment. The soluble fragment consists of three complete HA_1 chains disulfide bonded to three HA_2 chains complete except for their membrane anchor regions. This soluble trimeric fragment has been crystallized and its structure has been determined at 3 Å resolution in the laboratory of Don Wiley at Harvard University. The influenza virus that he used was the Hong Kong 1968 strain, which caused the "Asian flu" pandemic disease.

The subunit structure is divided into a stem and a tip

The monomeric subunit is divided into a long, fibrous stemlike region extending outward from the membrane with a globular region at its tip (Figure 5.21). The globular region contains only residues of HA_1, while the stem contains some residues of HA_1 and all of HA_2.

Both polypeptide chains are unusually extended (Figure 5.21). The amino terminus of HA_1 is found at the base of the stem close to the viral membrane. The first 63 amino acids of HA_1 (pale red in Figure 5.21) reach in a nearly fully extended structure almost 100 Å along the length of the molecule before the first compact fold. These 63 residues form part of the stem region of the subunit. The globular tip is an eight-stranded distorted jelly roll structure (red) comprising residues 116 to 261, which are folded into a distorted barrel. The remaining 70 residues of HA_1 return to the stem region, running nearly anti-parallel to the initial stretch of 63 residues.

The major structural feature of the HA_2 chain (blue in Figure 5.21) is a hairpin loop of two α helices packed together. The second α helix is 50 amino acids long and reaches back 76 Å toward the membrane. At the bottom of the stem there is an unusual β sheet of five antiparallel strands. The central β strand is from HA_1, and this is flanked on both sides by hairpin loops from HA_2. About 20 residues at the amino terminal end of HA_2 are associated with the activity by which the virus penetrates the host cell membrane to initiate infection. This region, which is quite hydrophobic, is called the fusion peptide.

The hemagglutinin molecule is trimeric

The **hemagglutinin trimer** molecule is 135 Å long (from membrane to tip) and varies in cross-section between 15 Å and 40 Å. It is thus an unusually elongated molecule. The long fibrous stems of each subunit form the major subunit contacts (Figure 5.22). In particular, the three long HA_2 helices, one from each subunit, intertwine and pack against each other for part of their lengths forming a core 40 Å long stabilized by both hydrophobic residues and internal salt bridges. This region of each long helix has very few interactions within its own subunit and would, in a monomer, expose a hydrophobic surface 36 Å long. It therefore seems likely that this helix does not fold until the subunits are assembled into a trimer and presumably this region has a different structure in a single subunit.

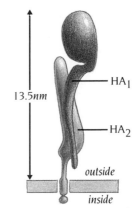

Figure 5.20 Schematic picture of a single subunit of influenza virus hemagglutinin. The two polypeptide chains HA_1 and HA_2 are held together by disulfide bridges.

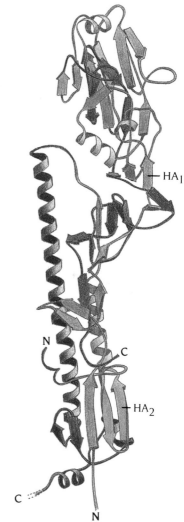

Figure 5.21 Schematic diagram of the subunit structure of hemagglutinin from influenza virus. The structure comprises about 550 amino acids arranged in two chains HA_1 (red) and HA_2 (blue). The first half of each chain has a lighter color in the diagram. The subunit is very elongated with a long stemlike region built up by residues from both chains and includes one of the longest α helices known in a globular structure, about 75Å long. The globular head is formed by residues only from HA_1. (Courtesy of Don Wiley, Harvard University.)

The receptor binding site is formed by the jelly roll domain

To initiate infection, the virus hemagglutinin binds to receptors of the target cell surface. These receptors contain sialic acid. Once bound to the receptor, the virus is then taken into the cell by endocytosis. The **receptor binding site** on the hemagglutinin molecule has been determined from experiments with mutants and from a determination of the structure of sialyllactose complexes with hemagglutinin.

The binding site is located at the tip of the subunit within the jelly roll structure. This site is a pocket at the top of the distorted barrel (Figure 5.23). Residues from β strands and loop regions at this end of the barrel are involved in forming this pocket. This is, therefore, another example of a binding site that is located in a pocket at one end of a barrel structure.

Hemagglutinin acts as a membrane fusogen

Influenza virus hemagglutinin is involved in two distinct steps in the infection of host cells. First, hemagglutinin binds to sialic acid residues of the carbohydrate side chains of cellular proteins projecting from the plasma membrane that the virus exploits as receptors. Although these cellular receptors have still to be identified, knowledge of the structure of the hemagglutinin molecule has shown where its binding sites for sialic acid are located. The virus is now bound to the plasma membrane. Bound viruses still attached to their membrane receptors are then taken into the cells by endocytosis. Proton pumps in the membrane of endocytic vesicles in which the bound virus is now contained cause an accumulation of protons and a consequent lowering of the pH inside the vesicle. The acidic pH (below pH6) somehow induces the hemagglutinin to act as a **membrane fusogen** and fulfill its second role, namely, to induce the fusion of the viral envelope membrane with the membrane of the endosome. This expels the viral RNA into the cytoplasm, where it can begin to replicate.

This fusogenic activity of influenza hemagglutinin can be exploited and manipulated. If, for example, the virus is bound to cells at a temperature too low for endocytosis and then the pH of the external medium is lowered, the hemagglutinin causes direct fusion of the viral envelope with the plasma membrane; infection is achieved without endocytosis. Similarly, artificial vesicles with hemagglutinin in their membrane and other molecules in their lumen can be caused to fuse with cells by allowing the vesicles to bind to the plasma membrane via the hemagglutinin and then inducing fusion by lowering the pH of the medium. In this way the contents of the vesicles are delivered to the recipient cell's cytoplasm. Unfortunately, elucidation of the structure of hemagglutinin has not explained how it acts as a membrane fusogen, which is the molecule's most interesting biological property. Indeed, knowledge of the molecular architecture has, if anything, deepened the mystery since we now know that the protein projects out from the viral envelope more than 100 Å. How do lipid bilayers fuse if they are 100 Å apart? The receptor site for the host membrane is at the tip of the molecule; the cleavage site from the viral membrane peptide is at the bottom of the molecule; and the fusion peptide, which brings the two membranes together, is in between, 100 Å from the top and 35 Å from the bottom. Clearly, at acidic pH the hemagglutinin must undergo a considerable, and still to be elucidated, conformational change.

Conclusion

Antiparallel β structures comprise the second major class of protein conformations. In these the antiparallel β strands are usually arranged in two β sheets that pack against each other and form a distorted barrel structure, the core of the molecule. Depending on the way the β strands around the barrel are connected along the polypeptide chain—in other words, depending on the topology of the barrels—they are divided into three main groups: up-and-down barrels, Greek key barrels, and jelly roll barrels.

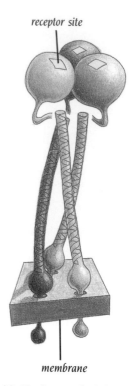

Figure 5.22 The hemagglutinin molecule is formed from three subunits. Each of these subunits is anchored in the membrane of the influenza virus. The globular heads contain the receptor sites that bind to sialic acid residues on the surface of eucaryotic cells. A major part of the subunit interface is formed by the three long intertwining helices, one from each subunit. (Adapted from I. Wilson et al., *Nature* 289: 368, 1981.)

(a)

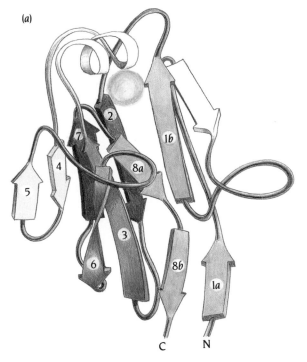

(b)

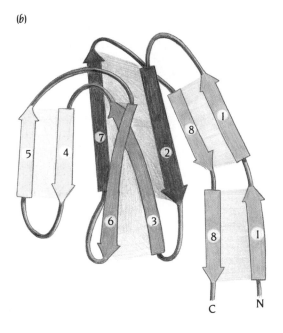

Up-and-down barrels are the simplest structures. Each β strand is connected to the next strand by a short loop region. Eight β strands arranged in this way form the core of a superfamily of proteins that include the plasma retinol-binding protein in mammals, biliverdin-binding proteins in insects, and β-lactoglobulin from milk. Members of this family as well as of the related P2 family with 10 β strands bind large hydrophobic ligands inside the barrel. The barrel seems to be particularly suited to act as a container for chemically quite diverse ligands. At least 20 members of this superfamily are presently known.

Most of the known antiparallel β structures, including the immunoglobulins and a number of different enzymes, have barrels that comprise at least one Greek key motif. An example is γ crystallin, which has two consecutive Greek key motifs in each of two barrel domains. These four motifs are homologous in terms of both their three-dimensional structure and amino acid sequence and are thus evolutionarily related. The jelly roll barrels are found in a variety of protein molecules, including viral coat proteins and the hemagglutinin protein from influenza virus. This structure looks complicated, but, in principle, it is very simple if one thinks of the analogy of wrapping a strip of paper around a barrel, like a jelly roll. The hemagglutinin receptor-binding domain forms such a jelly roll barrel of eight β strands, where the active site is at one end of the barrel.

The number of possible ways to form antiparallel β structures is very large. The number of topologies actually observed is small, and most β structures fall into these three major groups of barrel structures. The last two groups—the Greek key and jelly roll barrels—include proteins of quite diverse function, where functional variability is achieved by differences in the loop regions that connect the β strands that build up the common core region. In the up-and-down barrels the common function, to bind ligands, is provided by the interior of the barrel. Diversity in ligand binding is achieved by differences in the size of the barrel and in the amino acids that also participate in building up the common core.

The second protein in the membrane of influenza virus, neuraminidase, does not belong to any of these three groups of barrel structures. Instead, it forms a superbarrel of 24 β strands. In sets of 4, these β strands are arranged in six similar motifs that are the building blocks of the superbarrel. Each motif is a β sheet of 4 up-and-down-connected β strands. The enzyme active site is formed by loop regions at one end of the superbarrel.

Figure 5.23 The globular head of the hemagglutinin subunit is a distorted jelly roll structure (a). β strand 1 contains a long insertion, and β strand 8 contains a bulge in the corresponding position. Each of these two strands is therefore subdivided into shorter β strands. The loop region between β strands 3 and 4 contains a short α helix, which forms one side of the receptor binding site (yellow circle). A schematic diagram (b) illustrates the organization of the β strands into a jelly roll motif.

Selected readings

General

Chothia, C.; Janin, J. Relative orientation of close-packed β-pleated sheets in proteins. *Proc. Natl. Acad. Sci. USA* 78: 4146–4150, 1981.

Cohen, F.E.; Sternberg, M.J.E.; Taylor, W.R. Analysis and prediction of protein β-sheet structures by a combinatorial approach. *Nature* 285: 378–382, 1980.

Gilbert, W. Why genes in pieces? *Nature* 271: 501, 1978.

Godovac-Zimmerman, J. The structural motif of β-lactoglobulin and retinol-binding protein: a basic framework for binding and transport of small hydrophobic molecules? *Trends Biochem. Sci.* 13: 64–66, 1988.

Ptitsyn, O.B.; Finkelstein, A.V. Similarities of protein topologies: evolutionary divergence, functional convergence or principles of folding? *Q. Rev. Biophys.* 13: 339–386, 1980.

Ptitsyn, O.B.; Finkelstein, A.V.; Falk, P. Principal folding pathway and topology of all-β proteins. *FEBS Lett.* 101: 1–5, 1979.

Richardson, J.S. Handedness of crossover connections in β-sheets. *Proc. Natl. Acad. Sci. USA* 72: 1349–1353, 1975.

Richardson, J.S. β-sheet topology and the relatedness of proteins. *Nature* 268: 495–500, 1977.

Sawyer, L. Protein structure. One fold among many. *Nature* 327: 659, 1987.

Specific structures

Chothia, C. Conformation of twisted β-pleated sheets in proteins. *J. Mol. Biol.* 75: 295–302, 1973.

Chothia, C.; Janin, J. Orthogonal packing of β-pleated sheets in proteins. *Biochemistry* 21: 3955–3965, 1982.

Cohen, F.E.; Sternberg, M.J.E.; Taylor, W.R. Analysis of the tertiary structure of protein β-sheet sandwiches. *J. Mol. Biol.* 148: 253–272, 1981.

Colman, P.M.; Varghese, J.N.; Laver, W.G. Structure of the catalytic and antigenic sites in influenza virus neuraminidase. *Nature* 303: 41–44, 1983.

Daniels, R.S., et al. Fusion mutants of the influenza virus hemagglutinin glycoprotein. *Cell* 40: 431–439, 1985.

Edelman, G.M., et al. The covalent and three-dimensional structure of concanavalin A. *Proc. Natl. Acad. Sci. USA* 69: 2580–2584, 1972.

Edison, A.S. Propagation of an error: β-sheet structures. *Trends Biochem. Sci.* 15: 216–217, 1990.

Hardman, K.D.; Ainsworth, C.F. Structure of concanavalin A at 2.4 Å resolution. *Biochemistry* 11: 4910–4919, 1972.

Holden, H.M., et al. The molecular structure of insecticyanin from the tobacco hornworm *Manduca secta* L. at 2.6 Å resolution. *EMBO J.* 6: 1565–1570, 1987.

Huber, R., et al. Crystallization, crystal structure analysis and preliminary molecular model of the bilin binding protein from the insect *Pieris brassicae*. *J. Mol. Biol.* 195: 432–434, 1987.

Jones, T.A., et al. The three-dimensional structure of P2 myelin protein. *EMBO J.* 7: 1597–1604, 1988.

Lifson, S.; Sander, C. Antiparallel and parallel β-strands differ in amino acid residue preference. *Nature* 282: 109–111, 1979.

Lifson, S.; Sander, C. Specific recognition in the tertiary structure of β-sheets of proteins. *J. Mol. Biol.* 139: 627–639, 1980.

McLachlan, A.D. Repeated folding pattern in copper-zinc superoxide dismutase. *Nature* 285: 267–268, 1980.

McRee, D.E., et al. Crystallographic structure of a photoreceptor protein at 2.4 Å resolution. *Proc. Natl. Acad. Sci. USA* 86: 6533–6537, 1989.

Miller, L.; Lindley, P.; Blundell, T. X-ray analysis of the eye lens protein gamma-crystallin at 1.9 Å resolution. *J. Mol. Biol.* 170: 175–202, 1983.

Newcomer, M.E., et al. The three-dimensional structure of retinol-binding protein. *EMBO J.* 3: 1451–1454, 1984.

Papiz, M.Z., et al. The structure of β-lactoglobulin and its similarity to plasma retinol-binding protein. *Nature* 324: 383–385, 1986.

Pervaiz, S.; Brew, K. Homology of β-lactoglobulin, serum retinol-binding protein and protein HC. *Science* 228: 335–337, 1985.

Richardson, J.S., et al. Similarity of three-dimensional structure between the immunoglobulin domain and the copper, zinc superoxide dismutase subunit. *J. Mol. Biol.* 102: 221–235, 1976.

Richardson, J.S.; Getzoff, E.D.; Richardson, D.C. The β-bulge: a common small unit of nonrepetitive protein structure. *Proc. Natl. Acad. Sci. USA* 75: 2574–2578, 1978.

Sacchettini, J.C., et al. Refined apoprotein structure of rat intestinal fatty acid binding protein produced in *Escherichia coli*. *Proc. Natl. Acad. Sci. USA* 86: 7736–7740, 1989.

Sibanda, B.L.; Blundell, T.L.; Thornton, J. Conformation of β-hairpins in protein structures. A systematic classification with applications to modelling by homology, electron density fitting and protein engineering. *J. Mol. Biol.* 206: 759–777, 1989.

Sternberg, M.J.E.; Thornton, J.M. On the conformation of proteins: an analysis of β-pleated sheets. *J. Mol. Biol.* 110: 285–296, 1977.

Sternberg, M.J.E.; Thornton, J.M. On the conformation of proteins: hydrophobic ordering of strands in β-pleated sheets. *J. Mol. Biol.* 115: 1–17, 1977.

Summers, L., et al. X-ray studies of the lens specific proteins. The crystallins. *Pept. Prot. Rev.* 3: 147–168, 1984.

Tainer, J.A., et al. Determination and analysis of the 2 Å structure of copper, zinc superoxide dismutase. *J. Mol. Biol.* 160: 181–217, 1982.

Varghese, J.N.; Laver, W.G.; Colman, P.M. Structure of the influenza virus glycoprotein antigen neuraminidase at 2.9 Å resolution. *Nature* 303: 35–40, 1983.

Weiss, W., et al. Structure of the influenza virus haemagglutinin complexed with its receptor, sialic acid. *Nature* 333: 426–431, 1988.

Wiley, D.C.; Skehel, J.J. The structure and function of the hemagglutinin membrane glycoprotein of influenza virus. *Annu. Rev. Biochem.* 56: 365–394, 1987.

Wiley, D.C.; Wilson, I.A.; Skehel, J.J. Structural identification of the antibody-binding sites of Hong Kong influenza haemagglutinin and their involvement in the antigenic variation. *Nature* 289: 373–378, 1981.

Wilmot, C.M.; Thornton, J.M. Analysis and prediction of the different types of β-turns in proteins. *J. Mol. Biol.* 203: 221–232, 1988.

Wilson, I.A.; Skehel, J.J.; Wiley, D.C. Structure of the haemagglutinin membrane glycoprotein of influenza virus at 3 Å resolution. *Nature* 289: 366–373, 1981.

DNA Structures

6

DNA replication, DNA transcription, and the selection of genes that are to be transcribed from those that are to remain unexpressed all depend upon recognition of DNA by proteins. By a powerful combination of structural and genetic studies, we in recent years have begun to understand how these functions are achieved. These insights will be described in Chapters 7 and 8, with emphasis on the contribution of structural studies. Before tackling the structures of DNA-binding proteins and the complexes they form with DNA, however, we need to understand the structure of the **double-stranded, base-paired helical DNA** molecule on its own in order to see what possibilities it offers for the recognition by proteins of specific sequences.

The DNA double helix is different in A- and B-DNA

The DNA molecules of each cell contain all the genetic information necessary to ensure the normal development and function of the organism. This genetic information is encoded in the precise linear sequence of the nucleotide bases from which the DNA is built. DNA is a linear molecule. While its diameter is only about 20 Å, if stretched out its length can reach many millimeters. This means that concentrated solutions of DNA can be pulled into fibers in which the long thin DNA molecules are oriented with their long axes parallel.

Early diffraction photographs of such DNA fibers taken by Rosalind Franklin and Maurice Wilkins in London and interpreted by James Watson and Francis Crick in Cambridge revealed two types of regular or averaged DNA structures: **A-DNA** and **B-DNA**. The B-DNA form is obtained when DNA is fully hydrated as it is *in vivo*. A-DNA is obtained under dehydrated nonphysiological conditions. Improvements in the methods for the chemical synthesis of DNA have recently made it possible to study single crystals of short DNA molecules of any selected sequence. These studies have essentially confirmed the refined fiber diffraction models for A- and B-DNA and in addition have given details on individual variations from the averaged values. Furthermore, a new structural form of DNA, called **Z-DNA**, has been discovered.

Both A-DNA and B-DNA have the familiar shape of a right-handed helical staircase (Figure 6.1). The rails are the two antiparallel phosphate-sugar chains, and the rungs are purine-pyrimidine base pairs, which are hydrogen bonded to each other. In A-DNA there are an average of 10.9 base pairs per turn of the helix, which corresponds to an average helical-twist angle of 33.1° from one base pair to the next. The spacing along the helix axis from one base pair to the next is 2.9 Å. In B-DNA these values are 10.0° and 35.9° and 3.4 Å, respectively. There

Figure 6.1 Schematic drawing of B-DNA. Each atom of the sugar-phosphate backbones of the double helix is represented as connected circles within ribbons. The two sugar-phosphate backbones are highlighted by orange ribbons. The base pairs that are connected to the backbone are represented as blue planks. Notice that in B-DNA the central axis of this double helix goes through the middle of the base pairs and that they are perpendicular to the axis.

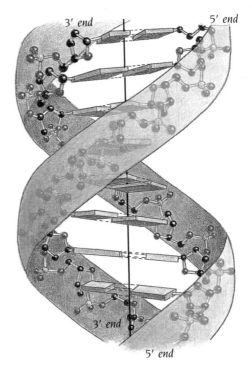

3' end   5' end

3' end

5' end

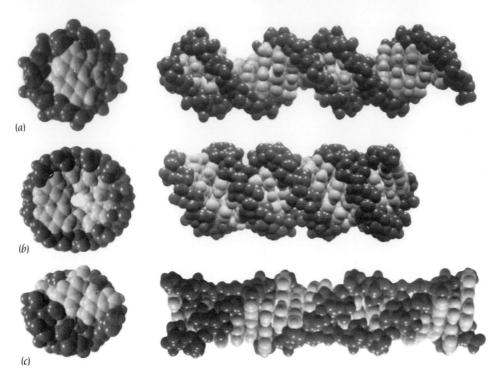

(a)

(b)

(c)

Figure 6.2 Three helical forms of DNA, each containing 22 nucleotide pairs, shown in both side and top views. The sugar-phosphate backbone is dark; the paired nucleotide bases are light. (a) B-DNA, which is the most common form in cells. (b) A-DNA, which is obtained under dehydrated nonphysiological conditions. Notice the hole along the helical axis in this form. (c) Z-DNA, which can form for selected DNA sequences under special circumstances. (Courtesy of Richard Feldmann.)

are, however, considerable variations in individual twist angles from the average values, and these variations are larger in A-DNA than in B-DNA. These variations are sequence dependent, and in B-DNA they might be important for specificity of interactions with proteins.

The DNA helix has major and minor grooves

The sugar-phosphate backbones are bulky and are on the edges of the helix, forming grooves within which the bases are exposed (Figure 6.2). These grooves are of two different widths, reflecting the asymmetrical attachment of the base pairs to the sugar rings of the backbone. Whereas in a regular helix the distance between the attachment points for the rungs would be the same at the front and the back of each step (Figure 6.3), in the DNA molecule each base-pair "rung" is effectively wider at one edge than at the other (Figure 6.4) so that the helical molecule has one narrower groove, known as the **minor groove**, and one wider groove, known as the **major groove** (Figure 6.5).

In B-DNA because the helical axis runs through the center of each base pair and the base pairs are stacked nearly perpendicular to the helical axis (Figures 6.1 and 6.5), the major and minor grooves are of similar depths. In A-DNA, on the other hand, the helical axis is shifted from the center of the bases into the major groove, bypassing the bases, and the base pairs are not perpendicular to this axis but are tilted between 13° and 19°. This arrangement makes the major groove very deep, extending from the surface all the way past the central axis

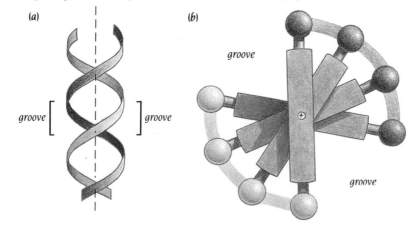

(a)

(b)

groove

groove

groove

groove

Figure 6.3 Schematic diagram illustrating that there are two similar grooves in a helical staircase. Four rungs are viewed from the top of the staircase in (b).

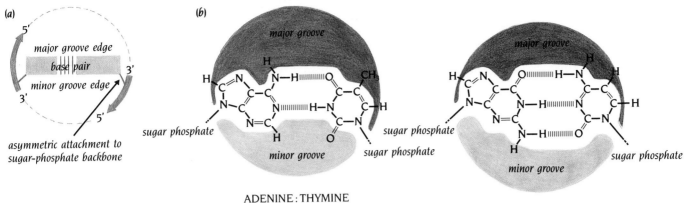

ADENINE : THYMINE

GUANINE : CYTOSINE

and part of the way out toward the opposite side, while the minor groove is shallow, scarcely more than a helical depression spiraling around the outside of the cylinder (Figure 6.5).

The edges of the base pairs form the floors of the two grooves. These edges are accessible from the outside and form the basis for sequence specific recognition of DNA. The edge of a base pair furthest from its attachment points to the sugar-phosphate backbones is the major groove edge; the one closest is the minor groove edge (Figure 6.4a).

Z-DNA forms a zigzag pattern

Z-DNA has a quite different structure (Figure 6.2). The helix is left-handed, and the sugar-phosphate backbone follows a zigzag path. The structure has been found for sequences with alternating G and C bases, such as CGCG and CGCGCG. Each cytosine has its sugar attached to the base in such a way that the pyrimidine ring swings away from the minor groove. This is the normal conformation for all four bases in A- and B-DNA. However, each guanine in Z-DNA has its sugar ring rotated 180° so that it bends inward toward the minor groove. The sugar-phosphate backbone of alternating C and G bases with their different conformations make a zigzag pattern around the helix. The shape of this helix is thin and elongated. It has a deep but quite narrow minor groove, whereas the major groove is pushed to the surface so that it is no longer a groove at all.

B-DNA is the preferred conformation in vivo

The specific protein–DNA interactions described in this book are all with DNA in its regular B-form, or, in some cases with distorted B-DNA. Whether or not Z-

Figure 6.4 The edges of the base pairs in DNA that are in the major groove are wider than those in the minor groove, due to the asymmetrical attachment of the base pairs to the sugar-phosphate backbone (a). These edges contain different hydrogen donors and acceptors for potential specific interaction with proteins (b).

Figure 6.5 Schematic diagram illustrating the major and minor grooves in A- and B-DNA. The sugar-phosphate backbone is represented by connected circles in color and the base pairs as blue planks. Four base pairs are shown from the top of the helix to highlight how the grooves are formed due to the asymmetric connections. The position of the helix axis is marked by a cross.

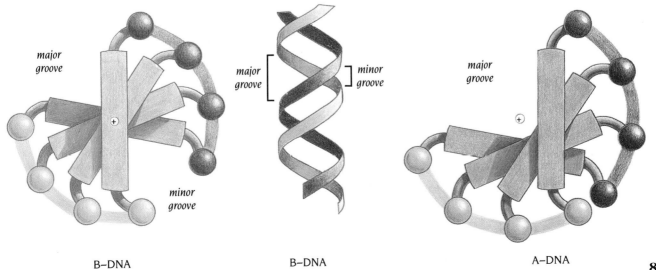

B–DNA

B–DNA

A–DNA

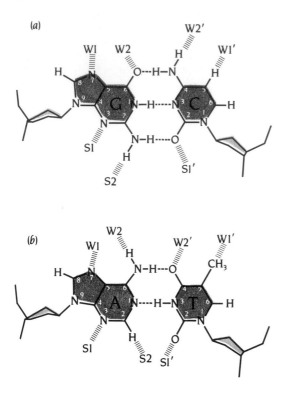

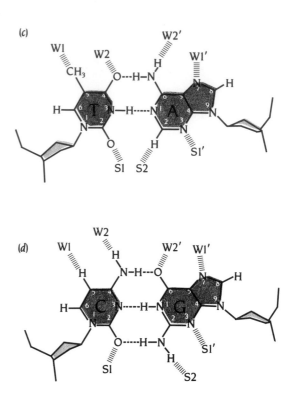

DNA occurs in nature is a matter of current research; there is growing evidence that certain GC-rich regions within long DNA molecules do adopt a Z-like conformation. Such an alternative structure could be a recognition signal for some important function of DNA. In biological systems DNA appears not to adopt the A conformation; however, double-stranded RNA does adopt this conformation preferentially *in vivo*.

Specific base sequences can be recognized in B-DNA

Let us now look at how specific sequences might in principle be recognized in B-DNA. The only regions where the bases are available for interaction are at the floor of the grooves. These are paved with nitrogen and oxygen atoms that can make hydrogen bonds with the side chains of a protein. The methyl group of thymine and the corresponding hydrogen in cytidine provide additional discriminatory recognition groups (Figure 6.4). These sites, which are represented in Figure 6.6 and schematically illustrated in Figure 6.7 in the form of a color code, form patterns that are different for the four possible Watson/Crick base pairs. If, for instance, we compare a GC and an AT base-pair reading from the purine to the pyrimidine ring, we can read the **recognition code** in the following way. For the GC pair, the major groove exposes a hydrogen-bond acceptor, G N7 (W1), another acceptor, G O6 (W2), a hydrogen-bond donor, C NH4(W2′), and, finally, a hydrogen atom at C5 (W1′). For the AT pair, the

Figure 6.6 The edges of the base pairs contain nitrogen and oxygen atoms that can make hydrogen bonds to protein side chains. An H atom in cytidine (C) and a methyl group in thymine (T) form additional sequence-specific recognition sites in DNA. W1, W2, W2′, and W1′ are the recognition sites at the edges of the base pairs in the major groove (W for wide) and S1, S2, and S1′ are those in the minor groove (S for small). The recognition sites are shown for all four base pairs: GC(a), AT(b), TA(c), and CG(d).

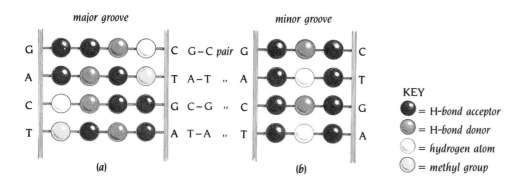

(a) (b)

KEY
= H-*bond acceptor*
= H-*bond donor*
= *hydrogen atom*
= *methyl group*

Figure 6.7 Color codes for the recognition patterns at the edges of the base pairs in the major (a) and minor (b) grooves of B-DNA. Hydrogen-bond acceptors are red; hydrogen-bond donors are blue. The methyl group of thymine is yellow, while the corresponding H atom of cytidine is white.

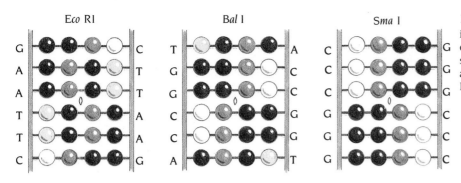

Figure 6.8 Sequence-specific recognition sites in the major groove of DNA for three restriction enzymes—Eco RI, Bal I, and Sma I. The DNA sequences that are recognized by these enzymes are represented by the color code defined in Figure 6.7.

corresponding groups are acceptor A N7, donor A NH6, acceptor T O4, and a methyl group at T5.

These patterns for potential hydrogen bonds are clearly quite different for the different base pairs in the major groove, and they could easily be recognized and distinguished by a protein molecule. This is not the case in the minor groove. If we look at the same two base pairs, we see that for the GC pair we have an acceptor G N3 (S1), a donor G NH2 (S2), and an acceptor C O2 (S1′). For the AT pair we find the following pattern: an acceptor A N3, a hydrogen atom at A2, and an acceptor T O2. In the minor groove there is only one discriminating feature between these two base pairs, a hydrogen-bond donor group in GC compared to a neutral hydrogen atom in AT. We can represent these recognition patterns by a color code as shown in Figure 6.7. Hydrogen-bond donors and acceptors of the base pairs have different colors as well as the H and CH3 groups of cytidine and thymine, respectively. Each of the four different base pairs is in this way represented by a unique color code, comprising four positions in the major groove and three in the minor groove. Clearly, the major groove is a much better candidate for sequence-specific recognition than the minor groove for two reasons. First, the major groove is wider than the minor, and the bases are thus more accessible to a protein molecule. Second, the pattern of possible hydrogen bonds from the edges of the base pairs to a protein are more specific and discriminatory in the major groove than in the minor (Figure 6.7).

Only a rather limited number of base pairs is needed to provide unique and discriminatory recognition sites in the major groove. This is illustrated in Figure 6.8, which gives the color codes for the hexanucleotide recognition sites of three different restriction enzymes—Eco RI, Bal I, and Sma I. It is clear that these patterns are quite different, and each can be uniquely recognized by specific protein–DNA interactions.

Bacteriophage repressor proteins provide excellent examples of sequence-specific interactions between side chains of a protein and bases lining the floor of the major groove of B-DNA. As we shall see, to fit the protein's recognition module into this groove it has to be made even wider; in other words, the B-DNA has to be distorted.

Conclusion

B-DNA is the most important conformation of DNA *in vivo*. It has a wide major groove and a narrow minor groove. Nitrogen and oxygen atoms at the edges of the base pairs are available at the bottom of the major groove to make hydrogen bonds with side chains of a protein. Such hydrogen bonds form the basis for sequence-specific recognition of DNA.

Selected readings

General

Dickerson, R.E. The DNA helix and how it is read. *Sci. Am.* 249(6): 94–111, 1983.

Dickerson, R.E., et al. The anatomy of A-, B- and Z-DNA. *Science* 216: 475–485, 1982.

Felsenfeld, G. DNA. *Sci. Am.* 253(4): 58–66, 1985.

Kennard, O.; Hunter, W.N. Oligonucleotide structure: a decade of results from single crystal x-ray diffraction studies. *Q. Rev. Biophys.* 22: 327–379, 1989.

Rich, A.A.; Nordheim, A.; Wang, A.H.-J. The chemistry and biology of left-handed Z-DNA. *Annu. Rev. Biochem.* 53: 791–846, 1984.

Saenger, W. Principles of Nucleic Acid Structure. Berlin: Springer, 1984.

Travers, A.A. DNA conformation and protein binding. *Annu. Rev. Biochem.* 58: 427–452, 1989.

Travers, A.A. DNA-protein interactions. London: Chapman and Hall, 1990.

Watson, J.D.; Crick, F.H.C. Genetic implications of the structure of deoxyribonucleic acid. *Nature* 171: 964–967, 1953.

Specific structures

Arnott, S.; Hukins, D.W.J. Optimised parameters for A-DNA and B-DNA. *Biochem. Biophys. Res. Comm.* 47: 1504–1509, 1972.

Franklin, R.E.; Gosling, R.G. Molecular structure of nucleic acids. Molecular configuration in sodium thymonucleate. *Nature* 171: 740–741, 1953.

Seeman, N.C.; Rosenberg, J.M.; Rich, A. Sequence-specific recognition of double helical nucleic acids by proteins. *Proc. Natl. Acad. Sci. USA* 73: 804–809, 1976.

Wang, A. H.-J., et al. Molecular structure of a left-handed DNA fragment at atomic resolution. *Nature* 282: 680–686, 1979.

Watson, J.D.; Crick, F.H.C. Molecular structure of nucleic acids. A structure for deoxyribose nucleic acid. *Nature* 171: 737–738, 1953.

Wilkins, M.H.F.; Stokes, A.R.; Wilson, H.R. Molecular structure of nucleic acids. Molecular structure of deoxypentose nucleic acids. *Nature* 171: 738–740, 1953.

Wing, R.M., et al. Crystal structure analysis of a complete turn of B-DNA. *Nature* 287: 755–758, 1980.

Structure, Function, and Engineering

Part 2

DNA Recognition by Proteins with the Helix-Turn-Helix Motif

7

Proteins that regulate transcription of DNA recognize specific DNA sequences through discrete DNA-binding domains within their polypeptide chains. These domains are in general relatively small, less than 100 amino acid residues. Many procaryotic DNA-binding domains contain a **helix-turn-helix motif** that recognizes and binds specific regulatory regions of DNA. The two α helices have the same orientation relative to each other and they are connected by a loop region of similar structure in all DNA-binding helix-turn-helix motifs. In eucaryotic proteins other DNA-binding motifs frequently occur, such as zinc fingers and leucine zippers, which will be described in Chapter 8. In this chapter we will discuss the functional properties of the helix-turn-helix motif and the way this motif is integrated into structurally different DNA-binding domains of procaryotic repressors and activators.

The mechanism of action of bacterial and bacteriophage repressors and activators is in principle very simple. Repressors bind tightly to the DNA at the promoter of a structural gene preventing the RNA polymerase from gaining access and hence blocking the initiation of transcription. Activators, on the other hand, work by binding next to the promoter and helping the polymerase to bind to the adjacent promoter, thereby increasing the rate of transcription of the gene. However, as we will see, the subtle regulation of these binding interactions can be quite complex.

The most thoroughly studied procaryotic regulator proteins belong to bacteriophage lambda and related phages. These phages code for two regulator proteins, namely, **repressor** and **Cro**. Repressor and Cro were given their names at an early date and subsequent studies have shown that, paradoxically, lambda Cro is a repressor protein, whereas lambda repressor has both repressor and activator functions; the names, however, have been kept for historical reasons. Repressor and Cro proteins of phage lambda operate a switch between two stable states of a bacterial cell. This system has been taken as a model for the kind of mechanism that may operate in the establishment of stable differentiated states during vertebrate embryogenesis. Both proteins contain the helix-turn-helix motif. Even though this motif was first observed in a different protein— CAP—that activates operons involved in sugar breakdown in *E. coli*, we will start our discussion with the two bacteriophage proteins since they are the best understood DNA regulatory proteins.

A molecular mechanism for gene control is emerging

Certain strains of *E. coli* can be stimulated by irradiation with a moderate dose of ultraviolet light to stop normal growth and start producing bacteriophages

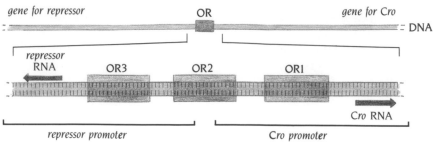

Figure 7.1 A region of DNA in the related bacteriophages lambda, 434, and P22 that controls the switch for synthesis of new phage particles. Two structural genes are involved in this switch; one coding for a repressor protein and one coding for the Cro protein. Between these genes there is an operator region (OR) that contain three protein binding sites—OR1, OR2, and OR3.

that eventually lyse the bacterium. Bacteria of these so-called lysogenic strains carry the DNA of the phage integrated into their own chromosomes, where it is dormant during normal cell growth; the phage DNA is replicated as an integral part of the bacterial chromosome, but the phage genes are not expressed. Ultraviolet light switches on the phage genes which then produce new phages and the cell eventually dies. We are now beginning to understand the molecular mechanism of this **genetic on-off switch** from results obtained by a powerful combination of genetic and x-ray structural studies.

Repressor and Cro proteins operate a procaryotic genetic switch region

Three related species of lysogenic bacteriophages have been studied, lambda, 434, and P22. A relatively small region of the phage genome contains all the genetic components of the on-off switch (see Figure 7.1). In each of the three species of phage this region comprises two structural genes coding for the two regulator proteins, Cro and repressor, that operate the switch and the **operator region** (OR) on which they act.

The two genes are transcribed in opposite directions from their two promoters, which occupy opposite ends of the operator region (see Figure 7.1). When RNA polymerase is bound to the right-hand promoter, *Cro* is switched on, along with the early lytic genes that lie to the right of *Cro*, and lysis results (Figure 7.2b). When the polymerase is bound to the left-hand promoter, repressor is switched on, and *Cro* and the lytic genes are repressed (Figure 7.2a), and the cell survives as a lysogenic strain.

Thus the lysis-lysogeny decision depends upon which of the two promoters in the operator region is able to bind polymerase, and that, in turn, depends upon the binding of the Cro and repressor proteins to three binding sites—OR1, OR2, and OR3—in the operator. These binding sites are situated in the middle

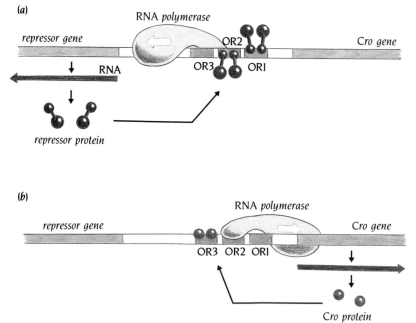

Figure 7.2 (a) The situation in the lysogenic bacterium. Repressor (red) and RNA polymerase (yellow) bound to the switch region in a lysogenic strain of *E. coli*. The repressor binds to OR1 and OR2, thereby turning off synthesis of Cro. The repressor also works as an activator for its own synthesis by facilitating RNA-polymerase binding to the repressor promoter through its binding to OR2. (b) The situation in the lytic phase. Synthesis of the Cro protein turns off synthesis of the repressor, since Cro binds to OR3 and blocks RNA-polymerase binding to the repressor promoter. Transcription of phage genes to the right can now occur.

of the operator in such a way that OR1 and OR2 overlap the promoter of the *Cro* gene and OR2 and OR3 overlap the promoter for the repressor gene (see Figure 7.1). Both Cro and repressor bind as dimers to all three sites, but Cro binds with the highest affinity to OR3, and when it is bound, it blocks the access of RNA polymerase to the promoter for the repressor gene; conversely, repressor binds with the highest affinity to OR1, and when it is bound, it blocks the access of the polymerase to the *Cro* promoter.

Thus, in the lysogen, where the phage genes to the right of *Cro* are switched off, repressor is bound to OR1, and it is also bound to OR2, where instead of blocking the access of the polymerase to its own promoter, it helps the polymerase to bind, thus producing more repressor. In the lysogen, repressor dimers are more or less continuously bound to OR1 and OR2 because once a molecule of repressor is bound at its high affinity site at OR1, it helps a second molecule to bind at OR2 through a cooperative interaction between the two molecules. Thus repressor acts both to repress Cro synthesis and to activate its own synthesis.

Cro, by contrast, acts purely as a repressor. When it is bound to its high affinity site at OR3, it prevents repressor synthesis by obstructing the access of polymerase to the left-hand promoter. In the absence of repressor, RNA polymerase can bind to the *Cro* promoter, and Cro can be synthesized along with the early phage genes to its right.

Thus, for a lysogen to switch over to the production of phage particles, repressor must be released from OR1 for long enough to allow Cro to be synthesized and bind to OR3. This is brought about by ultraviolet radiation, which causes changes in one of the bacterial proteins, the Rec A gene product, converting it from a DNA-binding protein to an enzyme that can cleave polypeptide chains. The modified Rec A protein cuts the repressor molecules in two pieces so that they can no longer form stable dimers. These repressor fragments have a much lower affinity for OR1 and OR2, and thus after some time the *Cro* promoter site becomes free to bind RNA polymerase. The *Cro* gene is then transcribed along with the early genes for viral replication. The newly formed Cro protein molecules bind to OR3, blocking the access of RNA polymerase, and synthesis of more repressor molecules is prevented (Figure 7.2b). The switch has been flipped, and the machinery of the cell is converted to the synthesis of phage particles.

The x-ray structure of the complete lambda Cro protein is known

How do repressor and Cro recognize the specific operator regions and achieve this subtle differential binding to the switch regions? The sequences of OR1, OR2, and OR3 in lambda are similar but not identical (see Figure 7.3). They are also partly **palindromic**, especially at their ends. The palindromic nature of these sequences is important because it means that the two halves of each binding site are related by an approximate twofold symmetry axis. The twofold axis of symmetry of the DNA provides a **dimeric protein** with two identical recognition sites, one for each subunit. Both Cro and repressor proteins are dimers, and the palindromic parts of OR1, OR2, and OR3 provide them with two almost but not quite identical recognition sites.

		1	2	3	4	5	6	7	8	9	10	11	12	13	14	15	16	17	
OR1	5′	T	A	T	C	A	C	C	G	C	C	A	G	T	G	G	T	A	3′
										•									
	3′	A	T	A	G	T	G	G	C	G	G	T	C	A	C	C	A	T	5′
OR2	5′	T	A	A	C	A	C	C	G	T	G	C	G	T	G	T	T	G	3′
										•									
	3′	A	T	T	G	T	G	G	C	A	C	G	C	A	C	A	A	C	5′
OR3	5′	T	A	T	C	A	C	C	G	C	A	A	G	G	G	A	T	A	3′
										•									
	3′	A	T	A	G	T	G	G	C	G	T	T	C	C	C	T	A	T	5′

Figure 7.3 The nucleotide sequences of the three protein-binding regions OR1, OR2, and OR3 of the operator of bacteriophage lambda. Palindromic base pairs that are most frequent at the two ends are green, and the pseudo twofold symmetry axis is indicated by a red dot.

(a)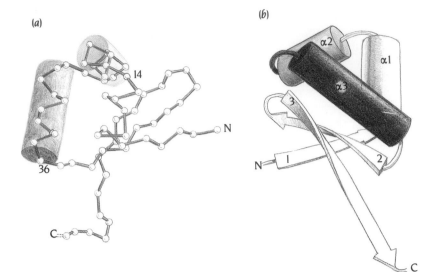

(b)

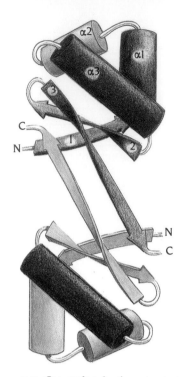

Figure 7.4 The DNA-binding protein Cro from bacteriophage lambda contains 66 amino acid residues that fold into three α helices and three β strands. (a) A plot of the C_α positions of the first 62 residues of the polypeptide chain. The four C-terminal residues are not visible in the electron density map. (Adapted from Anderson et al., *Nature* 290: 755, 1981.) (b) A schematic diagram of the subunit structure. α helices 2 and 3 that form the helix-turn-helix motif are colored blue and red, respectively. The view is different from that in (a). (Adapted from D. Ohlendorf et al., *J. Mol. Biol.* 169: 757, 1983.)

Figure 7.5 Cro molecules from bacteriophage lambda form dimers both in solution and in the crystal structure. The main dimer interactions are between β strands 3 from each subunit. In the diagram one subunit is green and the other is brown. α helices 2 and 3, the helix-turn-helix motifs, are colored blue and red, respectively, in both subunits. (Adapted from D. Ohlendorf et al., *J. Mol. Biol.* 169: 757, 1983.)

A first glimpse into the structural details of this system was obtained in 1981 when the group of Brian Matthews at Eugene, Oregon, determined the crystal structure of the Cro protein of phage lambda at 2.8 Å resolution. Cro is a small protein that forms stable dimers in solution. Each subunit is a single polypeptide chain of 66 amino acid residues with a very simple structure. It folds into three α helices and three strands of antiparallel β sheet and belongs therefore to the α + β class of structures, with the three α helices in a loop region between β strands 1 and 2 (Figure 7.4).

The α helices are not, however, packed against each other in the usual way as described in Chapter 3. Instead, α helices 2 and 3, residues 15–36, form a unique arrangement that in 1981 had only been observed once before in all the then known protein structures, in a different bacterial DNA-binding protein, the catabolite gene-activating protein CAP, whose structure and function is described later in this chapter.

Dimerization of pairs of Cro monomers depends primarily on interactions between β strand 3 from each subunit (Figure 7.5). These strands, which are at the carboxy end of the chains, are aligned in an antiparallel fashion and hydrogen bonded to each other so that the three-stranded β sheets of the monomers form a six-stranded antiparallel β sheet in the dimer (Figure 7.6).

The x-ray structure of the DNA-binding domain of the lambda repressor is known

Superficially, the lambda repressor protein is very different from lambda Cro. The polypeptide chain of the subunit is much larger, 236 amino acids, and is composed of two domains that can be released as separate fragments by mild proteolysis. In repressor the domain responsible for dimerization is separate from the domain with DNA-binding functions; the C-terminal domains form the strong subunit interactions that hold the dimer together, while the N-terminal domains of 92 residues bind specifically to operator DNA. Although the N-terminal domains can form dimers on their own, they do so only weakly, and for this reason they also bind the operator more weakly than does the intact repressor molecule. This feature is crucial for the switch from lysogeny to the

Figure 7.6 The three β strands of each subunit in lambda Cro are aligned in the dimer so that a six-stranded antiparallel β sheet is formed as shown in this topology diagram. The β strands from different subunits have different colors.

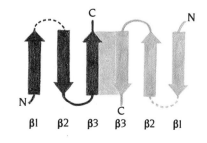

β1 β2 β3   β3 β2 β1

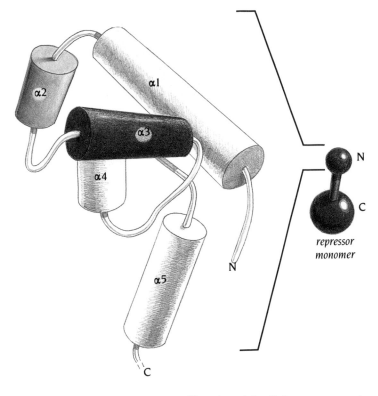

Figure 7.7 The N-terminal domain of lambda repressor, which binds DNA, contains 92 amino acid residues folded into five α helices. Two of these, α2 (blue) and α3 (red) form a helix-turn-helix motif with a very similar structure to that of lambda Cro shown in Figure 7.5. The complete repressor monomer contains in addition a larger C-terminal domain. (Adapted from C. Pabo and M. Lewis, *Nature* 298: 445, 1982.)

lytic cycle: the proteolytic cleavage, activated by ultraviolet light, separates the C-terminal domain from the N-terminal domain. The x-ray structure of the N-terminal DNA-binding domain of the lambda repressor was determined to 3.2 Å resolution in 1982 by Carl Pabo at Harvard University and revealed a structure with striking similarities to that of Cro, although the β strands in Cro are replaced by α helices in repressor.

The polypeptide chain of the 92 N-terminal residues is folded into five α helices connected by loop regions (Figure 7.7). Again the helices are not packed against each other in the usual way for α-helical structures. Instead, α helices 2 and 3, residues 33–52, form a helix-turn-helix motif with a very similar structure to that found in Cro.

In spite of the absence of the C-terminal domains, the DNA-binding domains of lambda repressor form dimers in the crystals (Figure 7.8), as a result of interactions between the carboxy terminal helix number 5 of the two subunits that are somewhat analogous to the interactions of the carboxy terminal β strand 3 in the Cro protein. The two helices pack against each other in the normal way with an inclination of 20° between the helical axes. The structure of the C-terminal domain, which is responsible for the main subunit interactions in the intact repressor, remains unknown.

Both lambda Cro and repressor proteins have a specific DNA-binding motif

The specific arrangement of two α helices joined by a loop region that has been observed both in lambda Cro and repressor, as well as in CAP, constitute a helix-turn-helix DNA-binding motif (Figure 7.9). The orientation of the two helices and the conformation of the loop regions are very similar in these three molecules, and this particular helix-turn-helix structure is unique to DNA-binding proteins.

Since these protein structures were determined in the absence of DNA, there was until recently no experimental evidence that these α helices were involved in DNA binding. However, Brian Matthews in 1981 realized from the structure of Cro that the subunit interactions in the dimer have an important consequence. They cause the second α helix (α3)in the helix-turn-helix motifs of the two subunits to be at opposite ends of the elongated dimeric molecule, separated

Figure 7.8 The N-terminal domains of lambda repressor form dimers, in spite of the absence of the C-terminal domains, which are mainly responsible for dimer formation in the intact repressor. The dimers are formed by interactions between α helix 5 from each subunit. The different subunits are colored green and brown, except the helix-turn-helix motif, which is colored blue and red as in Figure 7.5. (Adapted from C. Pabo and M. Lewis, *Nature* 298: 446, 1982.)

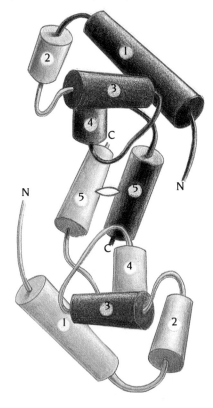

by a distance of 34 Å (Figure 7.5). This distance corresponds almost exactly to one turn of a B-DNA double helix. As a result, if the second α helix of one subunit binds into the major groove of B-DNA, the corresponding α helix of the other subunit can also bind into the major groove one turn further along the DNA molecule. This α helix (the second in the helix-turn-helix motif, colored red in the illustrations) was, therefore, called the **recognition α helix**, and Matthews proceeded to build a model of a possible Cro–DNA complex.

Model building predicts Cro–DNA interactions

Matthews was able to show, by **model building** on a graphics display, that the two recognition helices of the Cro dimer indeed fitted very well into the major groove of a piece of regular B-DNA as seen in Figure 7.10. The orientation of the two recognition α helices follows the orientation of the groove at both sites. Amino acid residues from these α helices can make contact with the edge of the base pairs in the major groove. These amino acids were thus assumed to be involved in recognizing specific operator regions.

   Approximately 10 base pairs are required to make one turn in B-DNA. The centers of the palindromic sequences in the DNA-binding regions of the operator are also separated by about 10 base pairs (see Figure 7.3). Thus if one of the recognition α helices binds to one of the palindromic DNA sequences, the second recognition α helix of the protein dimer is poised to bind to the second palindromic DNA sequence.

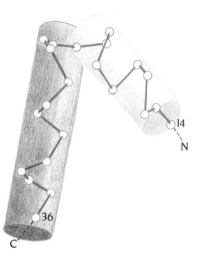

Figure 7.9 The DNA-binding helix-turn-helix motif in lambda Cro. C_α positions of the amino acids in this motif have been projected onto a plane and the two helices outlined. The second helix (red) is called the recognition helix because it is involved in sequence-specific recognition of DNA.

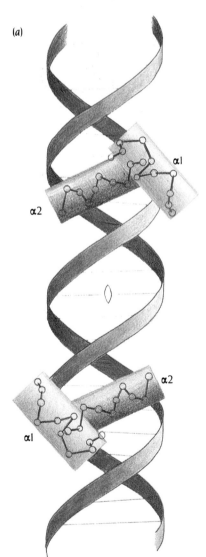

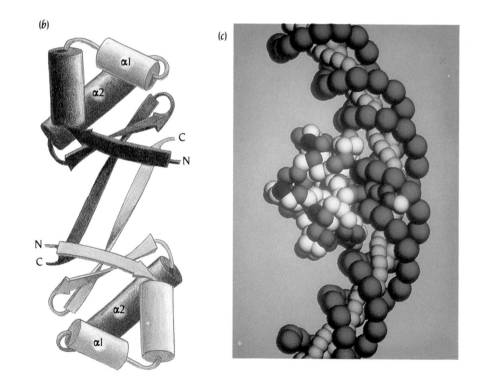

Figure 7.10 The helix-turn-helix motif in lambda Cro bound to DNA (orange) with the two recognition helices (red) of the Cro dimer sitting in the major groove of DNA. The binding model, suggested by Brian Matthews, is shown schematically in (a) with connected circles for the C_α positions as they were model built into regular B-DNA. A schematic diagram of the Cro dimer is shown in (b) with different colors for the two subunits. A schematic space-filling model of the dimer of Cro bound to a bent B-DNA molecule is shown in (c). The sugar-phosphate backbone of DNA is red, and the bases are yellow. Protein atoms are colored red, blue, green, and white. [(a) Adapted from D. Ohlendorf et al., *J. Mol. Evol.* 19: 113, 1983. (c) Courtesy of Brian Matthews.]

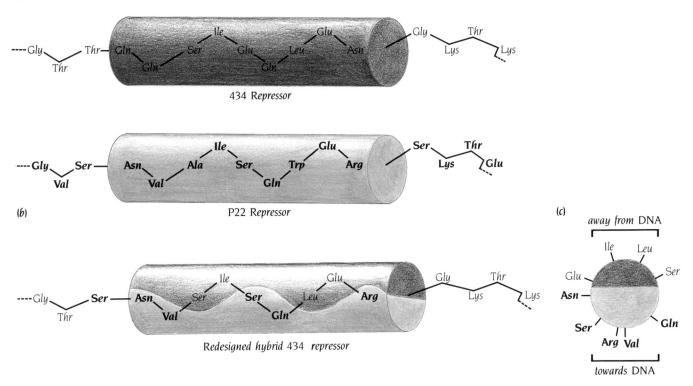

(a)

434 Repressor

P22 Repressor

(b)

Redesigned hybrid 434 repressor

(c)

away from DNA

towards DNA

Both the relative orientation of the α helices and the distance between them are determined by the way the dimer is formed through interactions between the subunits. Three features of the Cro structure therefore were claimed to be important for specific binding to operator DNA: (1) the presence of a helix-turn-helix motif that provides a recognition α helix, which binds in the major groove of B-DNA; (2) the specific amino acid sequence of this α helix, which recognizes different operator regions; and (3) the subunit interactions that provide the correct distance and relative orientation between the two recognition α helices of the dimer, thereby increasing the affinity between Cro and operator DNA.

This hypothetical DNA-binding mode of Cro was arrived at by intuition and clever model building. The validity of this model was considerably strengthened when the same features were subsequently found in the DNA-binding domains of the lambda-repressor molecule. The helix-turn-helix motif with a recognition helix is present in the repressor, and moreover the repressor DNA-binding domains dimerize in the crystals in such a way that the recognition helices are separated by 34 Å as in Cro.

Figure 7.11 The proposed DNA-binding surface of the recognition helix of bacteriophage 434 repressor was redesigned genetically to that of P22 repressor by changing six amino acid residues. The amino acid sequences of the recognition helices of the wild-type repressors are shown in (a) and that of the redesigned repressor in (b) and (c) viewed from the side and along the helix, respectively. The redesigned 434 repressor acquired all the DNA-binding properties of the P22 repressor. The six amino acid residues that were changed in the 434 repressor are shown in boldface in (b) and (c). (Adapted from R. Wharton and M. Ptashne, *Nature* 316: 602, 1985.)

Genetic studies agree with the structural model

The presence of this common helix-turn-helix motif poised for DNA binding in lambda Cro and repressor (as well as in CAP, as we shall see later) provided considerable stimulus for further genetic and structural studies of these and other procaryotic DNA-binding proteins. All the results obtained essentially supported the proposed mode of binding between these regulator proteins and DNA.

The most informative genetic experiments were made by the group of Mark Ptashne at Harvard University. In one experiment they redesigned the repressor from phage 434, replacing its recognition α helix with that of the corresponding α helix from Cro of phage 434. There were five amino acid differences between the two α helices. The redesigned repressor thereby acquired the differential binding properties of the 434 Cro protein. In another experiment the group changed selected amino acid residues in the recognition α helix of the 434 repressor to those that occur in the repressor of phage P22 (see Figure 7.11). The amino acids selected to be changed were those that in the structural model face the DNA and therefore presumably interact with it.

The consequences of these changes were impressive. The parental 434 repressor had no affinity for the P22 operator *in vivo*, but the redesigned 434 repressor controlled only P22 operators and not 434 operator regions *in vivo*. Purified redesigned 434 repressor bound specifically to P22 operator DNA *in vitro* and showed the same hierarchy of affinities for the three regions OR1, OR2, and OR3 as native P22 repressor.

These genetic experiments clearly demonstrated that the proposed structural model for operator DNA binding by these proteins was essentially correct. The second α helix in the helix-turn-helix motif is involved in recognizing operator sites as well as in the differential selection of operators by P22 Cro and repressor proteins.

The x-ray structure of DNA complexes with 434 Cro and repressor revealed novel features of protein DNA interactions

The general features of the model for DNA binding were experimentally confirmed when Stephen Harrison's group at Harvard University determined in 1987 the structure to 3.2 Å resolution of a complex of DNA and the DNA-binding domain of the 434 repressor. However, it also became evident, both from the structure of this complex and from further site-directed mutagenesis studies, that the selective recognition of the different operator regions by the 434 repressor depends mostly on other factors than the amino acid residues of the recognition helix. The complexity of the fine tuning of DNA regulation has been clearly demonstrated by Harrison's subsequent studies of complexes between different operator DNA regions and both 434 Cro and the DNA-binding domain of 434 repressor.

For purely practical reasons, the complexes that Harrison studied were modified versions of the natural complexes. Ideally, the complete repressor molecule should be bound to an operator region in the middle of a piece of DNA long enough to ensure that the DNA is maintained in the same structure as *in vivo*. Such a complex would, however, be difficult to crystallize. Instead, as a compromise, Harrison first studied the N-terminal DNA-binding domain of the repressor from phage 434, which comprises 69 amino acids, complexed with a piece of synthetic DNA containing 14 base pairs with a completely palindromic sequence. In other words, the DNA in the complex had a strict twofold symmetry analogous to the twofold symmetry of the dimeric repressor molecule. This synthetic DNA thus contains two identical halves, each of which, as we will see, binds one subunit of Cro or one repressor fragment. The three 14

		1	2	3	4	5	6	7	8	9	10	11	12	13	14	
ORI	5'	A	C	A	A	G	A	A	A	G	T	T	T	G	T	3'
	3'	T	G	T	T	C	T	T	T	C	A	A	A	C	A	5'
OR2	5'	A	C	A	A	G	A	T	A	C	A	T	T	G	T	3'
	3'	T	G	T	T	C	T	A	T	G	T	A	A	C	A	5'
OR3	5'	A	C	A	A	G	A	A	A	A	A	C	T	G	T	3'
	3'	T	G	T	T	C	T	T	T	T	T	G	A	C	A	5'
OL1	5'	A	C	A	A	G	G	A	A	G	A	T	T	G	T	3'
	3'	T	G	T	T	C	C	T	T	C	T	A	A	C	A	5'
OL2	5'	A	C	A	A	T	A	A	A	T	A	T	T	G	T	3'
	3'	T	G	T	T	A	T	T	T	A	T	A	A	C	A	5'
OL3	5'	A	C	A	A	T	G	G	A	G	T	T	T	G	T	3'
	3'	T	G	T	T	A	C	C	T	C	A	A	A	C	A	5'
Synthetic DNA	5'	A	C	A	A	T	A	T	A	T	A	T	T	G	T	3'
	3'	T	G	T	T	A	T	A	T	A	T	A	A	C	A	5'
		14'	13'	12'	11'	10'	9'	8'	7'	6'	5'	4'	3'	2'	1'	

Figure 7.12 There are six operator regions (OR1–OR3 and OL1–OL3), each of 14 base pairs, in bacteriophage 434. The palindromic base pairs of these regions are marked in green. Crystal structures have been determined of complexes between both 434 Cro and the repressor fragment with synthetic DNA fragments—one 14 base pairs long (a 14 mer), which is completely palindromic, and one 20 base pairs long (a 20 mer), which contains the sequence of OR1 in its middle region.

base-pair operator regions (OR) that the 434 repressor recognizes in the phage genome are not perfectly palindromic (Figure 7.12). The synthetic and the natural binding sites therefore are not identical in these operator regions.

In addition to the operator regions (OR1 and OR2) regulating lysogeny, the 434 repressor also controls a second set of operators, called OL, that regulate a different set of phage genes. Significantly, this second operator region also has three binding sites OL1, OL2, and OL3, each again 14 base pairs long. As Figure 7.12 shows, the synthetic DNA Harrison used to crystallize the repressor-DNA complex is very similar to OL2; it is 14 base pairs long (a 14 mer) with the only difference being an inversion of base pair 7 from A–T to T–A. Experiments with mutants had shown that this inversion did not alter the affinity of either intact repressor or the N-terminal DNA-binding fragment for the DNA.

The crystals of the complex described above diffracted only to medium resolution. However, by systematic variations of the length of the DNA fragment and its sequences at the ends, Harrison found one piece of DNA that gave crystals that diffracted to high resolution in complexes both with 434 Cro and the DNA-binding domain of 434 repressor. This DNA fragment, which is only partly palindromic, contains 20 nucleotides in each chain (a 20 mer) that are base paired such that the middle region is identical to operator region OR1 (Figure 7.12) and the 5′ ends contain one nonpaired nucleotide that is involved in packing of the fragments in the crystal.

By comparison of the crystal structures of these complexes with a further complex between the 434 repressor DNA-binding domain and a synthetic DNA containing the operator region OR3, Harrison has been able to resolve at least in part the structural basis for the differential binding affinity of 434 Cro and repressor to the different 434 operator regions.

The structures of 434 Cro and 434 repressor-binding domains are very similar

The 434 Cro molecule contains 71 amino acid residues that show 48% sequence identity to the 69 residues that form the N-terminal DNA-binding domain of 434 repressor. It is not surprising, therefore, that their three-dimensional structures are very similar (Figure 7.13). The main difference lies in two extra amino acids at the N terminus of the Cro molecule. These are not involved in the function of Cro. By choosing the 434 Cro and repressor molecules for his studies, Harrison has eliminated the possibility that any gross structural difference of these two molecules can account for their different DNA-binding properties.

The DNA-binding domain of 434 repressor also has significant sequence homology (26% identity) with the corresponding part of the lambda repressor and, consequently, a related three-dimensional structure (see Figures 7.8 and 7.13). Like its lambda counterpart, the subunit structure of the DNA-binding domain of 434 repressor, as well as that of 434 Cro, consists of a cluster of four α helices, with helices 2 and 3 forming the helix-turn-helix motif. The two helix-turn-helix motifs are at either end of the dimer and contribute the main protein–DNA interactions, while protein–protein interactions at the C-terminal part of the chains hold the two subunits together in the complexes. Both 434 Cro and repressor fragment are monomers in solution even at high protein concentrations, whereas they form dimers when they are bound to DNA. (It should be noted once again, however, that in the intact repressor the main dimerization interactions are believed to be formed by the C-terminal domain, which is not present in the crystal complex.)

The B-DNA conformation is distorted in the complexes

The real significance of Harrison's work with these protein-DNA complexes lies not in the structure of the protein domains but rather in the details of the structures of the bound DNA and in the protein–DNA interactions. The DNA in

Figure 7.13 The DNA-binding domain of 434 repressor. It is a dimer in its complexes with DNA fragments. Each subunit (green and brown) folds into a bundle of four α helices (1–4) that have a structure similar to the corresponding region of the lambda repressor (Figure 7.8) including the helix-turn-helix motif (blue and red). A fifth α helix (5) is involved in the subunit interactions, details of which are different from those of the lambda repressor fragment. The structure of the 434 Cro dimer is very similar to the 434 repressor shown here.

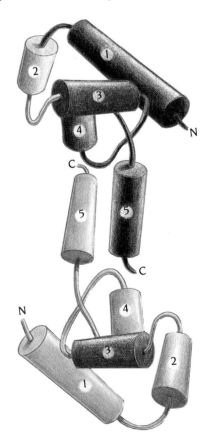

all the complexes is in the B-form but with significant **distortions**. Examination of the **local twist** between base pairs showed that the DNA was **overwound** (larger twist) at its center and underwound at the ends. The helical axis was also somewhat **bent** toward the recognition helices at the ends of the protein dimer. These distortions narrow the minor groove at the center and widen it at the ends, as is shown in Figure 7.14, which compares the observed structure of DNA complexed with 434 repressor fragments to that of regular B-DNA. Since short lengths of free DNA have no tendency to unwind at their ends, it is reasonable to believe that the conformational changes of the DNA complexed with 434 Cro and the 434 repressor fragment are a direct result of the protein–DNA interactions.

The synthetic DNA containing 14 base pairs (the 14 mer) and the OR1-containing 20 mer have very similar conformations in complexes with the repressor fragment. These conformations, however, are distinctly different in complexes with Cro. The overall bend and twist of the DNA are similar, but there is a significant difference in the local structure of two of the nucleotides in each half site of the operator. Binding of Cro and repressor fragment thus imposes different local DNA structures in their binding sites (Figure 7.15), which result from differences in the identity and conformations of various amino acid residues that interact with the DNA backbone.

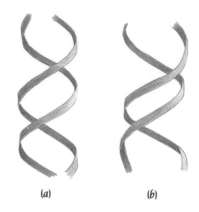

(a) (b)

Figure 7.14 (above) The changes of DNA structure from regular B-DNA (a) to a distorted version (b) when 434 Cro and repressor fragment bind to operator regions. The distortions essentially involve bending of DNA and overwinding of the middle regions. The diagram shows the sugar-phosphate backbones of DNA as orange ribbons viewed in the narrow groove in the middle region of the operator.

Conformational changes of DNA are important for differential repressor binding

The conformation of a synthetic DNA molecule containing the OR3 operator region (Figure 7.12) when complexed with the repressor fragment has provided an important clue to the molecular basis for the differential binding of repressor to different operators. All operator regions except OR3 have the sequence 5′-ACAA...at the ends of the DNA in each of the two binding sites (Figure 7.12). OR3 has this consensus sequence in its left-half site, but in the other half the sequence is instead 5′-ACAG.... Binding of the 434 repressor fragment to DNA containing the OR3 sequence induces different conformational changes in the two halves of the DNA. The left half has the same conformation as both halves have in other complexes with the repressor fragment, whereas the right half has the conformation that is found in complexes with Cro.

These results suggest that the repressor fragment is unable to impose upon the nonconsensus sequence 5′-ACAG...in OR3 the conformation of DNA that is required for tight binding between repressor and DNA. Remember that the repressor, as well as its DNA-binding fragment, binds more tightly to OR1 and OR2 than to OR3.

Sequence-specific protein–DNA interactions recognize operator regions

The protein–DNA interactions have been analyzed in detail at high resolution in the complex between the 434 repressor fragment and the OR1 containing 20 mer DNA. A pseudo twofold symmetry axis relates the two halves of this complex. The symmetry is not exact since the nucleotide sequence of the DNA

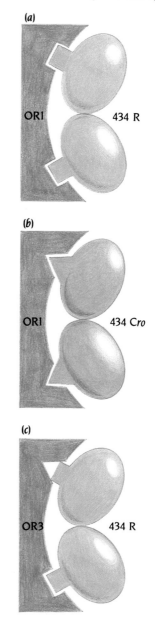

(a)

OR1   434 R

(b)

OR1   434 Cro

(c)

OR3   434 R

Figure 7.15 (right) Binding of 434 repressor fragment (a) and 434 Cro (b) to operator region OR1. This binding induces different structural changes in the region of the DNA that binds the proteins illustrated as different shapes of the binding regions of OR1. In the complex of operator OR3 with the 434 repressor fragment (c) the two half sites of OR3 are different: one is similar to OR1, with bound repressor, whereas the other has a different nucleotide sequence that adopts the Cro-type binding conformation on binding repressor. The binding surfaces of the DNA and repressor fragment do not complement each other as they do in the OR1 complex; consequently, the repressor fragment binds more weakly to OR3 than to OR1. DNA is schematically shown in orange and the proteins in blue or green.

is slightly different in the two halves (Figure 7.12). However, the interactions between one protein subunit and one half of the DNA are very similar to those between the second subunit and the other half of the DNA since most of the bases that interact with the protein are identical in the two halves. Details of the interaction are very similar to those in the complex with the palindromic synthetic 14 mer of DNA shown in Figures 7.16 and 7.17. The base pairs at one end of the DNA, 1–14′, 2–13′, etc. are called base pairs 1, 2, etc.

The protein dimer binds in such a way that the recognition α helices at opposite ends of the protein molecule are in the major groove of DNA as predicted, where they interact with base pairs at the end of the DNA molecule. Since these binding sites are separated by one turn of the DNA helix, it follows that at the center of the DNA molecule the narrow groove faces the protein. Therefore, there are no interactions between the protein and the bases of the DNA in this middle region of the operator.

Residues of the recognition α helix project their side chains into the major groove and interact with the edges of the DNA base pairs on the floor of the groove. Gln (Q) 28 forms two **hydrogen bonds** to N_6 and N_7 of A1 in base pair 1 (T14′–A1) (see Figure 7.17), and Gln 29 forms hydrogen bonds both to O_6 and N_7 of G13′ in base pair 2 (G13′–C2). At base pair 3 (T12′–A3) no hydrogen bonding to the protein occurs and direct contacts are all hydrophobic. The methyl groups of the side chains of Thr (T) 27 and Gln (Q) 29 form a **hydrophobic pocket** to receive the methyl group of T12′.

The first three base pairs in all six operator regions recognized by phage 434 repressor are identical (Figure 7.12). This means that interactions between these three base pairs and the two glutamine residues (28 and 29) cannot contribute to the discrimination between the six binding sites in the DNA; rather, these interactions provide a general recognition site for operator regions. This simple pattern of hydrogen bonds and hydrophobic interactions therefore accounts for the specificity of phage 434 Cro and repressor proteins for 434 operator regions. The role of Gln 29 is particularly important since it interacts with both base pairs 2 and 3. Its hydrogen bond to guanine specifies C–G at position 2 and the hydrophobic pocket formed by the hydrophobic part of its side chain together with residue 27, specifies A–T at base pair 3.

This general recognition function is crucial for the bacteriophage. When glutamines 28 and 29 are replaced by any other amino acid, the mutant phages are no longer viable. Moreover, 434 Cro protein has glutamine residues at these two positions in its recognition helix, as expected since it binds to the same set of operator regions.

Nonspecific protein–DNA interactions determine DNA conformation

It is apparent from the crystal structures of these protein–DNA complexes that the **differential affinities** of 434 repressor and Cro for the different operator

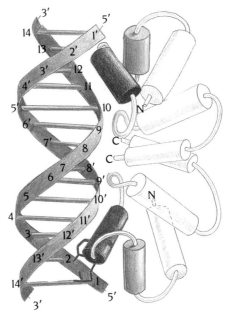

Figure 7.16 Overall view of the complex between 434 repressor fragment and a palindromic synthetic 14 mer of DNA (see Figure 7.12). The two binding sites of the repressor dimer to the DNA are identical. The recognition helices of the repressor are red, and the first helix of the helix-turn-helix motif is blue. (Adapted from J. Anderson et al., *Nature* 326: 846, 1987.)

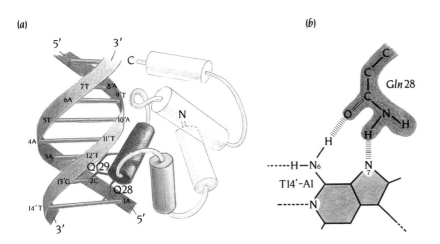

Figure 7.17 Sequence-specific protein–DNA interactions provide a general recognition signal for operator regions in 434 bacteriophage. In this complex between 434 repressor fragment and a synthetic DNA (a) there are two glutamine residues (28 and 29) at the beginning of the recognition helix in the helix-turn-helix motif that provide such interactions with the first three base pairs of the operator region. The side chain of Gln 28 forms two hydrogen bonds (b) to the edge of the adenine base of base pair T14′–A1 in the major groove of the DNA. (Adapted from J. Anderson et al., *Nature* 326: 846, 1987.)

regions are not determined by sequence-specific interactions between amino acid side chains of the recognition helix and base pairs in the major groove of DNA. Instead, they seem to be determined mainly by the ability of the DNA to undergo specific structural changes so that complementary surfaces are formed between the proteins and the DNA. Nonspecific interactions between the DNA sugar-phosphate backbone and the proteins are one important factor in establishing such structural changes.

The interface between the repressor fragment and DNA has complementary molecular surfaces over an extended area. Most interactions between one monomer of the repressor fragment and DNA occur within a single half site; in other words, each repressor subunit "sees" only one-half of the operator. In order to obtain complementary surfaces at both half sites and hence tight binding, it is essential that the structural change in the DNA coincides with the dimer organization of the repressor. It is because this coincidence does not occur in the binding of repressor fragment to OR3 that binding in this complex is relatively weak (see Figure 7.15).

In all complexes studied the protein subunit is anchored across the major groove with extensive contacts along two segments of the sugar-phosphate backbone, one to either side of the groove. Hydrogen bonds between the DNA phosphate groups and peptide backbone NH groups are remarkably prevalent in these contacts (Figure 7.18).

One of these interaction regions involves the loop after the recognition helix, residues 40–44 of the repressor (yellow in Figures 7.16 and 7.17a), where three main-chain NH groups form hydrogen bonds with phosphates 9′ and 10′. The side chain of Arg 43 in this loop projects into the minor groove and probably stabilizes its compression, by introducing a positive charge between the phosphate groups on opposite sides of the narrow groove. All residues in this loop, which are outside the helix-turn-helix motif, contribute to the surface complementarity between the protein and the sugar-phosphate surfaces of nucleotides 9′ and 10′.

These and other nonspecific interactions, which stabilize the appropriate DNA conformation, involve a large number of residues that are distributed along most of the polypeptide chain. In addition, other residues that are involved in the subunit interactions in the protein dimer ensure that each subunit is properly poised for binding to its half site in the operator DNA. Thus, the "unit" that is responsible for the differential binding to different operator DNA regions is really an entire binding domain, appropriately dimerized, and nearly all the protein–DNA contacts contribute to this specificity.

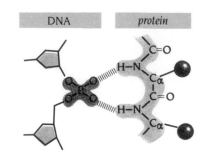

Figure 7.18 Nonspecific protein–DNA interactions are frequently formed by hydrogen bonds (striped) between backbone phosphate oxygen atoms of DNA (red) and main-chain NH groups of the protein (blue).

Local DNA structure modulates repressor binding

The ability of the protein–DNA contacts to accomplish proper changes in the DNA structure, in particular the overwinding in the central region, can be modulated by the actual nucleotide sequence of the DNA. The overwinding causes a narrowing of the minor groove in the central region of the operator. It is generally believed that A-T and T-A base pairs can more readily be accommodated in narrow grooves than G-C or C-G base pairs. This means that DNA with A-T, T-A base pairs in the central region of operator sequences should be able to adopt the conformation necessary for proper DNA–protein interaction more readily than DNA with G-C, C-G base pairs in these positions. This is confirmed by mutation experiments using synthetic operator regions; these show that it is possible to change the affinity between repressor and DNA by changing the base pairs in the middle region. Thus, if base pairs 7 and 8 are changed from T-A and A-T to G-C and C-G, the affinity for the repressor fragment is decreased fifteenfold. In other words, the base sequence in this region, by influencing local conformational changes in the DNA, alters the affinity of repressor for DNA. Here we have a case of local DNA structure, rather than direct sequence-specific DNA–protein interactions, modulating repressor binding.

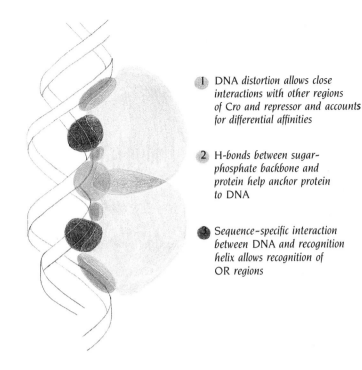

1. DNA *distortion allows close interactions with other regions of Cro and repressor and accounts for differential affinities*

2. H-*bonds between sugar-phosphate backbone and protein help anchor protein to* DNA

3. *Sequence-specific interaction between* DNA *and recognition helix allows recognition of* OR *regions*

Figure 7.19 Schematic diagram of the main features of the interactions between DNA and the helix-turn-helix motif in DNA-binding proteins.

The essence of phage repressor and Cro

The DNA-binding proteins Cro and repressor of the three phages lambda, P22, and 434 all recognize operator regions in DNA through a helix-turn-helix motif in their structures. The second α helix of this motif, the recognition helix, binds in the major groove of DNA. The proteins form dimers such that the two identical recognition helices are 34 Å apart, at the two ends of the elongated dimer, and bind to the ends of the operator DNA regions.

The first two amino acid residues of the recognition helix in these two proteins from phage 434 are glutamines, which form specific hydrogen bonds to the first two base pairs of the operator region. These interactions are identical in all operator regions of phage 434. A few residues on the recognition helix of the helix-turn-helix motif thus form the recognition signal of Cro and repressor proteins for the operator regions by specific interaction with the edge of the base pairs in the major groove of DNA as predicted by Matthews.

Such sequence-specific interactions do not, however, account for the differential affinity of Cro and repressor for the operator regions. These affinities are instead determined by the ability of the DNA to undergo specific structural changes. At least two factors are important to the ability of the complex to achieve the proper DNA conformation (Figure 7.19). Nonspecific protein–DNA interactions between the sugar-phosphate backbone of DNA and regions of the protein form a large interaction area of complementary surfaces that stabilizes the structural change in the DNA. These protein regions are not restricted to the helix-turn-helix motif but are spread over almost the entire polypeptide chain. The ability of these interactions to accomplish a structural change can be modulated by the sequence of DNA. A-T base pairs in this region facilitate the change whereas G-C base pairs do not.

Amino acid sequence relations identify helix-turn-helix motifs

Polypeptide chains that form a specific structural motif frequently show no or very low **sequence homology**, even when they are associated with the same specific function. Such motifs are therefore very difficult, if not impossible, to identify from amino acid sequences alone. One approach is to determine a number of amino acid sequences from related systems and hope that they will form a consistent group such that if protein A and B show no sequence homology, there is a protein C in the group that is homologous to both A and B. If one member of such a group is known to have a specific structural motif, then all members may have it.

Another approach is to ask the following question: Is it possible to use three-dimensional information about the motif to increase the probability of identifying amino acid sequences that form such a motif? The helix-turn-helix motif provides one of the few examples where both approaches have been successful.

Phage Cro and repressor proteins are evolutionarily related

The Cro and repressor proteins from phage lambda share a similar function that is performed by a similar structure: the helix-turn-helix motif. The remaining parts of their structures are, however, quite different. The repressor has a large C-terminal domain that is absent from Cro. The N-terminal DNA-binding domain of the repressor is entirely α helical, whereas Cro has a three-stranded antiparallel β sheet. We can now ask the following question: Are these similarities, and differences, reflected in the amino acid sequences to the extent that we can trace evolutionary relations between these two proteins?

A comparison of the amino acid sequences of lambda Cro and the N-terminal domain of the lambda repressor shows no statistically significant homology, even in the 20 residues that comprise the helix-turn-helix motif. From these sequences alone, therefore, it is not possible to draw any conclusions about possible evolutionary relationships between these two proteins.

This picture changed, however, when the amino acid sequences became available for the complete set of repressor and Cro proteins from all three related phages lambda, 434, and P22. This allowed a systematic search for sequence homologies using pairwise comparisons of Cro proteins and the N-terminal domains of the repressors. The results of this comparison are shown in simplified form in Table 7.1, where the percentage identities for these 15 different pairwise comparisons are listed.

Table 7.1

The results of pairwise amino acid sequence comparisons of Cro and repressor (R) proteins from phages lambda, 434, and P22

| | | Lambda | | 434 | | P22 | |
		Cro	R	Cro	R	Cro	R
Lambda	Cro	—	17	22	20	16	17
Lambda	R	17	—	17	26	20	14
434	Cro	22	17	—	48	17	30
434	R	20	26	48	—	15	30
P22	Cro	16	20	17	15	—	22
P22	R	17	14	30	30	22	—

An average of 65 amino acids were aligned in each comparison. The percentage identities are listed.

Some of these comparisons reveal obvious sequence homology; for example, the Cro and repressor proteins from phage 434 have 48% amino acid sequence identity. Other pairs clearly show no significant homology. A rigorous statistical analysis of these comparisons established that those pairs which have a sequence identity of 22% or more (Table 7.1) are significantly homologous. The pairs that do not show significant sequence homology can all be related to the homologous pairs. For example, the Cro and repressor from phage lambda have only 17% identity, which is not significant. However, lambda Cro is homologous to 434 Cro (22% identity), which, in turn, is homologous to 434 repressor (48% identity), which is homologous to lambda repressor (26% identity). This analysis showed that all six proteins form a consistent homologous set in which each member is evolutionarily related to all others. They have all evolved from a common ancestral gene.

The repressors may have evolved from an ancestral Cro or a single domain DNA-binding protein by recombinational addition of a C-terminal domain. In all three repressors the C-terminal domains are homologous to each other but quite different in sequence from the N-terminal domains. Alternatively, the one-domain Cro protein may have evolved from a repressor by loss of its C-terminal domain. This alternative is less likely since the function of the C-terminal domain is to form more stable dimers. The increased binding affinity conferred by dimerization probably evolved later and independently in Cro and repressor, since they form stable dimers in such different ways. Consequently, it is likely that the common ancestor to these DNA-binding domains was a small monomeric DNA-binding protein with the helix-turn-helix motif.

Cro and repressor have homologous sequences but partly different structures

The Cro and repressor proteins from phage lambda belong to a set of homologous and evolutionarily related proteins; they are, however, not similar in their three-dimensional structures. The only structural correspondence between them is the helix-turn-helix motif of 20 residues. The remaining regions are quite different with strands of β sheet in Cro and α helices in the repressor. Not even the additional α helix prior to the motif is formed by sequentially aligned regions (Figure 7.20).

We are thus confronted with an unusual situation: the amino acid sequences are more conserved than the three-dimensional structures. For example, the overall amino acid sequence identity between the whole lambda Cro (66 amino acids) and 434 repressor fragment (69 amino acids) is 20%. The homology is not significantly higher in the structurally conserved helix-turn-helix motif than in the regions where the two protein structures are different. For these related proteins, which are small and have no obvious core in their subunit structures, the regions outside the functional DNA-binding motif have been subject to changes during evolution at the structural level, including changes of secondary structure.

Sequence comparison using strong stereochemical constraints identifies helix-turn-helix motifs

To preserve the precise three-dimensional structure of a motif, there are always some crucial positions that can only be occupied by a few amino acids with special properties. The presence of other amino acids will distort the structure. In particular, invariant glycine residues are often found in such motifs.

The three-dimensional structure of the DNA-binding helix-turn-helix motif is strictly conserved. It comprises 20 residues (Figure 7.21). The first helix is formed by residues 1–7, the loop by residues 8–11, and the second helix, the recognition helix, by residues 12–20. In some motifs there are additional residues at the beginning of the first helix or at the end of the second helix, but the central 20 residues are always present.

Comparison of the three-dimensional structures of the helix-turn-helix motif in Cro and repressor of lambda and in CAP protein of *E. coli* revealed **stereochemical constraints** in several positions. It was then possible to define some conditions that the amino acid sequence of such a motif must fulfill to avoid structural distortions.

Figure 7.20 Comparison of elements of secondary structure in the polypeptide chains of Cro (a) and repressor (b) from phage lambda. The chains have been aligned by amino acid sequence comparisons with the other members of the homologous set of repressors and Cro proteins. Red regions with a black stripe are the helix-turn-helix motifs, other α helices are blue, and strands of β sheets are green.

(a)   1   66

(b)   1   92   C-terminal domain

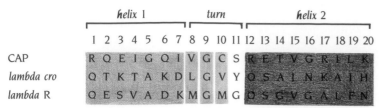

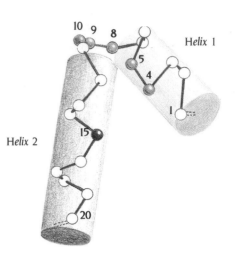

Figure 7.21 (a) Amino acid sequences for the helix-turn-helix motif of CAP, lambda Cro, and lambda repressor. (b) Schematic diagram of the helix-turn-helix motif. The positions subject to constraints in the amino acid sequence are colored. The C_α positions of each residue have been projected onto a plane, and the helical regions are outlined.

1. Residue 9 must be a glycine (G). The structure of the loop is such that this residue has a conformation corresponding to that of a residue in a left-handed α helix. This is common for glycine but occurs very rarely with other amino acids because of steric interference of the β carbon atom with the main chain.

2. Residues 4 and 15 are completely buried within the protein, and residues 8 and 10 are almost buried. All four of these residues should be hydrophobic or possibly weakly polar. In particular, residues 4 and 15 should not be charged.

3. The helical residues 3–7 and 15–20 are unlikely to be proline. Proline can be accommodated at the amino terminus of a helix but not within the body due to its inability to form a hydrogen bond from its main-chain nitrogen.

4. Residue 5 should not have a branched side chain. This side chain is wedged between the two helices, and a branched side chain would change the relative alignment of the helices.

Brian Matthews's model building of the binding of the helix-turn-helix motif to B-DNA also showed that the most likely candidates for amino acid side-chain interactions with nucleotide bases in the major groove were residues 11–13, 16–17, and 20. This prediction was validated when Harrison showed that in the 434 repressor–DNA complex Gln 28 and 29 (corresponding to residues 12 and 13 of the helix-turn-helix motif) were indeed involved in specific interactions with the DNA.

Having defined these stereochemical restrictions, Matthews used them in a search through all the known amino acid sequences of DNA-binding proteins. He first compared the sequences of fragments of these proteins with a master set of six helix-turn-helix sequences (most of them listed in class 1 of Table 7.2) and obtained a list of fragments with weak sequence homology to all members of this set. He then applied the stereochemical restrictions as a mask on these fragments and extracted those that fulfilled these restrictions. The sequences he extracted in this way are listed in the second class of Table 7.2, except CAP and *lac* R which belonged to the master set.

How valid is this method? In 1983, when Matthews identified these putative helix-turn-helix motifs, the three-dimensional structure of the *trp* repressor had not been determined. When this was achieved to 2.6 Å resolution by Paul Sigler's group at the University of Chicago in 1985, the helix-turn-helix motif in the structure corresponded exactly to the sequence of the fragment identified by Matthews. As this example shows, stereochemical constraints for a specific motif are, when properly used, a powerful complement to normal sequence comparisons when there is only a weak sequence homology. There is the danger, however, that sequences may be included that fortuitously obey the stereochemical constraints without having the three-dimensional structure from which the constraints were derived.

Sequence comparisons without using stereochemical constraints do not identify helix-turn-helix motifs unambiguously

What happens when sequence homology searches for DNA-binding regions are made without applying stereochemical constraints? In one such study the sequence of the CAP helix-turn-helix motif was compared with the sequences of procaryotic DNA-binding proteins. This simple homology search, in addition to extracting several of the sequences listed in classes 1 and 2 of Table 7.2 also extracted the set of sequences in class 3. These latter sequences when they were compared pairwise exhibited about as much homology to CAP and to each other as did the sequences in classes 1 and 2, namely, three to six identities.

Table 7.2

Sequence alignment of presumed helix-turn-helix DNA-binding motifs

	1	2	3	4	5	6	7	8	9	10	11	12	13	14	15	16	17	18	19	20
Class 1																				
lambda Cro	Q	T	K	T	A	K	D	L	G	V	Y	Q	S	A	I	N	K	A	I	H
lambda R	Q	E	S	V	A	D	K	M	G	M	G	Q	S	G	V	G	A	L	F	N
P22 Cro	Q	R	A	V	A	K	A	L	G	I	S	D	A	A	V	S	Q	W	K	E
P22 R	Q	A	A	L	G	K	M	V	G	V	S	N	V	A	I	S	Q	W	Q	R
434 Cro	Q	T	E	L	A	T	K	A	G	V	K	Q	Q	S	I	Q	L	I	E	A
434 R	Q	A	E	L	A	Q	K	V	G	T	T	Q	Q	S	I	E	Q	L	E	N
Class 2																				
CAP	R	Q	E	I	G	Q	I	V	G	C	S	R	E	T	V	G	R	I	L	K
trp R	Q	R	E	L	K	N	E	L	G	A	G	I	A	T	I	T	R	G	S	N
lac R	L	Y	D	V	A	E	Y	A	G	V	S	Y	Q	T	V	S	R	V	V	N
gal R	I	K	D	V	A	R	L	A	G	V	S	V	A	T	V	S	R	V	I	N
Mat a1	K	E	E	V	A	K	K	C	G	I	T	P	L	Q	V	R	V	W	C	N
lex R (28–47)	R	A	E	I	A	Q	R	L	G	F	R	S	P	N	A	A	E	E	H	L
Class 3																				
lex R (141–160)	N	G	Q	V	V	V	A	R	I	D	D	E	V	T	V	K	R	L	K	K
ara C	A	K	L	L	L	S	T	T	R	M	P	I	A	T	V	G	R	N	V	G
lys R	L	T	E	A	A	H	L	L	H	T	S	Q	P	T	V	S	R	E	L	A
tnp R	E	A	K	L	K	G	I	K	F	G	R	R	R	T	V	D	R	N	V	V
Class 4																				
Mat a1	K	E	E	V	A	K	K	C	G	I	T	P	L	Q	V	R	V	W	C	N
Mat α 2	L	E	N	L	M	K	N	T	S	L	S	R	I	Q	I	K	N	W	V	S
ftz	R	I	D	I	A	N	A	L	S	L	S	E	R	N	I	K	I	W	F	Q
Antp	R	I	E	I	A	H	A	L	C	L	T	E	R	Q	I	K	I	W	F	Q
Rad10	R	G	R	S	V	L	F	L	T	L	T	Y	H	K	L	Y	V	D	Y	I
ERCC-1	Q	S	T	C	A	L	F	L	S	L	R	Y	H	N	L	H	P	D	Y	I

Residue positions subject to stereochemical constraints are colored. In class 1 are Cro and repressor proteins from the related phages lambda, P22 and 434. In class 2 are examples of other procaryotic DNA-binding proteins (and one from yeast) where the sequence regions listed here fulfill the stereochemical constraints for the motifs. Class 3 are examples of suggested DNA-binding regions from procaryotic proteins that do not fulfill the stereochemical constraints. In class 4 are some examples of suggested DNA-binding regions from eucaryotic proteins that to some extent fulfill the stereochemical constraints. The proteins in classes 2–4 are as follows:

CAP = catabolite activating protein
trp R = tryptophan repressor from *E. coli*
lac R = lactose repressor from *E. coli*
gal R = galactose repressor from *E. coli*
lex R = rec A repressor from *E. coli*
ara C = arabinose repressor from *E. coli*
lys R = lysine repressor from *E. coli*
tnp R = transposon tn3 repressor from *E. coli*

Mat a1 and α2 = yeast mating-type regulatory protein
ftz = *Drosophila fushi tarazu* homeo box protein
Antp = *Drosophila antennapedia* homeo box protein
Rad10 = DNA repair protein from yeast
ERCC-1 = DNA repair protein from human

Three to six identities in a segment of 20 residues are far from significant. To illustrate this point, we extracted sequence fragments that showed identities to the sequence of the helix-turn-helix motif of the *trp* repressor shown in Table 7.2 using the sequence databases available at the time of writing this book. We obtained about 10,000 fragments that showed five identities, 1,000 that showed six identities, about 100 with seven, 10 with eight, and 1 with nine identities. This last was hemocyanin, an oxygen carrier in mollusks with no known DNA-binding function. Thus sequence homology alone among these fragments is not a valid criterion of a DNA-binding motif, unless the homology is very extensive or there is significant sequence homology within other regions of the proteins.

By contrast with the fragments in the first two classes, the fragments in class 3 do not obey the stereochemical restrictions that are required for a helix-turn-helix motif. None of them has a glycine at position 9. Residue 5 has a branched side chain in two of the cases. Residues 8 and 10 are both charged in one case and so on. The region (residues 141–160) of the rec A repressor from *E. coli* (lex R) that was extracted by the simple search for homology is in the C-terminal domain, which has significant amino acid sequence homology to the C-terminal domains of the lambda 434 and P22 repressors, but the latter are not involved in DNA binding. When one uses stereochemical constraints, it is, however, possible to identify a different region, residues 28–47, in the N-terminal domain of the rec A repressor as a putative helix-turn-helix motif. This region is listed in class 2 of Table 7.2. We conclude, therefore, that despite sequence homologies the fragments, listed in class 3, are probably not DNA-binding helix-turn-helix motifs.

The sequences shown in class 4 in Table 7.2 are all from eucaryotic DNA-binding proteins, and to some extent they fulfill the stereochemical restrictions of the procaryotic helix-turn-helix motif. There are significant sequence homologies among some members of this set. Notice that the yeast mating-type regulatory protein (Mat a1) sequence was identified as having a helix-turn-helix motif and belonging to class 2. When we look closely at the sequences in class 4, we see that some of the stereochemical restrictions are violated at some sites. For example, the yeast DNA repair protein (Rad10) sequence has a valine residue with a branched side chain at position 5, while the human DNA repair protein (ERCC-1) sequence has a proline residue at position 17. Furthermore, in all cases, with the exception of Mat a1, residue 9 is not glycine but serine, threonine, or cysteine. These three amino acids have small side chains, however, and recent genetic experiments have shown that glycine at this position is not absolutely essential for the function of lambda repressor. Recent structure determinations have shown that the helix-turn-helix motif can accommodate serine or cysteine in position 9. It is clear that these class-4 sequences represent a borderline case. They may well have a structure closely resembling the procaryotic helix-turn-helix motif but that can only be established by examining the structure of these proteins. In fact, the homeo domain Antp has recently been shown to have the helix-turn-helix motif in its structure as we shall discuss further in Chapter 8. This analysis of sequence homologies cannot yield unambiguous answers but it provides useful working hypotheses.

DNA binding is regulated by allosteric control

The DNA-binding capacity of most repressors and activators is regulated by small molecules, such as sugars, amino acids, or cyclic AMP, which bind to a distinct site on the protein. This binding causes a conformational change which alters, and therefore regulates, the DNA binding site and its affinity for DNA. Jacques Monod, at the Pasteur Institute in Paris, introduced the term allosteric for these binding sites to emphasize that the small ligands, known as **allosteric effectors**, and the sites at which they bind are sterically quite distinct from the functional binding sites. We will discuss two examples of allosteric control of DNA binding—the tryptophan (*trp*) repressor and the CAP protein, both from *E. coli*.

The trp *repressor forms a helix-turn-helix motif*

The *trp* repressor controls the operon for the synthesis of L-tryptophan in *E. coli* by a simple **negative feedback** loop. In the absence of L-tryptophan, the repressor is inactive, the operon is switched on and L-tryptophan is produced.

Figure 7.22 The subunit of the *trp* repressor. The subunit contains 107 amino acid residues that are folded into six α helices. Helices 4 (blue) and 5 (red) form the DNA-binding helix-turn-helix motif. (Adapted from R. Schevitz et al., *Nature* 317: 782, 1985.)

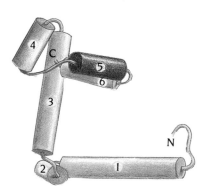

As the concentration of L-tryptophan increases, it binds to the repressor and converts it to an active form so that it can bind to the operator region and switch off the gene.

The structure of the *trp* repressor has been determined both with and without bound tryptophan to 1.8 Å resolution. The *trp* repressor is a dimer, like the other proteins we have discussed in this chapter. The polypeptide chain has 107 amino acids and is folded into six α helices (Figure 7.22). The structure is therefore α type like the lambda repressor, but the arrangement of the helices in the subunit is quite different. The six helices do not pack to form a regular structure with a hydrophobic core, and it is doubtful if monomers with the arrangement of helices shown in Figure 7.22 would be stable alone. Stability is conferred by the dimerization because the two subunits fit together to give a functional molecule that, in contrast to its subunits, has a compact globular form (Figure 7.23).

The helices at the N-terminal regions of the two polypeptide chains are intertwined and make extensive contacts in the central part of the molecule (Figure 7.23) to form a stable core. This core supports two "heads," one from each polypeptide chain, comprising the last three helices (Figure 7.23). In the middle of the subunit chain α helix 3 is quite long and forms the main link between the core and the head.

In these heads α helices 4 and 5 form the helix-turn-helix motif, which was accurately predicted by Matthews to be present in this structure. Helix 5 should thus be the recognition helix for DNA binding, and this is supported by both mutant studies and structural studies of a repressor-DNA complex.

A conformational change provides the molecular mechanism of the functional switch

What is the molecular mechanism by which the binding of tryptophan—the allosteric effector—confers DNA-binding capacity on this molecule? The structures of the *trp* repressor with and without bound tryptophan provided a simple and satisfying answer. Tryptophan binding causes a **conformational change** of the molecule that alters the orientations of the recognition helices. Two tryptophan molecules bind to the dimeric repressor in two identical cavities between the heads and the core (Figure 7.24b). They are wedged between the long connecting α helices (3) and the recognition helices of the heads. When tryptophan is bound in these cavities, the recognition helices are properly

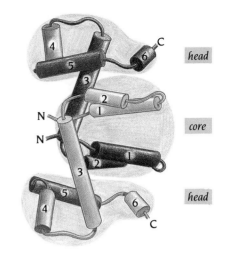

Figure 7.23 The α helices of the N-terminal region of the *trp* repressor are involved in subunit interactions and form a stable core in the middle of the dimer. α helices 4–6, which include the helix-turn-helix motif, form two "head" regions at the two ends of the molecule. α helix 3 connects the core to the head in both subunits. (Adapted from Schevitz et al., *Nature* 317: 782, 1985.)

Figure 7.24 Schematic diagrams of docking the *trp* repressor to DNA in its inactive (a) and active (b) forms. When L-tryptophan, which is a corepressor, binds to the repressor, the "heads" change their positions relative to the core to produce the active form of the repressor, which binds to DNA. The structures of DNA and the *trp* repressor are outlined. The positions of the DNA recognition helices (5) and the helices (3) that connect the core with the heads are indicated. The approximate position of the side chain of residue 77 is marked as a green ball (see text for the significance of this residue). (Adapted from Zhang et al., *Nature* 327: 591, 1987.)

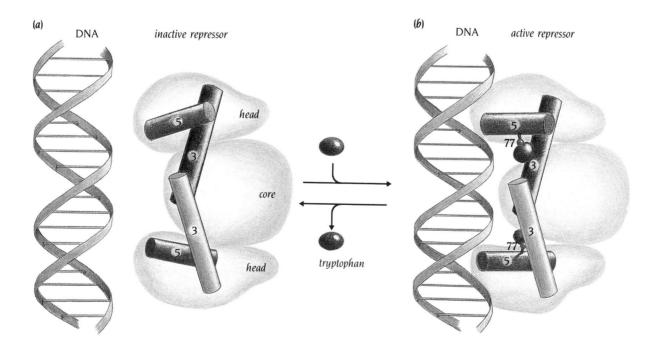

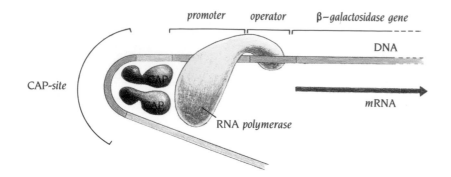

Figure 7.25 Catabolite gene activating protein, CAP, is a DNA-binding protein that assists RNA polymerase to bind more effectively to certain promoters and thereby CAP enhances the rate of initiation of RNA synthesis. A preliminary x-ray structure determination in the laboratory of Tom Steitz, Yale University, of a complex between CAP and a DNA fragment has shown that the CAP dimer induces sharp bends in DNA. *In vivo* CAP binding might induce a loop in DNA so that regions far from the operator site can interact with RNA polymerase. This might result in tighter binding of the enzyme to the DNA. (Adapted from T. Steitz, *Q. Rev. Biophys.* 23:229, 1990.)

poised 34 Å apart for binding DNA in the major groove, in a manner similar to that of 434 repressor. In the absence of tryptophan the position of the heads changes; they tilt inward removing the empty cavity, and in particular the recognition helices fold in toward the core (Figure 7.24a). This makes the distance between the two DNA binding sites of the repressor molecule too short by 5–6 Å to allow them to fit into the major groove. Specific DNA binding is abolished, and the repressor is inactive.

The elegant genetic studies by the group of Charles Yanofsky at Stanford confirm this mechanism. The side chain of Ala 77, which is in the loop region of the helix-turn-helix motif, faces the cavity where tryptophan binds. When this side chain is replaced by the bulkier side chain of Val, the mutant repressor does not require tryptophan to be able to bind specifically to the operator DNA. The presence of a bulkier valine side chain at position 77 maintains the heads in an active conformation even in the absence of bound tryptophan. The crystal structure of this mutant repressor, in the absence of tryptophan, is basically the same as that of the wild-type repressor with tryptophan. This is an excellent example of how ligand-induced conformational changes can be mimicked by amino acid substitutions in the protein.

CAP is a positive control element

CAP, catabolite gene activating protein, is a DNA-binding protein that assists RNA polymerase in binding effectively to certain promoters. In other words, it is a **positive control protein** that acts by allowing more frequent initiation of RNA synthesis. Alone, CAP is a nonspecific DNA-binding protein like the trp repressor, but it is converted by binding cyclic AMP to a form that binds strongly to specific operator regions as shown in Figure 7.25.

CAP controls a number of operons, all of which are involved in the breakdown of sugar molecules and one of which is the *lac* operon. When the level of the breakdown products of lactose is low, the concentration of cyclic AMP in the cells increases and CAP is switched on, binds to its specific operators, and increases the rate of transcription of adjacent operons.

The polypeptide chain of CAP folds into two domains

The CAP molecule, the structure of which was determined to 2.5 Å resolution in 1981 in the laboratory of Tom Steitz at Yale University, comprises two identical polypeptide chains of 209 amino acid residues (Figure 7.26). Each chain is folded into two domains that have separate functions (Figure 7.26b). The N-terminal domain binds the allosteric effector molecule, cyclic AMP; in addition, it provides all the subunit interactions that form the dimer. These are mainly interactions between two long helices, one (helix C) from each subunit. The C-terminal domain contains the helix-turn-helix motif that binds DNA. The structure of CAP thus illustrates an important principle common to many allosteric proteins. The allosteric and functional binding sites are often separated into different domains. In some allosteric proteins they reside in different subunits but whether they are in different domains or different subunits is more a reflection of genomic organization than structural requirements.

(a)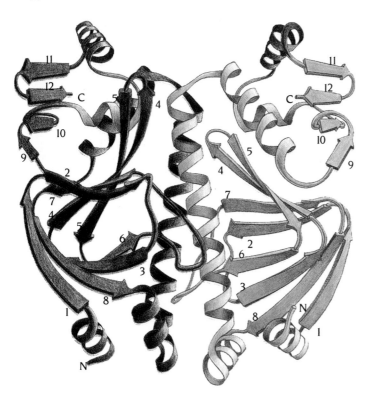

(b)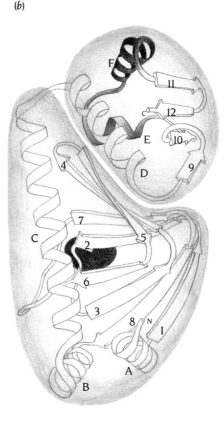

The major structural feature of the N-terminal domain of CAP, residues 1–135, is an eight-stranded jelly roll β barrel. This class of barrel structure was discussed in Chapter 5 using influenza virus hemagglutinin as an example. In contrast to other jelly roll barrels whose interiors are filled with hydrophobic side chains, the barrel in CAP contains a pocket that forms a major part of the cyclic AMP binding site (Figure 7.26b).

One cyclic AMP molecule is bound to each subunit. The other side of the binding pocket, away from the barrel, is lined with residues from the long α helix that connects the jelly roll structure with the DNA-binding domain (Figure 7.26b). This binding site is close to the subunit interface and deeply buried in the structure of the dimer. Each cyclic AMP molecule is in contact with amino acids from both subunits.

The C-terminal domain, residues 136–209, is built up from a β sheet of four antiparallel β strands and, in addition, three α helices (Figure 7.26b). The structure of this domain is thus of the α + β type, similar to lambda Cro. The arrangement of the secondary structure elements along the polypeptide chain is, however, somewhat different from that of Cro. One α helix (D) is at the N-terminal end of the domain and the other two (E and F) form the helix-turn-helix motif between β strands 10 and 11 in the molecule, but strands 2 and 3 in the C-terminal domain. The CAP dimer is formed in such a way that in the presence of cyclic AMP the two recognition helices F, one from each subunit, are arranged 34 Å apart as they are in Cro and the repressors (Figure 7.27).

A molecular mechanism has been suggested for the switch in CAP

The structure of CAP in the absence of cyclic AMP is unknown. Tom Steitz, nonetheless, has been able to suggest possible mechanisms for the allosteric transition mediated by the binding of cyclic AMP from a nonspecific DNA-binding protein to a specific DNA-binding protein. Since cyclic AMP makes no contact with the DNA-binding domains, it cannot induce a conformational change in them by direct interaction such as that which occurs in the *trp*

Figure 7.26 (a) Schematic diagram of the CAP molecule, which is a dimer. Each subunit (green and brown) contains 207 amino acids that fold into two domains, one of which contains the DNA-binding helix-turn-helix motif, colored blue and red in the diagram. The β strands are numbered (1–12) from the N terminus. (Adapted from T. Steitz and I. Weber, in *Biological Macromolecules and Assemblies*, eds. F. Jurnak and A. McPherson. Vol. 2. New York: Wiley, 1985.)

(b) Schematic diagram of the structure of one subunit of CAP. The β strands (1–12) and α helices (A–F) are labeled from the N terminus. The DNA-binding helix-turn-helix motif is colored blue and red with the recognition helix red. The binding site for cyclic AMP is purple. (Adapted from D. McKay and T. Steitz, *Nature* 290: 744, 1981.)

107

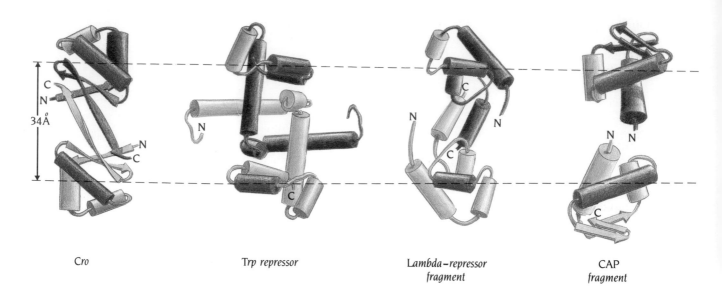

| Cro | Trp repressor | Lambda–repressor fragment | CAP fragment |

repressor. It is quite possible, however, that cyclic AMP induces a change in the relative orientation of the two subunits, since it binds close to the subunit interaction area. Such a change could be relayed to the DNA-binding domain and could change the relative position and orientation of the recognition helices of the two helix-turn-helix motifs in a similar way to the *trp* repressor. Such a change would have analogous effects on the DNA-binding properties of the CAP dimer.

Met *and* arc *repressors belong to a family of β-sheet DNA-binding proteins*

Many procaryotic regulatory proteins have DNA-binding domains whose amino acid sequences do not satisfy the steric requirements of helix-turn-helix motifs; therefore, additional structural motifs for specific DNA binding must exist. One such is exemplified by the *met* repressor in *Escherichia coli.* As long ago as 1959 the French molecular biologists Francois Jacob and George Cohen discovered, by genetic experiments, the regulatory system in *E. coli* that controls the biosynthesis of methionine. Beside being the amino-terminal amino acid at the beginning of all newly synthesized polypeptide chains in *E. coli*, methionine is also the precursor to S-adenosylmethionine, which is the chief donor of methyl groups in a variety of biochemical pathways. The *met* repressor uses S-adenosylmethionine as co-repressor to control its own gene as well as structural genes for enzymes involved in the synthesis of both methionine and S-adenosylmethionine.

The *met* repressor polypeptide chain is 104 amino acids long and forms stable dimers in solution. Its gene has been cloned, the repressor protein overexpressed, and the x-ray structure of the repressor–co-repressor complex has been determined to 1.7 Å resolution in the laboratory of Simon Phillips at Leeds University, England.

The dimeric *met* repressor molecule is formed by two highly intertwined monomers (Figure 7.28) analogous to the dimeric *trp* repressor (Figure 7.22). The subunit structures of the *met* and *trp* repressors are different, however. Each subunit of the *met* repressor is folded into three α helices and one β strand, which together account for about 50% of the polypeptide; the remainder of the polypeptide chain consists of turns or regions of irregular structure. Within the dimer the subunits are arranged in such a way that the β strands, one from each subunit, form a two-stranded antiparallel β sheet (Figure 7.28). This β sheet forms a protrusion at the surface of the molecule. A similar structure has been deduced for the *arc* repressor, based on NMR studies in the laboratory of Robert Kaptein at Utrecht University. The *arc* repressor of *Salmonella* bacteriophage P22 is involved in the switch between lysis and lysogeny.

Figure 7.27 Comparison of the positions of the recognition helices in the dimeric structures of Cro, *trp* repressor and the DNA-binding domains of CAP and lambda repressor. The helix-turn-helix motifs are colored blue and red with the recognition helix in each case red.

Figure 7.28 Schematic diagram of the *met* repressor molecule. One subunit of the dimeric molecule is blue, and the other is green. The main features of each subunit are from its amino end: a flexible loop region that changes its structure when the repressor binds to DNA, a β strand that in the dimeric molecule forms a two-stranded antiparallel β sheet with the corresponding β strand from the other subunit, and three α helices, αA, αB, and αC. (Adapted from J.B. Rafferty et al., *Nature* 341: 705–710, 1989.)

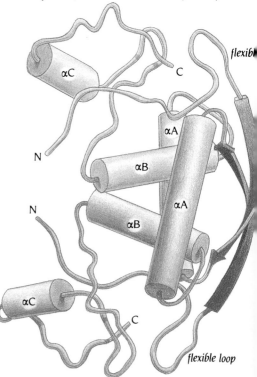

Simon Phillips has also determined the x-ray structure to 2.8 Å resolution of a complex between the *met* repressor, its corepressor, and a synthetic DNA fragment containing 18 base pairs. The DNA sequence of this fragment includes two copies of the consensus sequence 5′–A G A C G T C T–3′ (the *met* box) found in the operator regions for *met* repressor. The repressor–DNA complex consists of two dimeric repressor molecules, each of which binds to a *met* box in the duplex DNA fragment.

The structure of the *met* repressor dimer is essentially unchanged by binding to DNA with one important exception. Eight amino acid residues between the N terminus of the polypeptide chain and the beginning of the β strand form a flexible loop (Figure 7.28) that changes its conformation when the repressor binds to DNA. The DNA fragment is essentially in the B-DNA form but with kinks in the middle of the two *met* box sequences.

There are three main interaction areas between one operator region of DNA and one dimeric repressor molecule (Figure 7.29). One of these is formed by the two-stranded β sheet of the repressor that binds in the major groove of DNA. Side chains from the β strands interact with base pairs within the operator sequence. These interactions form the main sequence-specific protein–DNA interactions in the *met* repressor–DNA complex. The kink in the DNA structure narrows the major groove in this region so that the interaction between protein and DNA is very tight.

Two other regions of the *met* repressor, the flexible loop and the amino end of helix αB (Figure 7.29) also interact with DNA. These interactions are mainly hydrogen bonds between main-chain NH groups of the repressor and phosphate groups of DNA. These interactions contribute to the stability of the repressor–DNA complex. The *met* repressor apparently recognizes its operator sequence mainly by interactions that involve amino acid side chains from the β sheet of the dimeric repressor molecule. Such a two-stranded β sheet therefore constitutes another DNA-binding motif, which so far has been recognized only in the *met* and *arc* repressors but which might well prove to be of more general occurrence.

Conclusion

Many proteins that switch off or on gene expression in bacteria are dimeric molecules, and the DNA sequences that they specifically recognize are palindromic at their ends. The twofold symmetry of the protein is therefore matched by twofold symmetry at the ends of the recognition sequence.

The monomeric subunits have a helix-turn-helix motif that functions as the specific DNA-binding region. The second helix of this motif is the recognition helix, and side chains of its residues form hydrogen bonds and hydrophobic contacts with nucleotide bases at the bottom of the major groove in B-DNA. These interactions form the recognition signal of Cro and repressor proteins for the operator regions. In the dimeric protein molecules recognition helices in each of the two helix-turn-helix motifs, at opposite ends of the elongated molecule, are separated by 34 Å, corresponding to one turn of B-DNA. When one recognition helix binds to the DNA major groove, the second, 34 Å away, is positioned to bind into the major groove one helical turn along the DNA.

Binding of 434 Cro and repressor to operator regions imposes specific and different structural changes of the DNA. The differential affinity of these proteins for the operator regions is determined by the ability of the DNA to undergo structural changes. The different DNA conformations are stabilized by nonspecific protein–DNA interactions between the sugar-phosphate backbone of DNA and regions of the protein that form a large interaction area of complementary surfaces. The ability of these interactions to accomplish a structural change can be modulated by the actual sequence of DNA in the middle of the operator regions.

Analysis of the mode of binding of the helix-turn-helix motifs to DNA has revealed a set of stereochemical constraints that the motif must fulfill. These can now be used to predict the presence of this DNA-binding motif in other DNA-binding proteins whose structure is not yet known.

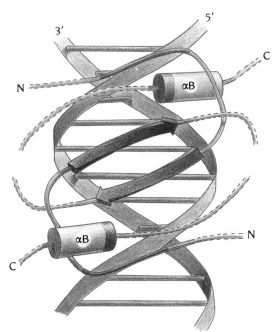

Figure 7.29 Simplified diagram of the *met* repressor–DNA structure illustrating the regions of the dimeric *met* repressor that contact DNA. The two-stranded β sheet of the repressor is bound in the major groove of DNA, where it forms the sequence-specific interactions. The flexible loop regions near the N termini of the polypeptide chains, and the amino ends of helices αB interact with the phosphate backbone of DNA. DNA is orange, one subunit of the dimeric repressor is blue, and the other subunit is green. Regions of the protein that interact with DNA have darker color. Regions of the repressor far removed from the DNA are absent or dotted. (Adapted from unpublished diagrams, courtesy of S. Phillips.)

Some of these procaryotic DNA-binding proteins are activated by the binding of an allosteric effector molecule. This event changes the conformation of the dimeric protein, causing the helix-turn-helix motifs to move so that they are 34 Å apart and able to bind to the major groove. In the CAP protein the domain that binds the allosteric effector—cyclic AMP—is distinct from the DNA-binding domain. This illustrates the modular nature of many allosteric proteins. The structure of a second distinct class of DNA binding proteins exemplified by the *met* and *arc* repressors has recently been determined. These are dimeric molecules in which a two-stranded β sheet binds in the major groove of B-DNA. Side chains of the β strands interact with base pairs in the operator sequence.

Selected readings

General

Anderson, W.F., et al. Proposed α-helical super-secondary structure associated with protein-DNA recognition. *J. Mol. Biol.* 159: 745–751, 1982.

Brennan, R.G.; Matthews, B.W. The helix-turn-helix DNA-binding motif. *J. Biol. Chem.* 264: 1903–1906, 1989.

Brennan, R.G.; Matthews, B.W. Structural basis of DNA-protein recognition. *Trends Biochem. Sci.* 14: 286–290, 1989.

Harrison, S.C.; Aggarwal, A.K. DNA recognition by proteins with the helix-turn-helix motif. *Annu. Rev. Biochem.* 59: 933–969, 1990.

Ptashne, M. Repressors. *Trends Biochem. Sci.* 9: 142–145, 1984.

Ptashne, M. A genetic switch: gene control and phage lambda. Palo Alto, CA: Blackwell, 1986.

Ptashne, M.; Johnson, A.D.; Pabo, C.O. A genetic switch in a bacterial virus. *Sci. Am.* 247: 128–140, 1982.

Saint-Girons, I., et al. Methionine biosynthesis in enterobacteriaceae: biochemical, regulatory, and evolutionary aspects. *CRC Crit. Rev. Biochem.* 23: S1-S42, 1988.

Sauer, R.T., et al. Homology among DNA-binding proteins suggests use of a conserved super-secondary structure. *Nature* 298: 447–451, 1982.

Steitz, T.A. Structural studies of protein-nucleic acid interaction: the sources of sequence-specific binding. *Q. Rev. Biophys.* 23: 205–280, 1990.

Steitz, T.A., et al. Structural similarity in the DNA-binding domains of catabolite gene activator and Cro repressor proteins. *Proc. Natl. Acad. Sci. USA* 79: 3097–3100, 1982.

Specific structures

Aggarwal, A.K., et al. Recognition of a DNA operator by the repressor of phage 434: a view at high resolution. *Science* 242: 899–907, 1988.

Anderson, J.E.; Ptashne, M.; Harrison, S.C. A phage repressor-operator complex at 7 Å resolution. *Nature* 316: 596–601, 1985.

Anderson, J.E.; Ptashne, M.; Harrison, S.C. Structure of the repressor-operator complex of bacteriophage 434. *Nature* 326: 846–852, 1987.

Anderson, W.F., et al. Structure of the Cro repressor from bacteriophage λ and its interaction with DNA. *Nature* 290: 754–758, 1981.

Boelens, R., et al. Complex of *lac* repressor headpiece with a 14 base-pair *lac* operator fragment studied by two-dimensional nuclear magnetic resonance. *J. Mol. Biol.* 193: 213–216, 1987.

Breg, J.N., et al. Structure of arc repressor in solution: evidence for a family of β-sheet DNA-binding proteins. *Nature* 346: 586–589, 1990.

Bushman, F.D., et al. Ethylation interference and x-ray crystallography identify similar interactions between 434 repressor and operator. *Nature* 316: 651–653, 1985.

Hochschild, A.; Ptashne, M. Homologous interactions of λ repressor and λ Cro with the λ operator. *Cell* 44: 925–933, 1986.

Jordan, R.S.; Pabo, C.O. Structure of the λ complex at 2.5 Å resolution: details of the repressor-operator interactions. *Science* 242: 893–899, 1988.

Kaptain, R., et al. A protein structure from nuclear magnetic resonance data. *J. Mol. Biol.* 182: 179–182, 1985.

Koudelka, G.B.; Harrison, S.C.; Ptashne, M. Effect of non-contacted bases on the affinity of 434 operator for 434 repressor and Cro. *Nature* 326: 886–891, 1987.

Koudelka, G.B., et al. DNA twisting and the affinity of bacteriophage 434 operator for bacteriophage 434 repressor. *Proc. Natl. Acad. Sci. USA* 85: 4633–4637, 1988.

Lawson, C.L.; Sigler, P.B. The structure of *trp* pseudorepressor at 1.65 Å shows why indole propionate acts as a *trp* "inducer." *Nature* 333: 869–871, 1988.

Lim, W.A.; Sauer, R.T. Alternative packing arrangements in the hydrophobic core of λ repressor. *Nature* 339: 31–36, 1989.

McKay, D.B.; Steitz, T.A. Structure of catabolite gene activator protein at 2.9 Å resolution suggests binding to left-handed B-DNA. *Nature* 290: 744–749, 1981.

Mondragón, A.; Wolberger, C.; Harrison, S.C. Structure of phage 434 Cro protein at 2.35 Å resolution. *J. Mol. Biol.* 205: 179–188, 1989.

Mondragón, A., et al. Structure of the amino-terminal domain of phage 434 repressor at 2.0 Å resolution. *J. Mol. Biol.* 205: 189–200, 1989.

Ohlendorf, D.H., et al. The molecular basis of DNA-protein recognition inferred from the structure of Cro repressor. *Nature* 298: 718–723, 1982.

Ohlendorf, D.H., et al. Comparison of the structures of Cro and λ repressor proteins from bacteriophage λ. *J. Mol. Biol.* 169: 757–769, 1983.

Otwinowski, Z., et al. Crystal structure of *trp* repressor/operator complex at atomic resolution. *Nature* 335: 321–329, 1988.

Pabo, C.O.; Lewis, M. The operator-binding domain of λ repressor: structure and DNA recognition. *Nature* 298: 443–447, 1982.

Pabo, C.O., et al. Conserved residues make similar contacts in two repressor-operator complexes. *Science* 247: 1210–1213, 1990.

Rafferty, J.B., et al. Three-dimensional crystal structures of *Escherichia coli* met repressor with and without corepressor. *Nature* 341: 705–710, 1989.

Schevitz, R.W., et al. The three-dimensional structure of *trp* repressor. *Nature* 317: 782–786, 1985.

Vershon, A.K., et al. The bacteriophage P22 arc and mnt repressors. *J. Biol. Chem.* 260: 12124–12129, 1985.

Warwicker, J.; Engelman, B.P.; Steitz, T.A. Electrostatic calculations and model building suggest that DNA bound to CAP is sharply bent. *Proteins* 2: 283–289, 1987.

Weber, I.; Steitz, T.A. The structure of a complex of catabolite gene activator protein and cyclic AMP refined at 2.5 Å resolution. *J. Mol. Biol.* 198: 311–326, 1987.

Wharton, R.P.; Brown, E.L.; Ptashne, M. Substituting an α helix switches the sequence-specific DNA interactions of a repressor. *Cell* 38: 361–369, 1984.

Wharton, R.P.; Ptashne, M. Changing the binding specificity of a repressor by redesigning an α helix. *Nature* 316: 601–605, 1985.

Wolberger, C., et al. Structure of phage 434 Cro/DNA complex. *Nature* 335: 789–795, 1988.

Zhang, R.-G., et al. The crystal structure of *trp* aporepressor at 1.8 Å shows how binding tryptophan enhances DNA affinity. *Nature* 327: 591–597, 1987.

Structural Motifs of Eucaryotic Transcription Factors

8

The regulation of transcription in eucaryotes is in general much more complex and currently less well understood than the rather simple switch mechanisms for the regulation of procaryotic genes, examples of which we discussed in Chapter 7. In eucaryotes, often complex sets of regulatory elements control the initiation of transcription of structural genes and there is more than one class of RNA polymerases. Two of these polymerases, RNA polymerase I and RNA polymerase III, transcribe the genes encoding transfer RNAs and ribosomal RNAs, respectively; genes that code for the messenger RNAs of proteins are transcribed by RNA polymerase II, and it is these genes that will be our principal focus. Upstream of the RNA polymerase II initiation site there are different combinations of specific DNA sequences, each of which is recognized by a corresponding site-specific DNA-binding protein. These proteins are called **transcription factors**, and each combination of DNA sequence and cognate DNA-binding protein (transcription factor) constitutes a **control module**. The essence of transcriptional regulation in eucaryotes is to use different combinations of this set of control modules to achieve the specific regulated expression of each gene. With over 50 modules to select from, the number of combinations is very high. The upstream regulatory sequences, the DNA part of each control module, can be divided into two main categories, those in the **upstream promoter element** and the **enhancer elements** (Figure 8.1). The upstream promoter element is usually 100 to 200 base pairs long, and this segment of DNA is relatively close to the site of initiation of transcription. Within each upstream promoter element there are DNA sequences specifically recognized by several different transcription factors, but one sequence that is rich in A-T base pairs, the so-called TATA box, is found in most, if not all, of these promoter elements. The TATA box is recognized by one of the transcription factors.

Enhancer elements are short DNA sequences that occur further upstream from the initiator site than the upstream promoter element. Enhancers contain specific sequences recognized by cognate transcription factors. The remarkable feature of enhancers is their distance from the promoters they control. They are often a few 1000 base pairs 5' to the promoter but may be 20,000 base pairs or more distant. This means that in eucaryotes transcription is regulated by a series of DNA sequences, each specifically recognized by a DNA-binding protein, dispersed over very long stretches of DNA 5' to the structural gene. Efficient gene expression, therefore, depends upon a series of interactions between a set of DNA-binding proteins and their corresponding DNA sequences and can take place only in cells in which the appropriate DNA-binding proteins are present. Each particular gene in every cell of the organism has the same DNA control

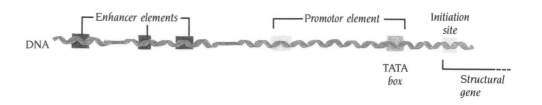

Figure 8.1 The transcriptional elements of a eucaryotic structural gene extend over a large region of DNA. The upstream regulatory sequences can be divided into two main regions: (1) those in the promotor element close to the polymerase initiation site and (2) the enhancer elements. One sequence, the so-called TATA box, is found in almost all promoter elements.

sequences, but not every cell has the complete or the same set of DNA-binding proteins. Cell-specific gene expression and all other regulatory events at the level of transcription depend on the complement of transcription factors present at any one time in each cell.

Transcription factors can be divided into two broad classes. The general factors are required for the expression of all structural genes transcribed by RNA polymerase II and are, therefore, ubiquitous. The specific transcription factors are present in only a restricted range of cell types, or only one cell type, and are responsible for the expression of cell-type specific proteins. The presence or absence of such transcription factors and their activation or inactivation therefore determine most of the functions of eucaryotic cells, from the changes they undergo during development and differentiation in a multicellular organism to their response to environmental changes, such as heat shock.

Transcription factors have two functionally different domains

The polypeptide chains of activating transcription factors are usually divided into two functionally different regions, one that binds to specific DNA sequences and another that activates transcription. The two functions are often embodied in separate domains of the protein. Three different types of activating regions have been found: those that have a net negative charge, those that contain a large number of proline residues, and those that contain a preponderance of glutamine residues. How they function to increase the initiation of transcription is not well understood, but it is believed that they act to enhance the binding of RNA polymerase to the promoter by contributing to the formation of a stable initiation complex at the TATA box. When the transcription factor is part of a DNA control module many hundreds of base pairs away from the promoter site (Figure 8.2), the intervening DNA is thought to loop out to enable the activating region of the protein to participate in the formation of the transcriptional initiation complex. It is not known whether the activator interacts directly with RNA polymerase itself, or with TATA-binding proteins, or perhaps with some third unknown protein that might act as an intermediary between the transcription factor and the initiation complex. But whichever the target, the interaction is presumed to modify the function of RNA polymerase and thereby enhance transcription.

Although we still know next to nothing about the interactions of transcription factors with each other, with RNA polymerase, or with some still to be

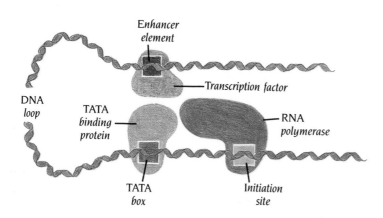

Figure 8.2 Schematic model for transcriptional activation. Interactions between one or several transcription factors and RNA polymerase, or the TATA-binding protein, causes the DNA to bend and loop out. Transcription factors can in this model modulate the function of RNA polymerase even though they bind to activating DNA elements far away from the initiation site in the nucleotide sequence.

114

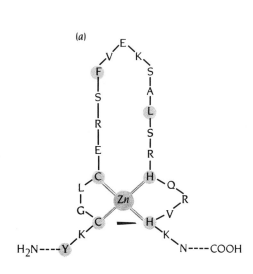

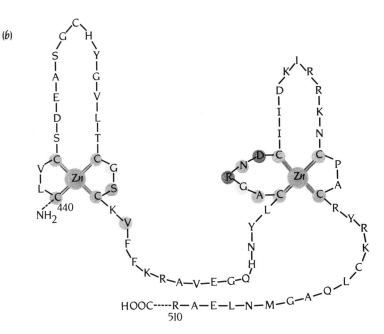

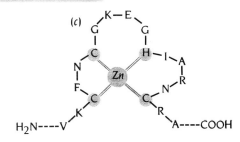

(a) Cys-Cys-His-His family

(b) Cys-Cys-Cys-Cys family

(c) Cys-Cys-His-Cys family

Figure 8.3 Amino acid sequences of zinc fingers from three different zinc finger families. The three-dimensional structures of these three polypeptides complexed to zinc have been determined by two-dimensional NMR techniques. These structures are shown in Figures 8.5, 8.7, and 8.10. (a) One zinc finger of the Cys-Cys-His-His family from a protein of unknown function, encoded by an embryonic gene, *Xfin*, from *Xenopus laevis*. Residues colored blue and yellow are invariant in most zinc fingers of this family. (b) The DNA-binding domain (residues 440–510) of the glucocorticoid receptor contains two zinc fingers of the Cys-Cys-Cys-Cys family. The colored residues in the second zinc finger are important for formation of receptor dimer. (c) One zinc finger of the Cys-Cys-His-Cys family from the DNA-binding gag protein p55 of retrovirus HIV.

identified components of transcription initiation complexes, we have made considerable progress toward understanding the structural basis of the recognition of specific DNA sequences by transcription factors. Genes that encode specific transcription factors are now being cloned and sequenced at a rapid pace. By comparing the deduced amino acid sequences of these proteins it has become apparent that their DNA-binding regions are built up from a very limited number of structural motifs. NMR methods recently have been used to determine the three-dimensional structures of three of these motifs: **zinc fingers**, **leucine zippers**, and **helix-turn-helix** motifs. About 80% of the known transcription factors have DNA-binding regions that contain one or another of these three motifs.

Three different families of zinc fingers have been observed

The zinc finger motif was first described in 1985 in the laboratory of Aaron Klug at the MRC Laboratory of Molecular Biology in Cambridge, where it was inferred from an analysis of the amino acid sequence of the transcription factor TFIIIA from *Xenopus laevis*. This factor, which regulates transcription of ribosomal 5S RNA, can be isolated in large quantities from the oocytes of this frog. Each oocyte contains about 20,000 molecules of TFIIIA complexed with 5S RNA to form 7S ribonucleoprotein particles.

The amino acid sequence of the 344 residue polypeptide chain of TFIIIA contains nine repeated sequences of about 30 residues each. The repeats are not identical in sequence but each contains 2 Cys residues at the amino end and 2 His residues at the carboxy end (Figure 8.3a). The last Cys residue and the first

3'

5'

Figure 8.4 Schematic representation of a hypothetical model for the interaction of double-helical DNA with tandemly repeated zinc fingers in transcription factors. The DNA-binding zinc fingers are represented as blue helices, DNA is orange. The protein lies on one face of the DNA helix with successive fingers pointing into the major groove alternately in "front" and "behind." (Adapted from L. Fairall et al., *J. Mol. Biol.* 192: 577, 1986.)

His residue are separated by a region of 12 residues that contains two invariant hydrophobic side chains. Since the protein contains intrinsic zinc atoms and its transcriptional activity is dependent on the presence of these zinc atoms, Klug suggested that the Cys and His residues are ligands to a zinc atom and that the loop between these residues forms the DNA-binding region (Figure 8.3a). Each of the nine repeats in TFIIIA is therefore called a zinc finger. As we shall discuss below, variations of this classic zinc finger (Figure 8.3b and c) have been found in other transcription factors.

Klug and his colleagues suggested that each zinc finger binds into the major groove of DNA, and they proposed a model (Figure 8.4) for DNA binding of the tandemly repeated zinc fingers in the complete transcription factor. In this model the protein lies on one face of the DNA helix with successive fingers pointing into the major groove alternately from opposite directions. Two fingers bind to each double helical turn of 10 base pairs in B-DNA. If all 9 fingers of the transcription factor TFIIIA bound to DNA, they would cover about 45 base pairs. This agrees quite well with experimental data.

Since this model of the zinc fingers in TFIIIA was proposed, zinc fingers have been found in other proteins from many different species. Some of these other zinc finger proteins are known to be transcription factors, but zinc fingers have also been found in proteins whose functions remain to be determined. TFIIIA has 9 zinc fingers, but other proteins have anything from 1 to more than 30 zinc finger motifs within their sequence. Furthermore, it is not only the number of zinc fingers that is variable; analysis of the amino acid sequences of these motifs has led to the recognition of three distinct but related families of zinc fingers. The variation occurs in the ligands to the zinc atom. The classic zinc finger of TFIIIA (Figure 8.3a) has two Cys and two His residues bound to zinc. Some proteins, exemplified by the DNA-binding domain of the glucocorticoid receptor in Figure 8.3b contain zinc fingers with four Cys ligands to zinc; others, such as a small DNA-binding protein in retroviruses (Figure 8.3c), have one His and three Cys residues bound to zinc. As we shall see, the three-dimensional structures of these three zinc finger families differ extensively from each other.

The zinc finger is a motif that is repeated in tandem to recognize DNA sequences of different lengths. Each finger is based on a similar framework, and each interacts with a small number of base pairs. The strength of the interaction can be varied by changes in the sequence of both the protein and the DNA and by varying the length of the spacing between the fingers. These changes allow a high level of specificity in recognition, and this modular design offers a large number of combinatorial possibilities for specific recognition of DNA.

The classic zinc finger has two cysteine and two histidine ligands bound to zinc

Many attempts have been made to crystallize the transcription factor TFIIIA, but so far no crystals have been obtained that diffract to high resolution. It has been possible, however, to determine the solution structure of individual zinc fingers of other members of the same family using NMR techniques. The first of these structures was determined in the laboratory of Peter Wright at the Scripps Clinic, La Jolla, from a chemically synthesized peptide corresponding to one zinc finger of a different protein from *Xenopus laevis*, encoded by an embryonic gene called *Xfin* (X for *Xenopus*, fin for fingers). The sequence of this gene established that the corresponding protein of 1363 amino acids contains

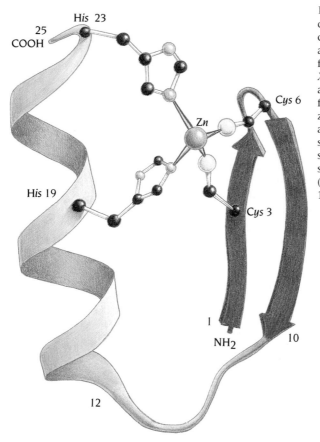

Figure 8.5 Schematic diagram of the three-dimensional structure of a 25-residue chemically synthesized peptide with an amino acid sequence corresponding to one of the zinc fingers in an embryonic protein, *Xfin*, from *Xenopus laevis*. The structure is built up from an antiparallel β hairpin motif (residues 1 to 10) followed by a helix (residues 12 to 24). The four zinc ligands Cys 3, Cys 6, His 19, and His 23 anchor one end of the helix to one end of the β sheet. Models, quite similar to the observed structure, were predicted from amino acid sequences of members of this zinc finger family. (Adapted from M.S. Lee et al., *Science* 245: 635, 1989.)

no less than 37 tandemly repeated zinc fingers. All these fingers belong to the classic zinc finger family with the characteristic pattern of two cysteine and two histidine residues separated by a region of 12 residues, the fingertip, containing two invariant hydrophobic side chains. The amino acid sequence of zinc finger number 31 was chosen for the chemical synthesis of a 25-residue-long peptide (Figure 8.3a) because it corresponds closely to a consensus sequence derived from the 148 known zinc finger members of this family, and chemical synthesis was the easiest way to produce the peptide.

The peptide requires zinc for the formation of a single folded conformation in aqueous solution. In the absence of zinc it does not bind DNA, whereas in the presence of zinc the folded conformation binds to DNA, but nonspecifically. Presumably, several fingers are required for sequence-specific binding to DNA.

The structure (Figure 8.5) confirms that a zinc finger belonging to this family is an independent folding unit; it can be described as a "miniglobular" protein with a hydrophobic core and with polar side chains on the surface. Residues 1 to 10 form an antiparallel hairpin motif where the zinc ligand Cys 3 is within the first β strand and the second zinc ligand, Cys 6, is in the tight turn between the β strands. The hairpin is followed by a helix, residues 12 to 24, of about three and a half turns. The remaining two zinc ligands, His 19 and His 23, are part of this helix. The helix is distorted between these zinc ligands to form a so-called 3_{10} helix, in which hydrogen bonds occur between every third residue of the helix instead of every fourth residue as in the normal α helix.

The hydrophobic residues Phe 10 and Leu 16 (Figure 8.3a), which are invariant in most zinc fingers of this family, form part of an interior hydrophobic core between the helix and the β strands. The zinc atom is also buried in the interior of the protein, and it is very unlikely that it participates in DNA binding. Instead, it seems to be necessary for the formation of the "finger" structure. Apparently, the few hydrophobic interactions in the core are not sufficient to provide a stable structure for this short polypeptide. Instead, the two ends of the molecule are held together at their proper places by the binding of side chains to the zinc atom.

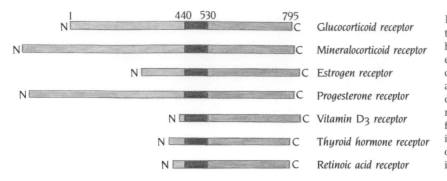

N———————————C Glucocorticoid receptor

440 530 795

Figure 8.6 Evolutionarily related members of the receptor family of transcription factors that have four cysteine residues bound to zinc in each of the two zinc fingers. The DNA-binding domains (red) have highly homologous amino acid sequences, whereas the ligand-binding domains (blue) are more variable. Residue numbers of the domain boundaries are given for the glucocorticoid receptor. Exchange of individual domains between different members of the family suggests that they can function as interchangeable modules.

How does an individual zinc finger of this family bind to DNA? There is some evidence from site-directed mutations that the residues from the helix are involved in specific DNA recognition. On the exposed surface of the helix there are several basic residues, Lys 13, Arg 18, Arg 21, and Lys 24, as well as polar side chains, Ser 14, Ser 17, and Gln 20, which could participate in both specific and nonspecific DNA binding. It therefore seems probable that this helix binds in the major groove of DNA in much the same way as the recognition helix in the helix-turn-helix motif of DNA-binding proteins, described in Chapter 7.

Two zinc fingers in the glucocorticoid receptor form one DNA-binding domain

The **glucocorticoid receptor** is a member of a family of nuclear transcription factors that also includes the thyroid hormone receptor, the retinoic acid receptor, the vitamin D3 receptor and different steroid hormone receptors (Figure 8.6). All members of this family contain a highly conserved DNA-binding domain that consists of about 70 residues and that binds to activating elements of DNA, called **hormone-response elements**. In addition, they all contain a variable C-terminal ligand-binding domain; some of them also have a large N-terminal domain, probably involved in transcriptional activation.

Protein fragments, produced from recombinant DNA in *Escherichia coli* and containing the glucocorticoid receptor DNA-binding domain exhibit sequence-specific DNA binding to glucocorticoid response elements. These protein fragments contain two zinc atoms that are required for proper folding and DNA binding. The three-dimensional solution structure of one such fragment (residues 440 to 510 of glucocorticoid receptor) has been determined by NMR methods in the laboratory of Robert Kaptein, University of Utrecht.

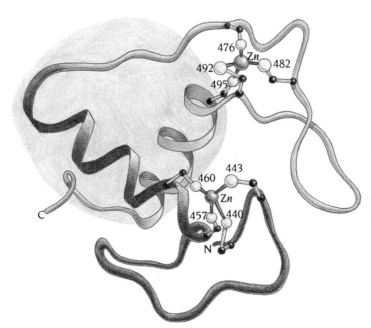

Figure 8.7 Schematic diagram of the three-dimensional structure of the DNA-binding domain of the glucocorticoid receptor. The two zinc fingers, defined from the amino acid sequence (Figure 8.3b) are colored red and green, respectively, and the region that joins the fingers is blue. Each zinc finger has one α helix and contains a zinc atom bound to four cysteine residues. The two α helices are part of a compact globular core. The zinc atoms and the residues between the zinc ligands form protrusions from this globular core. (Adapted from T. Härd et al., *Science* 249: 157, 1990.)

The amino acid sequence of this fragment (Figure 8.3b) indicated that there are two zinc finger regions, each binding a zinc atom through four cysteine residues instead of two Cys and two His residues. In the three-dimensional structure these two zinc fingers (red and green in Figure 8.7) are not separated into discrete units but are interwoven into a single globular domain with extensive interactions between the two finger units. The only regions of secondary structure of this domain are two α helices, one from each zinc finger. These two α helices in combination with the region that joins the two fingers (blue in Figure 8.7) form a compact core with a hydrophobic interior. The two zinc atoms and the protein regions between the zinc ligands form protrusions from this globular core. It is apparent from a comparison of the structure of a classic zinc finger (Figure 8.5) with that of the zinc fingers in the glucocorticoid receptor (Figure 8.7) that the three-dimensional structures of the zinc fingers from these two families are quite different. This difference should be reflected in different modes of binding to DNA.

Kaptein has suggested a model for the binding of the glucocorticoid receptor fragment to DNA, based on its three-dimensional structure and genetic and biochemical data. The hormone-receptor response elements have a partly palindromic DNA sequence indicating that the receptor binds as a dimer, as with the procaryotic repressors described in Chapter 7. Dimeric binding has received strong support from biochemical data. First, mutations have shown that the residues between the first and second zinc ligand in the second zinc finger (residues 477 to 481) are important for dimer formation. Second, residues at the N terminus of the α helix in the first zinc finger (residues 458, 459, and 462) are important for discriminating between different response elements; these residues are therefore probably in direct contact with DNA. Third, it has been shown that the receptor binds into the major groove of DNA.

The model (Figure 8.8) bears a striking resemblance to the binding of Cro and repressor to procaryotic DNA (compare Figure 7.16). The α helix in the first zinc finger (red) functions as a recognition helix that fits into the major groove of DNA. In the dimer these two recognition helices (red cylinders) have the correct orientation and distance between them to fit into two successive major grooves on one side of the DNA double helix. The second zinc finger (green) does not participate in DNA binding but is instead responsible for formation of a proper dimer. This model for DNA binding of the hormone-receptor zinc fingers is thus quite different from the model suggested for DNA binding of successive zinc fingers in the classic zinc finger family (compare Figures 8.4 and 8.8).

Yeast transcription factor Gal 4 contains a binuclear zinc cluster in its DNA-binding domain

The DNA-binding domains of some transcription factors from yeast and fungi contain six invariant cysteine residues with the following consensus sequence, where X is any amino acid:

$$- Cys - X_2 - Cys - X_6 - Cys - X_6 - Cys - X_2 - Cys - X_6 - Cys -$$

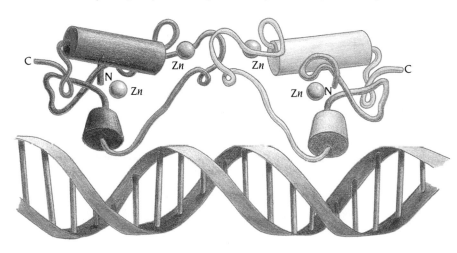

Figure 8.8 A hypothetical model of the DNA-binding domain of the glucocorticoid receptor bound to DNA as a dimer. The individual subunits have light and dark colors, respectively. The α helices of the first zinc finger of each subunit (red cylinders) are positioned in successive major grooves on one face of the DNA double helix (orange). The second zinc finger (green) of each subunit is involved in dimer formation. (Adapted from T. Härd et al., *Science* 249: 157, 1990.)

The lengths of the regions between the cysteine residues are strictly conserved, but their sequences vary. One of these factors is Gal 4 from *Saccharomyces cerevisiae*, which is required for transcriptional activation of the genes coding for galactose-metabolizing enzymes. Gal 4 is a protein of 881 amino acids in which the DNA-binding domain has been localized to the N-terminal 62 amino acids. The corresponding protein fragment of 62 amino acids has been produced from recombinant DNA and shown to require two zinc atoms per protein molecule to bind specifically to Gal 4 regulatory elements. The three-dimensional structure of this fragment is not yet known, but the arrangement of ligands to the metal atoms has been deduced from a solution NMR study of its cadmium complex. Cadmium and zinc protein complexes are assumed to have very similar structures (no exception to this assumption has yet been found), and the specific shifts produced by the ^{113}Cd isotope in protein NMR spectra can therefore be used to probe the environment of zinc in the protein.

The cadmium NMR experiments revealed an unusual ligand arrangement around the metal atoms (Figure 8.9). The two metal atoms are coupled by two bridging cysteine residues into a **binuclear cluster** with a short distance between the metal atoms. A tetrahedral coordination of the metal atom is completed by the remaining four cysteine residues, two for each metal atom. This arrangement of zinc atoms has never before been observed in protein structures; binuclear iron clusters, however, are present in several proteins, one of which is described in Chapter 3 (Figure 3.3). The presence of zinc clusters in eucaryotic transcription factors is yet another example of how biological systems recruit unusual chemistry to perform specific functions.

Retroviral zinc fingers have one histidine and three cysteine residues bound to zinc

During the assembly and budding stages of the life cycle of **retroviruses** a large polypeptide is produced from the viral gag gene and forms a complex with viral RNA. After the complex has been transported to the cell wall for budding, the gag polypeptide is proteolytically cleaved to give, among other proteins, a low molecular weight nucleic acid binding protein. In all retroviruses this protein contains either one or two regions with the following consensus sequence:

$$- Cys - X_2 - Cys - X_4 - His - X_4 - Cys -$$

Site-directed mutagenesis experiments in these regions in intact gag proteins from different retroviruses have shown that the region binds zinc and that it is intimately involved in recognition and binding to retroviral RNA during viral replication.

The three-dimensional structure of a chemically synthesized, 18-residue-long peptide with a sequence (Figure 8.3c) corresponding to one of the zinc fingers in the HIV gag protein has been determined by solution NMR methods in the laboratory of Michael Summers, University of Maryland. The structure of the zinc complex of this peptide (Figure 8.10) shows that the zinc atom is indeed coordinated to the side chains of the three cysteines and the histidine. Residues 1 to 9 form a distorted antiparallel hairpin structure, which contains the first two zinc ligands, Cys 3 and Cys 6. This part of the structure, including the positions of the zinc ligands, is quite similar to the corresponding part of the "classic" zinc finger structure (Figure 8.5). The remaining regions are, however, quite different. The loop between the second and the third zinc ligand, the fingertip, is much shorter, 4 residues rather than 12. Consequently, this loop does not form an α helix but an extended chain so that His 11 is brought in contact with the zinc atom. Finally, the carboxy end of this stubby zinc finger is positioned adjacent to the turn of the hairpin by Cys 16, which is the fourth zinc ligand. It is not known how this zinc finger interacts with RNA.

The structural studies of zinc fingers described in this chapter reflect the rapid methodological advances in structural biology during recent years. Zinc fingers are identified from sequence motifs within large structural genes that are now cloned and sequenced at a rapid pace. Corresponding peptides are synthesized chemically, and the three-dimensional structures of their zinc complexes

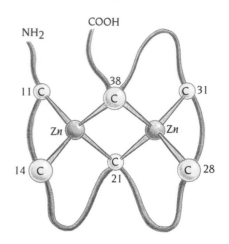

Figure 8.9 Schematic diagram of the binuclear zinc cluster in the DNA-binding domain (residues 1 to 62) of yeast transcription factor Gal 4. The two zinc atoms are close together and are bridged by Cys 21 and Cys 38. Cysteine residues 11 and 14 complete the tetrahedral coordination of one zinc atom, and cysteines 28 and 31, the other zinc atom. Loops that connect the nonbridging cysteines are three residues long, whereas all loops from bridging to nonbridging cysteines are six residues long. Binuclear zinc clusters have been observed only in this family of proteins.

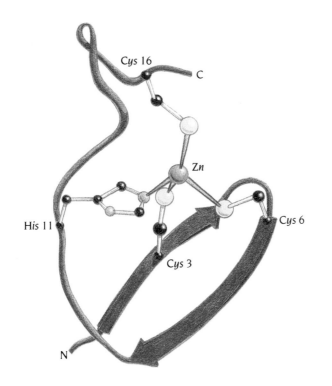

Figure 8.10 Schematic diagram of the three-dimensional structure of a complex between zinc and a synthetic peptide with an amino acid sequence corresponding to one of the zinc fingers in the gag protein of retrovirus HIV. The zinc atom is coordinated to three cysteine residues and one histidine. The first half of the structure (residues 1 to 9) is similar to the first 10 residues of the "classic" zinc finger, but the second half is quite different. (Adapted from M. Summers et al., *Biochemistry* 29: 329, 1990.)

$$H_2N-V-K-C-F-N-C-G-K-E-G-\boxed{H}-I-A-R-N-C-R-A-COOH$$
$$1 \quad\quad 3 \quad\quad\quad 6 \quad\quad\quad\quad\quad 11 \quad\quad\quad\quad 16 \quad 18$$

are determined by NMR techniques. These new technologies since 1985 have allowed the identification of numerous zinc fingers, which are divided into at least three different families, each with rather different structures. The common theme of these structures is that a zinc atom stabilizes a scaffold that holds one region of the peptide in proper position for sequence-specific interaction with DNA. The different families have different combinations of cysteine and histidine residues with different lengths of the spacer regions between them. These cysteine and histidine residues provide the ligands that bind to zinc and thereby stabilize the scaffolds.

Monomers of homeodomains bind to DNA through a helix-turn-helix motif

The zinc finger motifs are a eucaryotic solution to the problem of producing a three-dimensional scaffold to allow a protein to interact specifically with DNA. Zinc finger proteins have not been found in procaryotes. However, the helix-turn-helix solution to this problem that is used in procaryotes, as discussed in Chapter 7, is also exploited in eucaryotes in **homeodomain** proteins.

The **homeobox** is a DNA sequence of 180 base pairs within the coding region of certain structural genes. The homeobox was first discovered in the genome of the fruitfly *Drosophila* during studies of mutations that cause bizarre disturbances of the fly's body plan, so-called homeotic transformation. In the mutation *Antennapedia*, for example, legs grow from the head in place of antennae. Such homeotic mutations cause a whole set of cells to be misinformed as to their location in the organism and consequently to make a structure appropriate to another region. Homeoboxes have since been found in many different genes from vertebrates as well as invertebrates, and there are varying degrees of DNA sequence homology between different members of this superfamily.

Homeoboxes code for homeodomains, regions 60 amino acids long that most probably function as the DNA-binding regions of transcription factors.

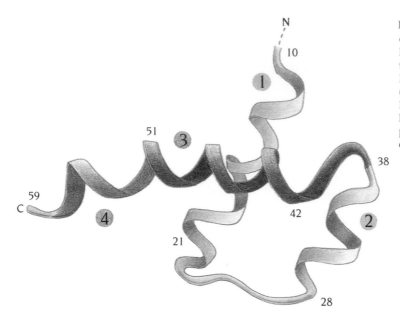

Figure 8.11 Schematic diagram of the three-dimensional structure of the *Antennapedia* homeodomain. The structure is built up from three α helices connected by short loops. Helices 2 and 3 form a helix-turn-helix motif (blue and red) similar to those in procaryotic DNA-binding proteins. Helix 3, the recognition helix, is longer than the corresponding helix in procaryotic repressors. (Adapted from Y.Q. Qian et al., *Cell* 59: 573, 1989.)

Nearly every one of the characterized homeobox genes in *Drosophila* is expressed only in characteristic subsets of embryonic cells, and almost every cell contains a unique combination of homeodomain proteins.

Comparisons of the amino acid sequences of homeodomains with the sequences of procaryotic DNA-binding proteins suggested that part of the homeodomain forms a helix-turn-helix motif similar to the DNA-binding motif described in Chapter 7 for Cro and repressor. This suggestion has recently been verified by determination of the NMR solution structures of the *Antennapedia* homeodomain and its complex with a DNA fragment in the laboratory of Kurt Wüthrich, ETH, Zurich, and by an x-ray structure determination of a DNA complex with a different homeodomain, an engrailed homeodomain from *Drosophila*, in the laboratory of Carl Pabo at Johns Hopkins University, Baltimore.

The structure of the homeodomain is built up from three α helices, connected by rather short loop regions (Figure 8.11). The three helices are formed by residues 10 to 21, 28 to 38, and 42–58. α helices 2 and 3 are part of the helix-turn-helix motif, in which the structure of the region 30 to 50 is virtually identical to the helix-turn-helix motif of Cro and repressor. The three-dimensional structure of the homeodomain motif is as similar to the procaryotic helix-turn-helix motifs as they are to each other. Surprisingly, the structural similarity is maintained even at the first residue of the turn, where the procaryotic glycine is replaced by a cysteine in the *Antennapedia* homeodomain. The turn has an unusual conformation that could normally be accommodated only by a glycine residue, and all procaryotic helix-turn-helix motifs have a glycine at this position, as described in Chapter 7. The *Antennapedia* homeodomain has the same "nonallowed" conformation despite having a cysteine at this crucial position. Presumably, the fold is sufficiently stable to force this residue to adopt an energetically unfavorable conformation.

Figure 8.12 Comparison of the three-dimensional structures of (a) the homeodomain, (b) the λ repressor DNA-binding domain, and (c) the trp repressor. The helix-turn-helix motif is blue and red with the recognition helix red. The helix-turn-helix motif is the only region of these structures that is similar. The remaining α helices (green) are all different in lengths and positions. The recognition helix of the homeodomain (a) is extended (light red) and kinked, in the *Antennapedia* homeodomain structure.

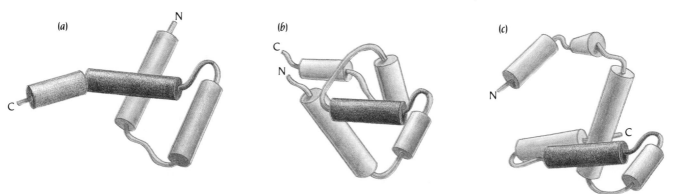

The helix-turn-helix motif is the only part of the homeodomain structure that is similar to procaryotic DNA-binding proteins (Figure 8.12). Helix 1 is arranged differently from the additional helices in λ or trp repressor. In addition, helix 3 is considerably longer than the recognition helices in the procaryotic repressors.

The homeodomain binds to DNA as a monomer, in contrast to procaryotic DNA-binding proteins containing the helix-turn-helix motif, which usually bind as dimers. Binding studies have shown that the homeodomain binds specifically to DNA fragments containing the sequence 5'-A-T-T-A-3' with an affinity of about 1 nanomolar. It also binds nonspecifically to DNA fragments with different sequences but with about 100 times lower affinity. Wüthrich has determined the NMR structure of a complex between the *Antennapedia* homeodomain and a DNA fragment containing 14 base pairs with the sequence 5'-G-A-A-A-G-C-C-A-T-T-A-G-A-G-3'. In order to avoid artificial dimer formation through S-S bridges between homeodomains, he used a mutant in which the single cysteine residue of the sequence was changed to a serine. Neither DNA binding nor the structure of the homeodomain was changed by the mutation.

The homeodomain-DNA complex has a large molecular weight, 17,800, which is at the upper limit for accurate structure determinations by current NMR techniques as will be discussed in Chapter 17. Consequently, no detailed information could be deduced about the interactions between specific amino acids and DNA. Such details were, however, obtained from Pabo's x-ray structure to 2.8 Å resolution of the engrailed homeodomain bound to a DNA-fragment also containing the 5'-A-T-T-A-3' nucleotide sequence.

Both complexes contain, as expected, one molecule of the homeodomain bound to the consensus sequence A-T-T-A of the double-stranded DNA fragment (Figure 8.13). The extended recognition helix α3 binds specifically in the major groove in the middle of the DNA fragment so that side chains in this helix can make specific contacts with DNA. All homeodomains contain four invariant residues, Asn 51, Arg 53, Trp 48 and Phe 49 in the middle of the long recognition helix. The first two invariant polar residues provide strong interactions with DNA; the other two, which are part of the hydrophobic core of the homeodomain, are important for accurate position of the recognition helix with respect to the remaining parts of the homeodomain structure. Asn 51 points into the major groove and makes a pair of hydrogen bonds with one of the adenine bases of the consensus sequence. Arg 53, in contrast, forms non-specific interactions with two phosphate groups of DNA. The fact that the only four invariant residues of homeodomains are in this region suggests that the extended recognition helix is crucial to the function of homeodomains and provides a major contribution

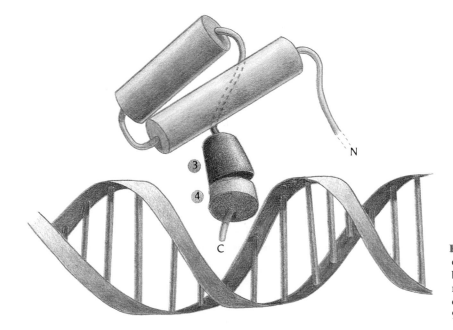

Figure 8.13 Schematic diagram of the three-dimensional structure of the homeodomain bound to a DNA fragment. The extended recognition helix is bound in the major groove of DNA. (Adapted from G. Otting et al., *EMBO J.* 9: 3085, 1990.)

to their tight, monomeric binding to DNA. In contrast, procaryotic helix-turn-helix motifs contain neither the extension of the recognition helix nor the four invariant residues.

A second interesting feature of the structure of the homeodomain-DNA complex is found in the N-terminal region of the protein. Arg 5, which is a highly conserved residue in many homeodomains, reaches into the minor groove and contacts a thymine base within the consensus sequence. This region of the protein has a fixed structure in the complex but is flexible in the absence of DNA.

In addition to these conserved specific interactions there are a number of specific as well as nonspecific protein-DNA interactions that involve protein side chains in the engrailed homeodomain that are not conserved in other homeodomains. Presumably these interactions are responsible for the specific recognition of the engrailed homeodomain by its cognate DNA response elements.

Leucine zippers provide dimerization interactions for some eucaryotic DNA-binding proteins

One unifying concept that emerges from the different DNA-binding motifs discussed so far is that these motifs provide three-dimensional scaffolds that match the contours of DNA. These scaffolds dictate proper positioning of the interacting protein surface against DNA, allowing interactions between amino acid side chains and the base pairs that constitute specific binding sites on DNA. In fact, it is the amino acid sequences of the scaffolds rather than those involved in DNA interactions that make it possible to identify a particular DNA-binding motif within the sequence of a DNA-binding protein. The same principle applies

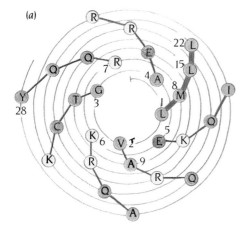

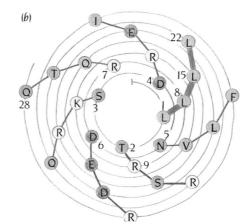

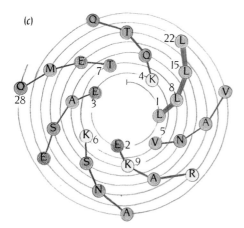

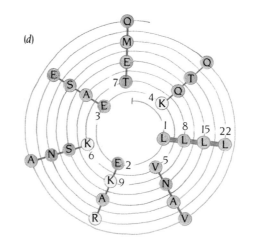

Figure 8.14 The amino acid sequences, represented as α-helical spirals with 3.6 residues per turn, of a region of 28 residues from (a) the eucaryotic DNA-binding proteins GCN 4 from yeast, (b) C/EBP from rat liver, and (c) the viral oncogene product v-jun. Positively charged residues are yellow, negatively charged, red; polar residues are blue, and hydrophobic, green. Residues with a spacing of 7 residues are connected. All three sequences show a characteristic heptad pattern of leucine residues (connected by thick bars), which has formed the basis for suggesting the leucine zipper motif as a scaffold for DNA recognition. (d) Representation of the v-jun sequence as a coiled-coil helical spiral with 3.5 residues per turn. In this helix the position of every seventh residue repeats at exactly the same face of the helix.

even more strongly to the third class of motifs in eucaryotic DNA-binding domains, the leucine zippers.

This motif was first recognized in the amino acid sequences of a yeast transcription factor, GCN 4, a mammalian transcription factor, C/EBP, and three nuclear transforming oncogene products, fos, jun, and myc, which act as transcription factors. When the sequences of these proteins are plotted on a helical wheel or spiral (Figure 8.14), a remarkable pattern of leucine residues emerges. In all five proteins there is a region of about 30 residues in which every seventh residue is a leucine. Since the periodic repeat of an α helix is 3.6, it follows that if the region folds into an α helix these leucine residues form a ridge on one side of the helix, positioned on every second turn of the eight-turn-long α helix (Figure 8.15). In the first model that was suggested on the basis of this pattern, it was assumed that this ridge forms a dimerization surface. In this model (Figure 8.15a) the two α helices, one from each polypeptide chain, were assumed to be antiparallel and the leucine residues to interdigitate to form a zipperlike structure, hence the name leucine zippers.

This model has subsequently been shown to be partly wrong. The dimerization concept is correct, the region is in α-helical conformation and the leucine residues are involved in dimer formation. The dimer is, however, built up from two parallel, not antiparallel, α helices, and the leucine residues are probably not interdigitated but rather adjacent, facing each other. Therefore the zipper analogy is not correct but the catchy name persists.

A chemically synthesized peptide with an amino acid sequence corresponding to the leucine zipper region of the yeast transcription factor GCN 4 forms stable dimers in solution. The secondary structure of this peptide has been determined by NMR methods; the whole region of 32 residues is α helical, and the NMR data indicated that the two helices of the dimer are parallel. This model has been confirmed by a preliminary x-ray structure determination by Tom Alber at Utah Medical Center of a similar synthetic peptide (Figure 8.15b).

The helical arrangement closely corresponds to a well-known structure based on heptad repeats of hydrophobic residues found in some α-helical fiber proteins such as keratin and fibrinogen. In keratin fibrils two long parallel α helices intertwine and form a so-called **coiled-coil** structure (Figure 8.16). The helical repeat in a coiled coil is reduced from 3.6 residues per turn in a regular α helix to 3.5 residues per turn, so that the interaction pattern of side chains between the helices repeats integrally every 7 residues. The currently accepted model of the leucine zipper is a short coiled coil of two parallel helices comprising about eight turns of each helix, in which four leucine residues from each helix are within the dimer interaction area.

How does the leucine zipper motif provide a scaffold for interaction of proteins with DNA? It has been shown that the yeast transcription factor GCN 4 and the nuclear oncogene products of *fos* and *jun* function as dimers and that the zipper region is responsible for dimer formation. Site-directed mutagenesis experiments on these transcription factors have also demonstrated that the DNA-binding residues are not within the zipper motif but in a region of the sequence immediately N terminal to the motif. These regions recognize palindromic sequences of DNA that form two directly abutted binding sites, one for each subunit of the dimeric protein molecules. A **scissors-grip** model (Figure 8.17) has been proposed for the recognition of DNA by this region in which the two polypeptide chains join to form a Y-shaped molecule. The stem of the Y corresponds to the zipper region, which forms a coiled pair of α helices. The bifurcating arms of the Y constitute the two DNA-binding regions, one from each polypeptide chain. They are suggested to be α helical and directly joined to the helices of the coiled stem. The bifurcation point contacts the major groove of DNA at the very center of the symmetric recognition site so that the

Figure 8.15 Schematic diagrams of α helices in which every seventh residue is a leucine, as found in the leucine zipper motifs in eucaryotic transcription factors. (a) The original leucine zipper model. Two α helices, one from each polypeptide chain, form the dimer interface. In this model the two α helices are antiparallel to each other. The leucine residues interdigitate to form a zipperlike arrangement. This model is not correct, but it has formed the basis for the name of the motif, the leucine zipper. In the correct model (b) the two α helices are arranged parallel to each other and the leucine residues face each other rather than interdigitate. [Figure (b) is adapted from a preliminary x-ray model of the GCN 4 leucine zipper; courtesy of T. Alber.]

Figure 8.16 Schematic diagram of the coiled-coil structure. Two α helices are intertwined and gradually coiled around each other. The α helices in the coiled coil are slightly distorted so that the helical repeat is 3.5 residues compared to 3.6 in a regular α helix. There is therefore an integral repeat of 7 residues along the helices that is reflected in the amino acid sequences of coiled-coil fiber proteins such as keratin and fibrinogen. Since a heptad repeat of leucine residues is found in the leucine zipper motif, the zipper is assumed to be in a coiled-coil conformation.

11nm

DNA-binding regions, the arms of the Y, are positioned to track along each half of the DNA recognition site in the major groove. Since these arms are fairly long, comprising about 20 amino acid residues, their α helices are suggested to be kinked so that they can follow the path of the major groove. It remains to be demonstrated by direct structure determination how accurate this model is.

Most of the three-dimensional structures of eucaryotic transcription factors described in this chapter have been determined by NMR techniques. They demonstrate the power of this method in the rapid structural analysis of small protein molecules. Information obtained in this way permits the construction of models for protein-DNA interactions of different transcription factors. These models form a sound basis for further studies using genetic and biochemical methods. The aim at the end of the day is to understand how simple molecular interactions can regulate the complex machinery of gene transcription so that one fertilized egg can develop into a complex organism that can respond to a large variety of input signals.

Conclusion

Eucaryotic transcription factors bind to modular DNA control elements widely dispersed within the large control regions of eucaryotic genes. The DNA-binding regions of different transcription factors are built up from a limited number of structural motifs. Of the known transcription factors, 80% contain one of three motifs: zinc fingers, leucine zippers, or helix-turn-helix motif. Structural information on these motifs has been obtained by NMR methods.

There are at least three different families of zinc finger motifs. The "classic" zinc finger is about 30 residues long and contains one zinc atom bound to 2 cysteine and 2 histidine residues in the order Cys-Cys-His-His. The DNA-binding region is probably an α helix that contains the 2 histidine residues bound to zinc. Zinc fingers are abundant among transcription factors and are usually found in several tandem repeats along a polypeptide chain.

The DNA-binding domains of the hormone-receptor family of transcription factors contain two closely interacting zinc fingers, in which each zinc atom is bound to 4 cysteine residues. One of the zinc fingers is directly involved in DNA binding through an α helix whereas the second zinc finger mediates the formation of receptor dimers. The two α helices of the dimer molecule bind in successive major grooves of the DNA, much as procaryotic repressor molecules bind.

Retroviral DNA-binding proteins contain a short stubby zinc finger of about 18 residues in which one zinc atom is bound to 3 cysteine and 1 histidine residue in the order Cys-Cys-His-Cys. The DNA-binding domain of transcription factor Gal 4 from yeast contains a zinc cluster where two zinc atoms are close together and are bridged by 2 cysteine residues. Four additional cysteine residues complete a tetrahedral coordination around each zinc atom.

Homeodomains contain the helix-turn-helix motif. They bind as monomers to specific nonpalindromic DNA sequences. The extended recognition helix of the motif binds in the major groove of DNA. Many side chains from this helix provide tight and specific protein–DNA interactions. The N-terminus of the homeodomain forms an extended chain that binds in the minor groove of DNA to the same nucleotides to which the recognition helix binds.

A heptad repeat of leucine residues is present in the leucine zipper motif in some eucaryotic transcription factors and in the products of the oncogenes *jun*, *fos*, and *myc*. The zipper region is α helical, and the leucine residues form the basis for a dimer interaction surface such that two parallel α helices form a coiled coil. A scissors-grip model has been suggested for the role of the zipper region in DNA recognition.

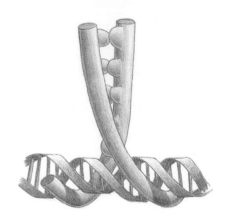

Figure 8.17 The scissors-grip model of DNA recognition by the leucine zipper motif. Two DNA-binding domains (blue) dimerize through their α-helical leucine zipper region to form a Y-shaped molecule. The stem of the Y, which is perpendicular to the axis of the DNA double helix, is formed by the coiled-coil zipper region. The leucine residues are green. The arms of the Y each form single α helices that contain the DNA recognition regions of the domains and which are positioned in the major groove of DNA. These helices are kinked so that they can follow the path of the major groove. The bifurcation point is in the center of the DNA recognition site, and each helix binds to one-half of the symmetric DNA recognition site. (Adapted from C.R. Vinson et al., *Science* 246: 911, 1989.)

Selected readings

General

Evans, R.M. The steroid and thyroid hormone receptor superfamily. *Science* 240: 889–895, 1988.

Gehring, U. Steroid hormone receptors: biochemistry, genetics and molecular biology. *Trends Biochem. Sci.* 12: 399–402, 1987.

Johnson, P.F.; McKnight, S.L. Eucaryotic transcriptional regulatory proteins. *Annu. Rev. Biochem.* 58: 799–839, 1989.

Klug, A.; Rhodes, D. "Zinc fingers": a novel protein motif for nucleic acid recognition. *Trends Biochem. Sci.* 12: 464–469, 1987.

Landschulz, W.H.; Johnson, P.F.; McKnight, S.L. The leucine zipper: a hypothetical structure common to a new class of DNA-binding proteins. *Science* 240: 1759–1764, 1988.

Levine, M.; Hoey, T. Homeobox proteins as sequence-specific transcription factors. *Cell* 55: 537–540, 1988.

Ptashne, M. How eucaryotic transcriptional activators work. *Nature* 335: 683–689, 1988.

Scott, M.P.; Tamkun, J.W.; Hartzell, G.W. The structure and function of the homeodomain. *Biochim. Biophys. Acta* 989: 25–49, 1989.

Struhl, K. Promoters, activator proteins, and the mechanism of transcriptional initiation in yeast. *Cell* 49: 295–297, 1987.

Struhl, K. Helix-turn-helix, zinc-finger, and leucine-zipper motifs for eucaryotic transcriptional regulatory proteins. *Trends Biochem. Sci.* 14: 137–140, 1989.

Struhl, K. Molecular mechanisms of transcriptional regulation in yeast. *Annu. Rev. Biochem.* 58: 1051–1077, 1989.

Specific structures

Affolter, M., et al. DNA-binding properties of the purified *Antennapedia* homeodomain. *Proc. Natl. Acad. Sci. USA* 87: 4093–4097, 1990.

Berg, J.M. Proposed structure for the zinc-binding domains from transcription factor IIIA and related proteins. *Proc. Natl. Acad. Sci. USA* 85: 99–102, 1988.

Green, D.M.; Berg, J.M. A retroviral Cys-Xaa$_2$-Cys-Xaa$_4$-His-Xaa$_4$-Cys peptide binds metal ions: spectroscopic studies and a proposed three-dimensional structure. *Proc. Natl. Acad. Sci. USA* 86: 4047–4051, 1989.

Härd, T., et al. Solution structure of the glucocorticoid receptor DNA-binding domain. *Science* 249: 157–160, 1990.

Kissinger, C.R., et al. Crystal structure of an engrailed homeodomain–DNA complex at 2.8 Å resolution: a framework for understanding homeodomain – DNA interactions. *Cell* 63: 579–590, 1990.

Klevit, R.E.; Herriott, J.R.; Horvath, S.J. Solution structure of a zinc finger domain of yeast ADR1. *Prot.: Struct. Funct. Gen.* 7: 215–226, 1990.

Kouzarides, T.; Ziff, E. The role of the leucine zipper in the fos-jun interaction. *Nature* 336: 646–651, 1988.

Lee, M.S., et al. Three-dimensional solution structure of a single zinc finger DNA-binding domain. *Science* 245: 635–637, 1989.

Lin, Y.-S., et al. How different eucaryotic transcriptional activators can cooperate promiscuously. *Nature* 345: 359–361, 1990.

Miller, J.; McLachlan, A.D.; Klug, A. Repetitive zinc-binding domains in the protein transcription factor IIIA from *Xenopus* oocytes. *EMBO J.* 4: 1609–1614, 1985.

Oas, T.G., et al. Secondary structure of a leucine zipper determined by nuclear magnetic resonance spectroscopy. *Biochemistry* 29: 2891–2894, 1990.

O'Neil, K.T.; Hoess, R.H.; DeGrado, W.F. Design of DNA-binding peptides based on the leucine zipper motif. *Science* 249: 774–778, 1990.

O'Shea, E.K.; Rutkowski, R.; Kim, P.S. Evidence that the leucine zipper is a coiled coil. Science 243: 538–542, 1989.

O'Shea, E.K., et al. Preferential heterodimer formation by isolated leucine zippers from fos and jun. *Science* 245: 646–648, 1989.

Otting, G., et al. Protein-DNA contacts in the structure of a homeodomain-DNA complex determined by nuclear magnetic resonance spectroscopy in solution. *EMBO J.* 9: 3085–3092, 1990.

Pan, T.; Coleman, J.E. GAL4 transcription factor is not a "zinc finger" but forms a Zn (II)$_2$ Cys$_6$ binuclear cluster. *Proc. Natl. Acad. Sci. USA* 87: 2077–2081, 1990.

Qian, Y.Q., et al. The structure of the *Antennapedia* homeodomain determined by NMR spectroscopy in solution: comparison with procaryotic repressors. *Cell* 59: 573–580, 1989.

Robertson, M. Homeoboxes, POU proteins and the limits to promiscuity. *Nature* 336: 522–524, 1988.

Ruiz Altaba, A.; Perry-O'Keefe, H.; Melton, D.A. *Xfin*: an embryonic gene encoding a multifingered protein in *Xenopus*. *EMBO J.* 6: 3065–3070, 1987.

Severne, Y., et al. Metal binding "finger" structures in the glucocorticoid receptor defined by site-directed mutagenesis. *EMBO J.* 7: 2503–2508, 1988.

Summers, M.F., et al. High-resolution structure of an HIV zinc fingerlike domain via a new NMR-based distance geometry approach. *Biochemistry* 29: 329–340, 1990.

Vinson, C.R.; Sigler, P.B.; McKnight, S.L. Scissors-grip model for DNA recognition by a family of leucine zipper proteins. *Science* 246: 911–916, 1989.

Weiss, M.A., et al. Folding transition in the DNA-binding domain of GCN4 on specific binding to DNA. *Nature* 347: 575–578, 1990.

DNA Polymerase Is a Multifunctional Enzyme

9

DNA synthesis is one of the most fundamental and complex enzymatic reactions in a living cell. The basic principles involved, namely, that each DNA strand serves through Watson/Crick base pairing as a template for the synthesis of its complement to produce two identical daughter copies of the DNA molecule, emerged immediately from the DNA structure. However, the biochemical details of these conceptually simple reactions, which involve several dozen different proteins, are still not fully worked out. At the heart of these reactions are **nucleotide polymerases** that catalyze the sequential addition of each correct nucleotide specified by the template to the nascent chain (Figure 9.1).

The first nucleotide polymerase to be isolated, **DNA polymerase I (Pol I)** of *E. coli*, was discovered and purified by Arthur Kornberg in 1956. Extensive

Figure 9.1(a) DNA replication. (a) Synthesis of new DNA proceeds from a replication fork where the strands of the parent DNA are separated. Both these strands serve as templates for replication of DNA during which the new DNA strands are always synthesized in the 5'-to-3' direction. One strand can therefore be synthesized as one continuous chain whereas the second strand, which uses the 5'-to-3' parent strand as template, must initially be made as a series of short DNA molecules, called **Okazaki fragments**. See **Figure 9.1(b)** on the following page.

(a)

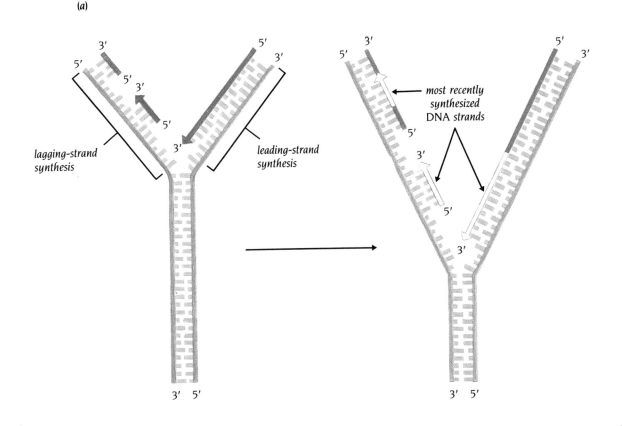

biochemical studies since then have shown that the enzyme has three functions within its single polypeptide chain: it not only polymerizes DNA but edits and repairs it as well. In this chapter we will examine the structure of a fragment of this enzyme, which is the only high-resolution structure of a nucleotide polymerase presently available, and we will ask what the structure tells us about how the enzyme's three functions are accommodated in a single polypeptide chain. The activities entailed in its three functions are as follows:

1. To polymerize DNA, the enzyme catalyzes the addition of deoxyribonucleoside 5′-triphosphates to the 3′-OH terminus of a preexisting DNA strand (Figure 9.2), the **primer strand**, but only if a **DNA template** is also present. The template directs the selection of nucleotides through **Watson/Crick base pairing**. An important feature of this reaction is that the enzyme adds a number of nucleotides without physically dissociating from the template.

2. The editing or proofreading function of the enzyme depends upon a **3′–5′ exonuclease activity** (Figure 9.3a). The nucleotide that is removed must have a free 3′-OH and it must not be part of a double helix. This allows the enzyme to remove the last nucleotide on a growing strand if it does not base-pair with the template. Coordination of the polymerase and 3′–5′ exonuclease activities enables the enzyme to synthesize DNA correctly *in vitro* given a template and a primer. The enzyme recognizes if it has added an incorrect nucleotide that does not fit the template and removes it by the 3′–5′ exonuclease activity. In other words, it can perform error-free, primed, template-directed DNA synthesis.

3. The enzyme also has a **5′–3′ exonuclease activity** that is quite different from the editing nuclease activity since it removes DNA or primer RNA ahead of the growing point of a DNA chain (Figure 9.3b). The bond to be cleaved must be in a double-helical region, and the activity is used to link Okazaki fragments and for repair of DNA.

Coordination of the polymerase and 5′–3′ exonuclease activities enables the enzyme to remove the RNA primer from **Okazaki fragments**. The polymerase can bind to the free 3′ end of an Okazaki fragment. Its 5′–3′ exonuclease activity cuts ribonucleotides in the RNA primer of the previous fragment and its polymerase activity fills the gap (Figure 9.1b). This process, in which the enzyme simultaneously polymerizes new DNA and degrades RNA ahead of the growing site continues until the primer is removed.

Coordination of all three activities accounts for the function of DNA polymerase I as a **repair enzyme** *in vivo*. Given a break in one of the strands of double-stranded DNA, the enzyme can remove nucleotides on both sides of the break and fill in correct nucleotides. A second enzyme, DNA ligase, is then required to seal or ligate the new stretch of DNA to the remainder of the chain.

What are the structural requirements of an enzyme that is able to perform all these functions within a single polypeptide chain? It must have active sites where the actual chemical events take place. The polymerase reaction is a nucleophilic attack on the 3′-OH terminus on the innermost phosphorus atom of the incoming deoxyribonucleoside triphosphate. Extending from this site must be binding sites for both primer and template. These sites must be constructed so that the enzyme can walk in a processive way along the template. There must also be active sites for the two nuclease activities either adjacent to

(b)

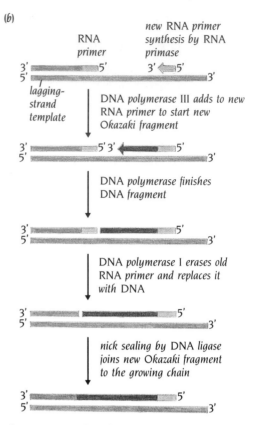

Figure 9.1(b) The Okazaki fragments require an **RNA primer** at the 5′ ends to initiate DNA synthesis, which is started by the enzyme DNA polymerase III. This enzyme adds nucleotides in the 5′-to-3′ direction until it encounters the RNA primer of the previous Okazaki fragment. A different enzyme, DNA polymerase I, is required to remove the ribonucleotide primer by its 5′–3′ nuclease activity and continue DNA synthesis by filling in the gaps with deoxyribonucleotides by its polymerase activity. A fragment of DNA polymerase I has been crystallized and its three-dimensional structure is described in this chapter.

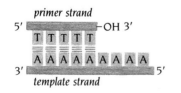

DNA polymerase can add nucleotides to the 3′ OH end of a primer strand when it is base paired

DNA polymerase is unable to add nucleotides to the 3′ OH end of a primer strand when it is not base paired

Figure 9.2 The polymerase reaction catalyzed by DNA polymerase I. This enzyme catalyzes the stepwise addition of a deoxyribonucleotide to the 3′-OH end of a polynucleotide chain (the primer strand) that is paired to a second template strand. The new strand therefore grows in the 5′-to-3′ direction. Because each incoming deoxyribonucleoside triphosphate must pair with the template strand in order to be recognized by the polymerase, this strand determines which of the four possible deoxyribonucleotides will be added.

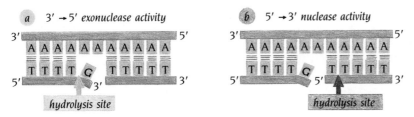

a 3′ → 5′ exonuclease activity

3′ _____ 5′
A A A A A A A A A A
T T T T G T T T T T
5′ _____ 3′
      hydrolysis site

b 5′ → 3′ nuclease activity

3′ _____ 5′
A A A A A A A A A A A
T T T T T G T T T T T
5′ _____ 5′ _____ 3′
        hydrolysis site

Figure 9.3 DNA polymerase I has two different nuclease activities that reside in different domains of the enzyme. One domain catalyzes hydrolysis of nucleotides at the 3′ end of DNA chains, a 3′-to-5′ exonuclease activity (a). To be removed, a nucleotide must have a free 3′-OH terminus and it must not be part of a double helix. The second domain hydrolyzes DNA starting from the 5′ end of DNA. This 5′-to-3′ nuclease activity can occur at the 5′ terminus or at a bond several residues away from it (b). The cleaved bond must be in a double-helical region in this reaction.

or separated from the polymerase active site. A combination of structural, biochemical and genetic studies has led to the conclusion that DNA polymerase I has three domains, each responsible for a separate enzymatic activity (Figure 9.4). The atomic structure of two of the domains, those with the polymerase and 3′–5′ exonuclease activities, has been determined.

The Klenow fragment of E. coli DNA polymerase I (Pol I) can be crystallized

E. coli DNA polymerase I is a single subunit protein comprising 928 amino acids in one polypeptide chain. One can think of the DNA polymerase gene as a fused operon in which three coding sequences specifying the three domains, each with a distinct enzymatic function, have fused so that they are always transcribed and translated in equimolar ratios. The gene has been cloned and sequenced. The polypeptide chain can easily be cleaved into two fragments by limited proteolysis. The larger C-terminal fragment of around 605 residues is called the **Klenow fragment** because its properties were first described by Hans Klenow in Copenhagen (Figure 9.4).

The Klenow fragment retains the polymerase and the 3′–5′ exonuclease activity. The smaller N-terminal fragment has only the 5′–3′ exonuclease activity. All three activities reside in separate domains in intact DNA polymerase I and the 5′–3′ exonuclease domain can be physically separated from the other two domains without impairing their activities.

The segment of the DNA polymerase I gene that codes for the Klenow fragment has also been cloned into an expression vector making available large quantities of the pure protein. This facilitated growth of large crystals that diffracted to high enough resolution for structure determination. The structure that was determined to 3.3 Å resolution by Tom Steitz's group at Yale University was actually of a complex between the Klenow fragment and an inhibitor of the 3′–5′ exonuclease activity, a deoxynucleoside monophosphate.

The Klenow fragment has two separate domains

The polypeptide chain of the Klenow fragment of Pol I is folded into two separate domains. The smaller N-terminal domain comprises residues 324–517 of the whole Pol I molecule, and the larger C-terminal domain comprises the

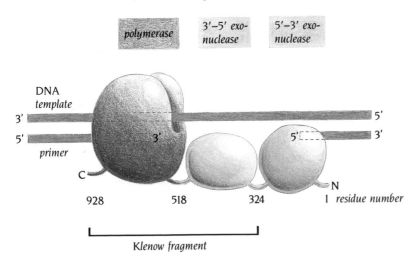

Figure 9.4 The polypeptide chain of DNA polymerase I from *E. coli* is divided into three domains, each having a specific catalytic activity. The N-terminal domain of about 300 amino acids has 5′–3′ nuclease activity and can be cleaved off without impairing the activities of the other two domains, the Klenow fragment. The C-terminal domain of the Klenow fragment has the polymerase activity, and the remaining 200 amino acids have 3′–5′ exonuclease activity.

131

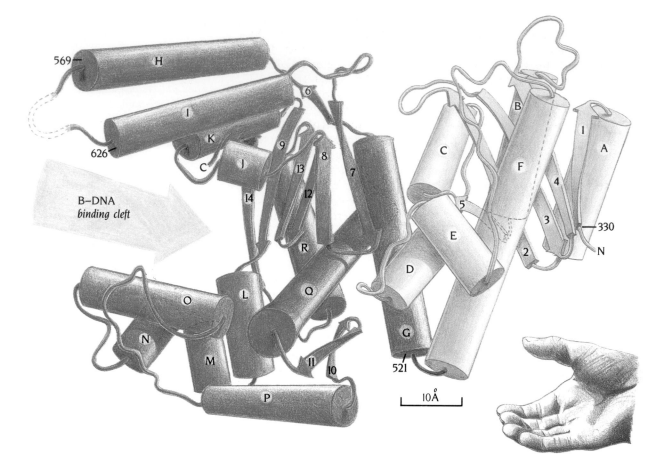

Figure 9.5 A schematic diagram of the three-dimensional structure of the Klenow fragment of *E. coli* DNA polymerase I. The large C-terminal domain (red) is an α + β structure with polymerase activity that has a cleft large enough to bind B-DNA. The N-terminal domain (green, residues 330–517), which has the 3′–5′ exonuclease activity, is of the α/β type. The beginnings and ends of β strands and α helices are numbered from the amino acid sequence of the complete DNA polymerase chain. α helices are labeled A to R and β strands 1 to 14 from the amino end of the Klenow fragment. The region between residues 569 (helix H) to 626 (helix I) is disordered in the crystal and not visible. The analogy of the structure of the large domain with a right hand is indicated. (Adapted from D.L. Ollis et al., *Nature* 313: 762, 1985.)

remaining residues, 518–928. Figure 9.5 shows a schematic drawing of the structure of the polypeptide chain. There is only one covalent connection between the two domains, and there are relatively few packing interactions. The domains, therefore, are only loosely linked in their three-dimensional structure.

The large domain has a large binding cleft

The large C-terminal domain has a unique folding motif of the α + β class that had not previously been found in any other structure. The most striking feature of this domain is a very deep cleft about 20–24 Å wide and 25–35 Å deep, big enough to bind B-DNA. A six-stranded antiparallel β sheet forms the bottom of this cleft, and large protrusions of α helices form its sides.

We can imagine the domain is a right hand (Figure 9.5) with the β sheet forming the palm and the fingers arranged as if to hold a rod. The curled fingers are represented by the more extended side of the cleft (Figure 9.5), which comprises a number of consecutive α helices (L–Q), that form a wall 50 Å long. The other wall is formed primarily by two α helices (H and I) that project out from the protein hanging over the crevice like a thumb. These two helices are connected by 50 residues that are disordered or flexible and therefore invisible in the crystal. The position of this flexible region would allow it to provide the fourth wall of the cleft and thus enclose the imaginary rod that the hand is holding.

The antiparallel β sheet does not have a barrel structure. Instead it is an open twisted β sheet. One side of the sheet faces the cleft, and the other side is packed against α helices as in other α + β structures. The β sheet has an unusual and so far unique topology (Figure 9.6).

The large domain has polymerase activity

When Tom Steitz saw the unusually big cleft in the large domain, he immediately realized that it might form the binding site for the DNA primer and

132

therefore might also be associated with the active site for the polymerase activity. He then asked the following question: Does the cleft have the right dimensions and properties to bind DNA? Model building with double-stranded B-DNA showed that it fits into the cleft with minor rearrangements of side chains and slight movements of the helical protrusions (Figure 9.7). The model is also consistent with the surface properties of the Klenow fragment. Theoretical calculations show that virtually all positive charges at the surface lie within the cleft and even follow a helical path along the cleft complementary to the phosphate groups of the DNA. The structure of the large domain of the Klenow fragment is thus consistent with the hypothesis that its large cleft binds the primer for DNA synthesis.

The next question—Is the polymerase active site located within the large domain?—was answered by cloning the segment of the gene coding for the C-terminal domain (residues 515–928) as a separate entity and then overexpressing it and purifying the polypeptide. Isolated domains of a protein, obtained either by mild proteolysis of the intact molecule or by genetic engineering, are often relatively insoluble. The separation tends to expose on the surface of the domains hydrophobic patches that are buried in the intact molecule; these newly exposed regions can induce aggregation followed by precipitation. Problems of this sort complicated the purification of the C-terminal domain of DNA polymerase I. Nevertheless, it was possible to show that this domain had polymerase activity but no measurable 3′–5′ exonuclease activity. In addition, biochemical experiments using labeled substrates for the polymerase reaction have shown that amino acids 758 and 766 (Figure 9.5) in helix O, within the large cleft, are part of the polymerase active site. In conclusion, all the amino acids necessary for DNA polymerase activity are within the large domain of the Klenow fragment.

The small domain has an α/β structure

The small domain is folded into an α/β structure with an open twisted mixed β sheet comprising five strands (Figure 9.5). Residues 330–420 form the five β strands and two connecting helices as shown in Figure 9.8a. These helices cover one side of the β sheet. The remaining residues of the domain (421–517) form a number of helices that cover the other side of the β sheet. The central β strand is antiparallel to the other four parallel strands. This is so far a unique topology

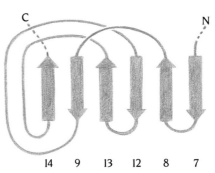

Figure 9.6 Topology diagram of the six-stranded antiparallel β sheet in the large domain of the Klenow fragment. The connection between β strands 9 and 12 makes a long excursion that builds up one side of the DNA-binding cleft and contains α helices L–Q as well as the antiparallel hairpin of β strands 11 and 12, as can be seen in Figure 9.5.

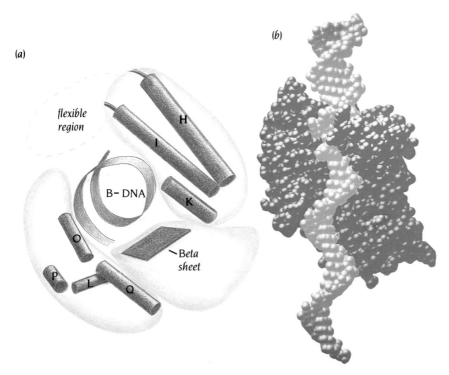

Figure 9.7 (a) Schematic diagram of the results of model building B-DNA into the cleft of the large domain of the Klenow fragment. The domain is divided into four parts, corresponding to the four walls surrounding the double-stranded DNA (orange). The numbering system of those helices that are outlined is the same as in Figure 9.5. The position of the flexible region is unknown but has been drawn to indicate that it could cover the fourth wall of the cleft. (b) Computer-generated space-filling diagram of a complex between the Klenow fragment and B-DNA viewed in a direction perpendicular to that in (a). (Courtesy of Tom Steitz.)

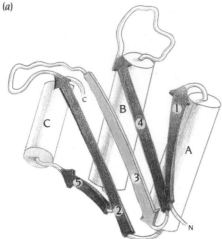

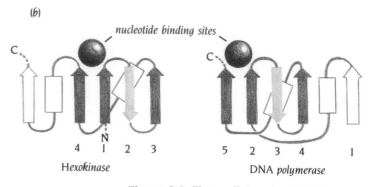

Figure 9.8 The small domain of the Klenow fragment that has the 3′–5′ exonuclease activity and that binds nucleotides is of the α/β type with one antiparallel β strand. (a) Residues 330–420 of DNA polymerase I form a five-stranded β sheet with two connecting helices and one helix, C, at the carboxy terminus. The antiparallel strand, number 3, is green, and the four parallel strands are red. (b) Comparison of topological diagrams of the mixed β sheets in hexokinase and DNA polymerase. Compare Figure 4.8c for a schematic diagram of the hexokinase domain. Colored β strands comprise the same topology in the two structures. Nucleotides (purple circles) are bound at the carboxy termini of the β strands at equivalent positions that can be predicted from the topology of the β strands.

for a mixed β sheet, but it is similar to the ATP-binding domain of hexokinase described in Chapter 4 (Figure 4.8c). β strands 1–4 of the hexokinase structure have the same topology as strands 2–5 in the small domain of the Klenow fragment (Figure 9.8b). In hexokinase one can correctly predict, using the rules described in Chapter 4, that the ATP binding site is in a crevice between the loops from the carboxy ends of β strands 1 and 4. Mononucleotides bind to the analogous position in the Klenow fragment, between β strands 2 and 5 (Figure 9.8b and Figure 9.9). We stress, however, that the similarities between these two structures are only topological; the actual structures are quite different. The β strands are twisted differently, and the helices on both sides of the β sheet have quite different lengths and orientation. The fact that the general rules for the position of binding sites in α/β structures are valid, even for this mixed β sheet of unique topology, greatly increases confidence in their generality.

The small domain has 3′–5′ exonuclease activity

The crystal structure showed that the deoxynucleoside monophosphate, dNMP, is bound to the small domain in the region of the structure described above (Figure 9.9). Since dNMP inhibits the 3′–5′ exonuclease reaction (presumably by product inhibition), it was assumed that the dNMP binding site in the crystal corresponds to the position of the 3′ terminus of the growing DNA chain that

Figure 9.9 (below) Deoxynucleoside monophosphates bind to the small domain in the Klenow fragment. (a) An overview of the nucleotide binding site. Base, sugar, and phosphate are colored blue, green, and red, respectively. β strands and binding regions are dark red, and the α helices D–G are yellow. The binding is stabilized by metal ions that bind both to protein atoms and the phosphate group. Only one of these metal ions is shown here. (b) Details of the nucleotide binding site including the two metal ions (yellow circles), which are bound to the protein by carboxy groups from Asp and Glu side chains. [(b) Courtesy of Tom Steitz.]

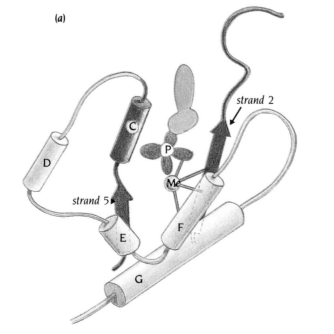

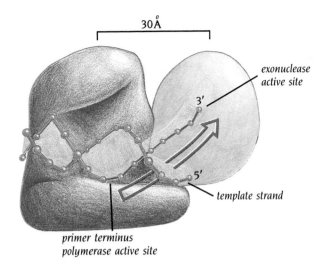

Figure 9.10 A schematic drawing showing how the Klenow fragment may bind to a DNA molecule when the 3′ terminus of the primer strand is in the exonuclease active site. The extent of movement necessary to bring the primer terminus from the polymerase active site to the exonuclease active site is indicated by an arrow. (Adapted from a diagram by Tom Steitz.)

is attacked in the 3′–5′ exonuclease reaction. This assumption implies, however, that the polymerase and exonuclease active sites are far apart (Figure 9.10). The distance between the polymerase active site in the large domain and the assumed exonuclease binding site in the small domain is, in fact, about 30 Å. Are these interpretations correct? Again, genetic experiments were used to test the structural model.

The dNMP binding site contains two bound divalent metal ions. Two mutants were constructed in which Asp and Glu residues that bind to the metal ions were changed to Ala. Both mutants had wild-type polymerase activity but no exonuclease activity. Crystallographic studies of these mutants showed that neither the large nor the small domain was perturbed. These experiments prove conclusively that the dNMP binding site is associated with exonuclease activity and that the exonuclease active site is distinct from the polymerase active site.

How the processivity and fidelity of DNA synthesis is achieved

These structural and genetic analyses of the Klenow fragment led to some provocative ideas on the mechanisms of **processivity**—how several nucleotides can be incorporated each time the enzyme binds to DNA—and on **fidelity**—how mismatched base pairs are edited out.

Processivity requires not only a tight DNA-enzyme complex to prevent dissociation between each step but also a relative movement between the enzyme and the template DNA, since the same active site is used for each successive addition of a nucleotide to the growing primer chain. Tom Steitz has suggested that when DNA is bound, the flexible region between helices H and I (Figure 9.5) folds across the top of the binding cleft, as indicated in Figure 9.7, allowing the protein to surround completely the DNA. This type of interaction would contribute to processivity by decreasing the rate of DNA dissociation relative to movement and addition of further nucleotides within the complex.

How does the relative movement occur? Steitz noticed that in the model of the enzyme complexed with B-DNA α helices J and K (Figure 9.5) are positioned partially into the major groove of DNA and may function like the thread of a nut to fix the exact position of the DNA major groove relative to the protein. Thus, during DNA synthesis the polymerase would be translocated in a spiral path along the DNA.

A different type of movement may occur during editing. If a mismatched nucleotide is incorporated by the polymerase active site, the DNA must move 30 Å relative to the protein to bring the mismatched pair into the site of the 3′–5′ exonuclease editing (Figure 9.10). From structural grounds it is very unlikely that protein conformational changes can bring the two active sites together; therefore, a relative movement of protein and DNA must occur. To account for a movement of 30 Å of DNA relative to the protein, Steitz has proposed that the

135

recognition of a mismatched pair by the polymerase active site triggers a sliding movement of the DNA over this large distance, which includes at least eight base pairs. Such a mechanism requires that the translocation step is slowed down or absent if a mismatched base has been incorporated and that the sliding movement stops when the mismatched base reaches the 3′–5′ exonuclease active site.

Comparison with other polymerases reveals evolutionary relationships

Can we learn anything about the structure of other DNA or RNA polymerases from the structure of the Klenow fragment? Do these polymerases comprise a family of structurally related enzymes based on the same structural motifs as *E. coli* DNA polymerase I? In the absence of more structural data the only method for detecting possible similarities in structure is to look for sequence similarities. Both the epsilon (ε) subunit of *E. coli* DNA polymerase III and bacteriophage T7 polymerase show sequence homology to the Klenow fragment, but no sequence homology has been detected with eucaryotic DNA polymerases.

E. coli DNA polymerase III is a complex enzyme that consists of several different polypeptide chains. The 3′–5′ exonuclease activity resides in one of these polypeptides, the ε subunit, which is small and shows significant sequence homology to the small domain of the Klenow fragment. The homologous regions surround the inhibitor binding site including the residues that interact with the nucleotide. In the case of *E. coli* Pol III, the domains of the enzyme are separated at the gene level, but the functions are brought together after translation by the formation of a noncovalent complex of polypeptide chains, while in Pol I all the domains are genetically fused together.

Bacteriophage T7 DNA polymerase is an interesting enzyme. It is the product of gene 5 of the phage, and the polypeptide chain specified by this gene carries both polymerase and exonuclease functions. The phage gene product, however, is inactive until it associates with a host protein, thioredoxin of *E. coli*. Thioredoxins are ubiquitous proteins involved in a variety of cellular reactions through their redox active disulfide bridge. This redox activity is irrelevant, however, to the activation of T7 DNA polymerase, since mutant thioredoxins, in which cysteine residues have been replaced by alanine and as a result have lost redox activity, activate the polymerase. In fact, it is still a mystery how thioredoxin activates the polymerase, although the three-dimensional struc-

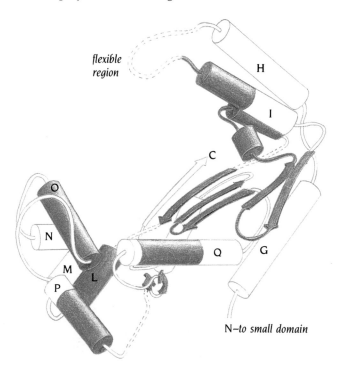

Figure 9.11 Schematic drawing of the large domain of the Klenow fragment showing the regions that have amino acid sequence homology with T7 DNA polymerase. Homologous regions are colored. Also indicated is the position of the flexible region whose sequence is highly homologous between the two polymerases. (Adapted from D.L. Ollis et al., *Nature* 313: 818, 1985.)

ture of thioredoxin is known (Figure 2.7a). The best guess is that the thioredoxin molecule provides some particular structural environment without which the phage enzyme is inactive.

Although the structure of T7 DNA polymerase has not been solved, we can be confident that its polymerase domain has a structure similar to the polymerase domain of the Klenow fragment. The reason for this prediction is the striking amino acid homology between the sequences of the two domains. Several segments of sequence, ranging in length from 9 to 45 residues, are homologous, and eight of these homologous segments constitute most of the structures that line the DNA binding cleft, including the flexible subdomain, in the Klenow fragment (Figure 9.11).

The regions between the homologous segments differ considerably in length in the two proteins. In the Klenow fragment these nonhomologous sections of the polypeptide chain are the segments that make excursions to the surface of the domain structure, and one imagines that insertions or deletions in these regions could be tolerated so long as the DNA binding cleft remains intact.

The extensive homology between the polymerase domains of the Klenow fragment and T7 DNA polymerase implies that the two enzymes have evolved from a common precursor, the likelihood being that the phage acquired the bacterial enzyme at some stage in evolution. Whether or not the domains that perform template-directed copying of DNA in all DNA polymerases have evolved from a common ancestor is an open question. If there was a common ancestor, the eucaryotic and procaryotic polymerases must have diverged so far that all traces of sequence homology have been lost.

DNA-binding proteins are constructed from modules

Some of the DNA-binding proteins discussed in this and the two preceding chapters illustrate the important principle that protein molecules frequently are constructed from **modules** that are linked together in a single polypeptide chain. Each module is folded into one domain that is responsible for a particular function.

DNA polymerases provide an outstanding example of this principle. In *E. coli* DNA polymerase I there are three modules in a single polypeptide chain (Figure 9.12), each with a separate function: 5'–3' nuclease activity, polymerase activity, and 3'–5' exonuclease activity. The gene for this enzyme can be regarded as a fusion product of three genes, each coding for a separate module, such that all three functions are synchronously controlled by the same operator region. In other systems we find some of these functions separated or even absent. Thus in the gene for DNA polymerase of bacteriophage T7 the module for the 5'–3' nuclease activity is absent. In *E. coli* DNA polymerase III the 3'–5' exonuclease activity is in one polypeptide chain, and the other activities are in a second chain. The active enzyme is a noncovalent complex between these two chains. In this case the modules are separated at the gene level but the functions are brought together after translation by forming a complex of different subunits.

A second and very intriguing example of the modular construction of proteins and the conservation of modules during evolution is provided by the CAP family of proteins. Cyclic AMP acts as an effector molecule not only in procaryotes but also in eucaryotes. High levels of cyclic AMP in mammalian cells result in activation of a protein kinase that is the first enzyme in a cascade of enzymatic reactions that results in activation of the phosphorylase-catalyzed breakdown of glycogen. Cyclic AMP binds to regulatory subunits in this protein

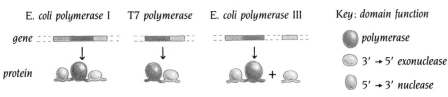

E. coli polymerase I   T7 polymerase   E. coli polymerase III   Key: domain function

gene

protein

polymerase

3' → 5' exonuclease

5' → 3' nuclease

Figure 9.12 Functional domains are combined in different ways to perform similar functions in different DNA polymerases.

137

kinase that have about 40% sequence homology to the cyclic AMP-binding domain of CAP from *E. coli*. From this sequence homology, and the known three-dimensional structure of CAP (Chapter 7), it has been possible to build models of these eucaryote regulatory subunits. These models have been correlated with functional properties such as the binding of analogues to cyclic AMP, and they are now used to target site-directed mutations in this system.

It appears, therefore, that a common ancestral protein, capable of binding cyclic AMP, has evolved into the appropriate receptor or transducer of cyclic AMP both in bacteria and mammals. The modular design of proteins has allowed this cyclic AMP module to switch on transcription of catabolic genes in bacteria by coupling to a DNA-binding domain and to activate enzymes for glycogen breakdown in mammals via a protein kinase subunit. A similar modular design has been found in the lac repressor. This classic repressor is a fairly large protein molecule with 360 amino acids. In spite of numerous attempts by many crystallographers no one has yet been able to obtain good crystals of this protein or of fragments of it. There is thus no direct information on the three-dimensional structure of the complete lac repressor. However, genetic and biochemical data have shown that its DNA binding site is in the N-terminal region whereas the sugar binds in the middle and C-terminal region of the polypeptide chain. Amino acid sequence homology and stereochemical considerations indicate that residues 5–26 of the lac repressor form a helix-turn-helix motif; this has been confirmed by a 2-D NMR structure determination of the corresponding peptide. Genetic experiments furthermore confirm that this region is involved in operator recognition.

There is also significant amino acid sequence homology between a region in the C-terminal part (especially residues 200–300) of the *lac* repressor and periplasmic sugar-binding proteins. These proteins are members of a family of proteins that are essential components of active transport systems for different carbohydrates, amino acids and ions in Gram-negative bacteria. The structure of the arabinose-binding protein was discussed in Chapter 4 as an example of an α/β protein. It is safe to predict that the lactose-binding module of the lac repressor has essentially the same structure as arabinose-binding protein. The lac repressor therefore provides another example of how domains with different functions, DNA binding and sugar binding, can combine and give rise to a new function, control of gene expression.

Conclusion

DNA polymerase I from *E. coli* catalyzes synthesis of DNA by several different chemical reactions. A combination of biochemical, genetic, and structural studies led to the conclusion that the enzyme has three domains, each responsible for a separate function. The N-terminal domain has an unknown structure that carries out the 5′–3′ exonuclease activity. The middle domain is of the α/β type and has the 3′–5′ exonuclease editing activity. The C-terminal domain has a unique fold of the $\alpha + \beta$ type that creates a deep cleft for binding DNA primer and template. The active site for polymerization is separated from the editing site by about 30 Å so that coordination of these two activities requires movement of DNA relative to the enzyme.

Selected readings

General

Joyce, C.M.; Steitz, T.A. DNA polymerase I: from crystal structure to function *via* genetics. *Trends Biochem. Sci.* 12: 288–292, 1987.

Kornberg, A. DNA Replication. *J. Biol. Chem.* 263: 1–4, 1988.

Kornberg, A. DNA Replication, 3rd ed. New York: W.H. Freeman, 1990.

Ogawa, T.; Okazaki, T. Discontinuous DNA replication. *Annu. Rev. Biochem.* 49: 421–457, 1980.

Specific structures

Derbyshire, U., et al. Genetic and crystallographic studies of the 3'–5'-exonucleolytic site of DNA polymerase I. *Science* 240: 199–201, 1988.

Freemont, P.S., et al. A domain of the Klenow fragment of *Escherichia coli* DNA polymerase I has polymerase but no exonuclease activity. *Prot.: Struct., Funct., Gen. 1:* 66–73, 1986.

Freemont, P.S., et al. Cocrystal structure of an editing complex of Klenow fragment with DNA. *Proc. Natl. Acad. Sci. USA* 85: 8924–8928, 1988.

Holmgren, A., et al. Three-dimensional structure of *E. coli* thioredoxin-S_2 to 2.8 Å resolution. *Proc. Natl. Acad. Sci. USA* 72: 2305–2309, 1975.

Huber, H.E., et al. Interaction of mutant thioredoxins of *Escherichia coli* with the gene 5 protein of phage T7. *J. Biol. Chem.* 261: 15006–15012, 1986.

Müller-Hill, B. Sequence homology between *lac* and *gal* repressor and three sugar-binding periplasmic proteins. *Nature* 302: 163–164, 1983.

Ollis, D.L.; Kline, C.; Steitz, T.A. Domain of *E. coli* DNA polymerase I showing sequence homology to T7 DNA polymerase. *Nature* 313: 818–819, 1985.

Ollis, D.L., et al. Structure of the large fragment of *Escherichia coli* DNA polymerase I complexed with dTMP. *Nature* 313: 762–766, 1985.

Steitz, T.A., et al. High-resolution x-ray structure of yeast hexokinase, an allosteric protein exhibiting a non-symmetric arrangement of subunits. *J. Mol. Biol.* 104: 197–222, 1976.

Warwicker, J., et al. Electrostatic field of the large fragment of *Escherichia coli* DNA polymerase I. *J. Mol. Biol.* 186: 645–649, 1985.

Weber, I.T.; McKay, D.B.; Steitz, T. Two helix DNA-binding motif of CAP found in *lac* repressor and *gal* repressor. *Nucleic Acids Research* 10: 5085–5102, 1982.

Weber, I.T., et al. Predicted structures of cAMP-binding domains of type I and II regulatory subunits of cAMP-dependent protein kinase. *Biochemistry* 26: 343–351, 1987.

Enzymes That Bind Nucleotides

10

Nucleotides play a central role in cellular metabolism, in which they are the major currency of energy exchange. They channel the energy released during the catabolism of food or captured from light during photosynthesis into the energy-requiring processes of the organism, such as synthesis of DNA, RNA, and proteins, active transport across membranes, and movement.

Chemically, a nucleotide consists of a base covalently linked to a sugar, which, in turn, is linked to at least one phosphate group (Figure 10.1a). Energy-transfer processes involving nucleotides can be divided into two distinct classes depending on which parts of the nucleotide engage in the energy transfer. The most familiar class involves high-energy phosphate bonds in triphosphates, such as ATP or GTP. The energy released by hydrolysis of these high-energy phosphate bonds is used to drive a large variety of metabolic processes. We will briefly discuss some kinases where phosphoryl groups are transferred between such nucleotides and metabolites.

The second class utilizes the base for electron transfer through hydrogen atoms in **oxidation-reduction (redox) processes**. The usual bases, A,T,G,C,U, that are present in RNA and DNA are not suitable for this purpose. Instead, two different bases are used, nicotinamide (Figure 10.1) and isoalloxazine (flavin) (Figure 10.2). Humans are unable to synthesize these molecules, which therefore have to be provided in the diet as vitamins. In nucleotides these two bases

Figure 10.1 (a) The basic unit of cofactors involved in energy transfer is a nucleotide that is built up from a base that is linked to a ribose sugar, a nucleoside, that in turn is linked to a phosphate group. There are two classes of energy-transfer mechanisms: one uses energy of phosphate hydrolysis; the other uses electron transfer through hydrogen atoms on the base. (b) Nicotinamide adenine dinucleotides, NAD and NADP, are used as cofactors in a variety of enzymes, frequently dehydrogenases, to shuttle electrons between different oxidation-reduction systems. NAD contains an adenine-ribose-phosphate AMP part coupled through a pyrophosphate bond to a nicotinamide-ribose-phosphate NMN part. NADP has an extra phosphate linked to the 2'-OH ribose group of the AMP part. The oxidized forms, NAD and NADP, accept electrons in the form of a hydride ion that binds covalently to the C4 atom of the nicotinamide moiety, giving the reduced forms NADH and NADPH. (c) These cofactors bind to dehydrogenases in an energetically unfavorable form, shaped like a boomerang, that is stabilized by interactions with the protein.

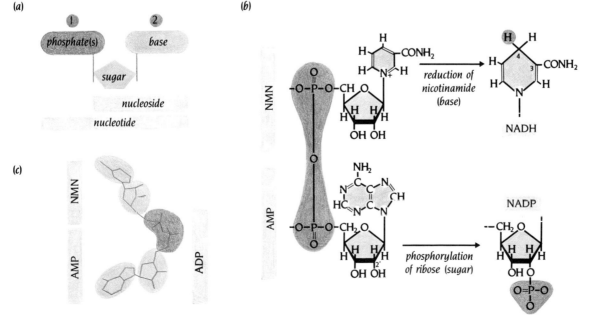

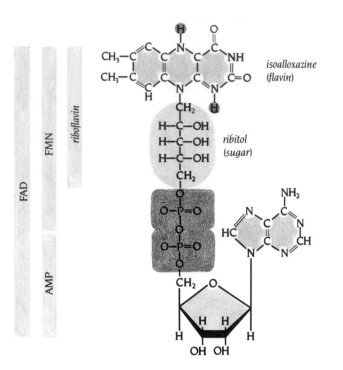

Figure 10.2 Certain redox-active enzymes, such as oxidases, use flavin nucleotides, FMN and FAD, as cofactors. The electron acceptor in these cofactors is a fused three-membered ring system, isoalloxazine, which is linked to a sugar, ribitol. The FMN cofactor is a mononucleotide where the riboflavin moiety (nucleoside) is coupled to a phosphate group at the end of the ribitol chain. This FMN moiety is coupled to an adenosine-monophosphate, AMP nucleotide, to form a complete FAD molecule. The reduced forms of the cofactors have accepted two hydrogen atoms (red) in their isoalloxazine ring system.

can be reduced by accepting hydrogen atoms at specific positions. The transfer of hydrogen from a substrate to the nucleotide base is catalyzed by the dehydrogenases and oxidases.

The electron transfer and the high-energy phosphate systems are coupled in the mitochondria through the process of oxidative phosphorylation, where the energy from such reduced nucleotides drives the synthesis of ATP.

Nicotinamide is present in the dinucleotide **NAD (nicotinamide adenine dinucleotide)** as well as in NADP, which has an extra phosphate group (Figure 10.1). An NAD molecule comprises adenine-ribose-phosphate–phosphate-ribose-nicotinamide. The first half, AMP, is linked to the second half, NMN (nicotinamide mononucleotide), through a pyrophosphate bond.

The flavin base is present in two redox-active nucleotides, **FMN (flavin mononucleotide)** and **FAD (flavin adenine dinucleotide)** (Figure 10.2). FMN comprises a flavin base linked to the linear sugar alcohol ribitol (instead of ribose), which is linked to a phosphate. The dinucleotide FAD consists of AMP linked to FMN by a pyrophosphate bond. These redox-active nucleotides are called coenzymes in contrast to ATP, which is also required for the activity of many enzymes but is not referred to as a coenzyme. This distinction is mainly semantic and has been kept for historical reasons.

The fundamental role of nucleotides in metabolism implies that nucleotide-binding enzymes must have arisen very early in the evolution of living organisms. This then raises the question of whether the multitude of different nucleotide-binding enzymes that exist today evolved independently of each other, or are they rather the descendants of one or a small number of primitive, ancestral nucleotide-binding proteins? What do we know about how enzymes bind redox-active nucleotide coenzymes, and does either the primary structure or the three-dimensional structure of these enzymes tell us anything about their evolution?

The structures of several NAD-dependent dehydrogenases are known

The **NAD-dependent dehydrogenases** comprise one of the largest and best-studied families of nucleotide-binding proteins: over 100 different members have already been identified. We will discuss the three-dimensional structures of four of these enzymes, those that catalyze oxidation of alcohol (LADH), lactate (LDH), malate (MDH), and glyceraldehyde-3-phosphate (GAPDH) (see Table 10.1).

Table 10.1
NAD-dependent dehydrogenases of known structure

Enzyme	Source	Group	Resolution (Å)	Year
Alcohol dehydrogenase (LADH)	horse liver	Carl Brändén Uppsala, Sweden	2.4	1973
Lactate dehydrogenase (LDH)	dogfish	Michael Rossmann Purdue University	2.0	1970
Malate dehydrogenase (MDH)	pig heart	Len Banaszak St Louis	3.0	1972
Glyceraldehyde-3-phosphate dehydrogenase (GAPDH)	lobster	Michael Rossmann Purdue University	2.9	1973
Glyceraldehyde-3-phosphate dehydrogenase (GAPDH)	B. stearothermophilous	Alan Wonacott London	2.2	1977

The chemical reactions catalyzed by these four enzymes are in principle very simple. An alcohol group of the substrate is oxidized by the transfer of a hydrogen atom from the carbon atom that binds the hydroxyl group to the oxidized form of the coenzyme, NAD (Figure 10.3). In addition, a proton is removed from the alcohol hydroxyl group. For GAPDH there is an additional phosphorylation reaction coupled to the dehydrogenation step.

All enzymatic reactions are reversible and in principle can go in both directions. Alcohol dehydrogenases from mammals and yeast provide illuminating examples of this principle. These enzymes have homologous amino acid sequences and catalyze the same reactions using the same mechanism. In mammals, ingested ethanol is removed from the body by oxidation to acetaldehyde, which is further oxidized to acetic acid in a subsequent enzymatic reaction. Certain individuals have a low alcohol tolerance due to mutations in these enzymes. In yeast, on the other hand, ethanol is produced from acetaldehyde during anaerobic fermentation of sugar molecules. Due to this subtle in vivo difference in the ability to influence the levels of ethanol and acetaldehyde, we, on the one hand, can produce alcohol-containing beverages in breweries and wineries and, on the other hand, can consume these drinks without toxic effects at moderate concentrations.

During dehydrogenation the hydrogen atom is directly transferred from the substrate to the coenzyme. These enzymes, therefore, not only must recognize and bind both NAD and the correct substrate, but also must position them in the active site sufficiently close and in the correct orientation to allow direct hydrogen transfer to take place (Figure 10.4).

All the NAD-dependent dehydrogenases use the same coenzyme, NAD, for similar chemical reactions, but the substrates differ in their size and shape and in the charge of the groups attached to the reactive alcohol. Are these functional similarities and differences reflected in the structures of these enzymes as we might expect them to be?

The dehydrogenase polypeptide chains are modular

Dehydrogenases are in general fairly large protein molecules; LDH and GAPDH are tetramers of four identical polypeptide chains, whereas MDH and LADH are dimers. The lengths of their polypeptide chains vary slightly, but they are all around 350 residues and the amino acid sequences are known. There is significant sequence homology between LDH and MDH; this is reflected in the three-dimensional structures of their subunits, which are essentially the same. Therefore, we will only discuss in detail one of these enzymes, LDH. No significant sequence homology can be detected, however, between LDH, GAPDH, and LADH, not even in small local regions of the polypeptide chains. From sequence comparisons alone one would, therefore, conclude that they are completely unrelated enzymes. But the three-dimensional structures tell a different story.

Figure 10.3 The chemical reactions catalyzed by the dehydrogenases LADH, LDH, MDH, and GAPDH are in principle quite simple. During oxidation a hydride ion (red) is transferred from the substrate to NAD. In addition a proton (blue) is released from the alcohol group of the substrate to the solvent. Different dehydrogenases have different specificities for the side chains R_1 and R_2 that give the enzymes their substrate specificities.

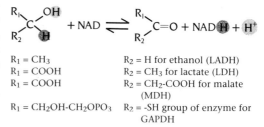

$R_1 = CH_3$ $R_2 = H$ for ethanol (LADH)
$R_1 = COOH$ $R_2 = CH_3$ for lactate (LDH)
$R_1 = COOH$ $R_2 = CH_2\text{-}COOH$ for malate (MDH)
$R_1 = CH_2OH\text{-}CH_2OPO_3$ $R_2 = $ -SH group of enzyme for GAPDH

(In the reaction catalyzed by GAPDH a covalent intermediate with the enzyme is first formed, which is exchanged with a phosphate group after the dehydrogenation reaction.)

143

The long polypeptide chains of these four dehydrogenases fold into two clearly separated domains. This is illustrated by the structural diagrams of LADH and GAPDH in Figure 10.5. The two domains have different functions, and each domain is therefore a functional module. One of them binds the coenzyme, NAD, and the second binds the substrate and provides the amino acids that are necessary for catalysis. We have discussed the concept that proteins are constructed in a modular fashion, with separate domains along the polypeptide chain fulfilling different functions, in previous chapters, but historically the structures of the dehydrogenases provided one of the first clear-cut examples of this principle.

In LDH and GAPDH the NAD-binding domains are formed from the N-terminal portion of the polypeptide chains, whereas this domain is formed by the C-terminal region of LADH (Figure 10.6). In other words, the functionally similar NAD-binding domains occur in different regions of the polypeptide chain in these dehydrogenases. This is reminiscent of the situation with the DNA-binding domains in CAP and lambda repressors that are respectively C terminal and N terminal.

The active site of these enzymes is in a cleft between the two domains. The substrate and the NAD binding sites in their respective domains are oriented so that the reactive part of the bound coenzyme, C4 of the nicotinamide ring, is in close proximity to the hydrogen atom to be transferred from the substrate. In LADH and GAPDH the domains are flexible and move closer to each other during catalysis so that the reactants are completely shielded from the solvent during hydrogen transfer.

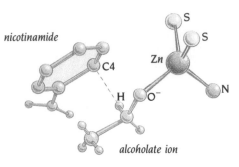

Figure 10.4 The nicotinamide ring of the coenzyme NAD and the substrate, ethanol are oriented for direct hydrogen transfer when they are bound to the enzyme LADH. In this enzyme a zinc atom in the catalytic domain binds the alcohol group of the substrate. This zinc is essential for catalysis in LADH both by binding the alcohol and by participating in the abstraction of the proton and, in addition, by polarizing the C-O bond that facilitates hydrogen transfer. The catalytic mechanisms are different in LDH and GAPDH, neither of which contains zinc.

The NAD-binding domains have similar structures

Despite the absence of amino acid sequence homology, the **NAD-binding domains** of LDH, GAPDH, and LADH have very similar three-dimensional

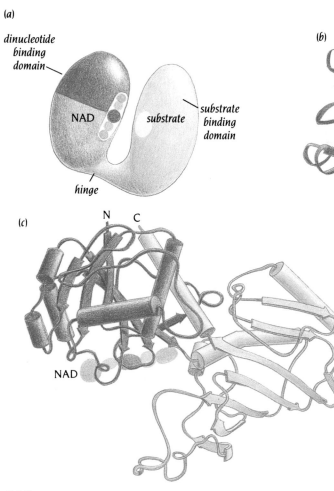

Figure 10.5 (a) The subunits of NAD-dependent dehydrogenases are folded into two separate domains, one of which, the dinucleotide-binding domain, binds the cofactor NAD. The second domain is responsible for the specificity of substrate binding and also provides the catalytic groups. The diagrams (b) and (c) illustrate schematically the domain organization of the subunit structures of liver alcohol dehydrogenase (b) and glyceraldehyde-phosphate dehydrogenase (c). The two grey balls in (b) represent zinc atoms. The catalytic zinc atom is close to the cofactor, NAD. [(b) Adapted from H. Eklund et al., *J. Mol. Biol.* 146: 561, 1981 and diagram (c) from G. Biesecker et al., *Nature* 266: 328, 1977.]

structures. The catalytic domains, on the other hand, have completely different structures in these three enzymes; indeed, each has a unique topology, which has not yet been observed in other proteins.

The NAD-binding domain is an open, parallel six-stranded β sheet with helices on both sides of the sheet (Figure 10.7a and c). This α/β structure is symmetrical because it is built from two halves with identical topology and similar structure (Figure 10.7b). Each half is formed by a pair of β-α-β motifs, as shown in Figure 10.7b. A crossover connection (yellow in Figure 10.7a) links the two halves, which are joined by hydrogen bonds between β strands 1 and 4 into a six-stranded β sheet. Each half of this symmetrical domain is called a **mono-nucleotide-binding motif** for two reasons. First, each half of the NAD-binding domain binds one of the two nucleotides in the dinucleotide NAD. Second, the half structure occurs in proteins that bind mononucleotides, for example, flavodoxin, which binds FMN.

Not only do the NAD-binding domains of LDH, GAPDH, and LADH have identical topology, but large parts of their actual structures are so similar that many of their main chain atoms superimpose within 2 Å. These superimposable parts are colored in figure 10.8; they comprise the complete β1-αA-β2 motif including the loop regions, the major parts of α helices B and D, and the remaining four β strands. The other regions of the domain are of different lengths and conformations, but the total lengths of the NAD-binding domains are about the same in all these enzymes, around 140 amino acids.

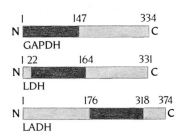

Figure 10.6 The NAD-binding domains (red boxes) of the three dehydrogenases GAPDH, LDH, and LADH are located at different positions within the polypeptide chains of these enzymes. In GAPDH and LDH they are in the N termini of the chains, whereas in LADH this domain is within the C-terminal region.

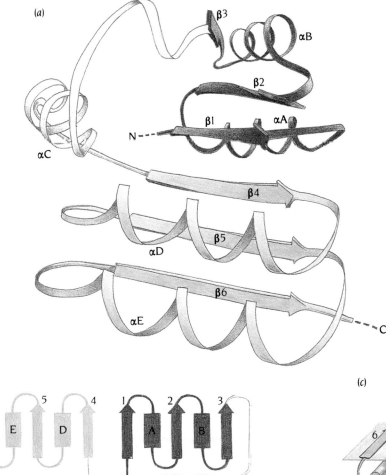

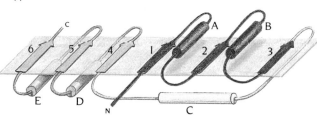

Figure 10.7 The NAD-binding domains have similar three-dimensional structures in spite of their completely different amino acid sequences. The structure is of the α/β type with an open twisted parallel β sheet in the middle surrounded by helices on both sides and is divided into two similar halves (red and green). These are called mononucleotide-binding motifs or Rossmann folds, after Michael Rossmann, who first pointed out that this was a frequently occurring motif in nucleotide-binding proteins. The diagrams show an idealized NAD-binding motif (a) and its topology diagrams (b and c).

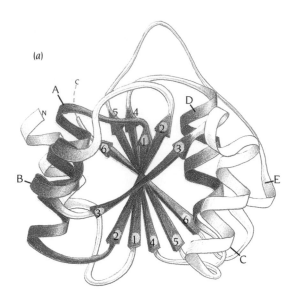

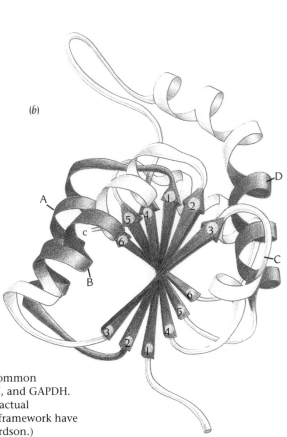

Figure 10.8 The six β strands and three of the four α helices form a common structural framework (red) in the NAD-binding domains of LADH, LDH, and GAPDH. The diagrams illustrate this framework viewed along the β sheet in the actual structures of LADH (a) and LDH (b). The regions outside this common framework have quite different structures in these two enzymes. (Adapted from J. Richardson.)

NAD binds in a similar way to each domain

In all three enzymes the regions of similar structure in the NAD-binding domains form the structural framework of the main chain that binds the NAD molecule. The actual interactions between the NAD and the protein mainly occur, however, through amino acid side chains. Two things are remarkable about this binding. First, the NAD molecules bind in almost identical positions in the three proteins, and the conformation of the NAD molecule itself is the same in all three cases. This conformation is not the most stable conformation for free NAD, as shown by theoretical calculations; nor is it the conformation found in crystals of NAD. It, therefore, follows that interaction with the protein forces NAD into an energetically unfavorable conformation. Second, NAD binding involves different combinations of side chains in different enzymes. In fact, comparisons of even the same enzyme, LADH, from different species have shown that these side chains vary between species almost as frequently as nonfunctional amino acids at the surface of the protein (Figure 10.9). In other words, there are several equally good ways of reaching a common structural goal, and this is reflected in the evolution of these enzymes.

The NAD-binding cleft is located outside the carboxy ends of β strands 1 and 4 (Figure 10.10). This is where the strand order is reversed. Thus, the binding cleft is precisely where it would be predicted in this α/β structure following the rule discussed in Chapter 4. As Figure 10.10 shows, the pyrophosphate group in

Figure 10.9 Amino acid sequences of alcohol dehydrogenases from different species. Residues that participate in NAD binding (orange) are not invariant but vary within these species almost as frequently as nonfunctional residues.

	46	55	199	230
Human α	C G T D D H V V S G	- - - - - -	G L . G G V G L S A I M G C K A A G A A R I I A V D I N K D K F A
π	C H T D A S V I D S	- - - - - -	G L . G G V G L S A V M G C K A A G A S R I I G I D I N S E K F V
Horse E	C R S D D H V V S G	- - - - - -	G L . G G V G L S V I M G C K A A G A A R I I G V D I N K D K F A
Pea	C H T D V Y F W E A	- - - - - -	G L . G A V G L A A A G G A R I S G A S R I I G V D L V S S R F E
Yeast 1	C H T D L H A W H G	- - - - -	G A A G G L G S L A V Q Y A K A M G Y . R V L G I D G G E G K E E

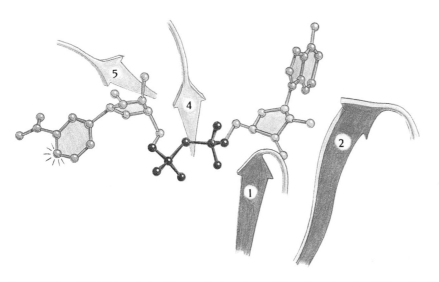

Figure 10.10 The coenzyme NAD is bound to the NAD-binding domain outside the carboxy edge of the parallel β sheet with the pyrophosphate group straddling the sheet and the two ends on opposite sides of the β sheet. The diagram shows the coenzyme viewed from the top of the four middle β strands of the domain. The loop regions from these strands form a crevice where the pyrophosphate group is bound.

the middle of NAD binds to the central region of the domain straddling the β sheet. The nucleosides (see Figure 10.1a) on the flanks of NAD bind outside the β strands to opposite sides of the β sheet. Adenosine binds to the first mononucleotide-binding motif and nicotinamide ribose to the second motif.

Hydride transfer to NAD is stereospecific

There are many enzymatic reactions in which the enzyme distinguishes between the **stereospecific isomers** of its substrate and binds only one of them. The dehydrogenases, however, are the classical example of another type of stereospecificity where the enzyme transfers a group (a hydrogen atom) in a stereospecific manner between two molecules. We now understand the structural basis of this stereospecificity. For many years the NAD-dependent dehydrogenases have been divided into two classes, A and B, depending on the **stereospecificity of hydrogen transfer** to the C4 atom of the nicotinamide ring (Figure 10.11). The ring is asymmetric because of the carboxamide substituent at its C3 atom. The two hydrogen atoms at C4 in the reduced form of NAD (above and below the ring in Figure 10.11) are therefore not equivalent. The two positions can be distinguished by using a deuterated substrate so that the enzyme transfers a deuterium atom to NAD instead of a hydrogen atom. LADH and LDH belong to **class A** because the hydrogen atom is transferred from the substrate to the position above the plane of the ring in Figure 10.11 whereas GAPDH belongs to **class B** and transfers hydrogen to the position below the ring.

Many theories have been proposed, and many experiments performed, to try to deduce a fundamental mechanistic or metabolic reason for the existence of these two classes of dehydrogenases. The structures of these enzymes provide a simple and trivial answer; this difference merely reflects the different structures of the catalytic domains. The stereochemistry is preserved for a specific enzyme within different species because the conformation of the enzyme is essentially preserved.

(a)

(b)

a pocket that allows only the 'A' form

a pocket that allows only the 'B' form

Figure 10.11 (a) Stereospecificity of hydrogen transfer to the nicotinamide part of the coenzyme NAD. In the A-form the transferred hydrogen is above the plane of the ring (red) and in the B-form it is below (green) when the carboxamide group is oriented, as shown in this diagram. (b) Structural basis for the stereospecificity of hydrogen transfer in dehydrogenases. Each enzyme has a pocket for binding the carboxamide group so that only one face of the nicotinamide ring is available for hydride transfer.

147

In all three enzymes the nicotinamide ring of the coenzyme is positioned in a cleft between the two domains close to the substrate that is bound to the catalytic domain. One side of the ring interacts with the structural framework of the NAD-binding domain, and the other side faces the substrate binding site (Figure 10.11). Interactions between the carboxamide group of NAD and the region of the protein that links the two domains determines if the A or the B side of the nicotinamide ring faces the substrate and therefore determines the stereospecificity of the transfer of the hydrogen atom. The conformational difference between the two states of NAD is a simple flip of the nicotinamide ring 180° around the glycosidic bond that links it to the ribose. This difference positions the carboxamide group in two different ways. Therefore, depending on the relative orientation in the enzyme of the two domains and the side chains in the region that joins them, the carboxamide group makes a better fit with the A side of the nicotinamide ring facing the substrate in some enzymes and with the B side in others. In this way the structure of the enzyme discriminates between the two stereochemical forms of NAD.

Are the NAD-binding domains evolutionarily related?

In Chapter 4 we discussed the evolution of α/β-barrel structures and reached the conclusion that they probably were not derived from a common ancestral gene. The barrel structures are similar, but their functions are not. The situation is different for the NAD-binding domains we have just discussed because both structure and function are preserved. The structural framework is similar as well as the conformation of bound NAD, which positions the nicotinamide ring close to the substrate. Yet, most of the side chains that interact with NAD can vary. This is clear from comparisons of the amino acid sequences of LADH from different species (Figure 10.9). Many different combinations of side chains can form this framework and preserve the function. The functionally important aspect of these domains is, therefore, the structural framework and not individual amino acid residues, except those that are necessary to preserve the structure. The essential lesson of these enzymes is that similar structural frameworks can be obtained from many different amino acid sequences.

It is, therefore, impossible to decide with any degree of confidence between the two possible evolutionary histories of the NAD-binding domains, namely, convergent evolution from different ancestral genes or divergent evolution from a common ancestor. It is, however, attractive to speculate that the dehydrogenases arose by the fusion of a gene for an ancestral NAD-binding protein with genes for primitive proteins able to bind different metabolites, such as alcohol, lactate, malate, glyceraldehyde-3-phosphate, and so on. The fusion of genes coding small proteins with single functions to generate a greater variety of more complex multifunctional proteins is a well-demonstrated evolutionary mechanism (recall DNA polymerase I) and occurs all the time in the somatic cells of the immune system, as will be discussed in Chapter 12.

The NAD-binding motif can be predicted from amino acid sequence

The absence of significant amino acid sequence homologies between even the NAD-binding domains of the different dehydrogenases means that we cannot say anything about the evolutionary histories of these proteins. We have remarked that the apparent nonconservation of the amino acid sequences of the proteins extends to the site of NAD binding. But although most of the residues at this site can vary, there are certain key invariant residues that make it possible to predict from the amino acid sequence the regions of the polypeptide chains that are involved in NAD binding. The reason for this is that in the NAD-binding domain, as in the calcium-binding and DNA-binding motifs discussed earlier, there are strong stereochemical constraints at specific positions in the polypeptide chain that must be respected to preserve the structure and function of the domain. The amino acids at these key sites are diagnostic.

(a)

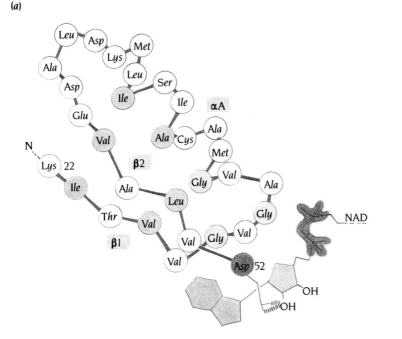

Figure 10.12 (a) Schematic diagram of the β1-αA-β2 motif in dinucleotide-binding proteins that has been used to derive a fingerprint to predict dinucleotide-binding regions in proteins of known amino acid sequence. The sequence shown here is from LDH. Hydrophobic side chains are required at certain positions (green) for packing the β strands against the α helix. Three invariant glycine residues (yellow) are required to form a tight loop and to bring the ADP part of the coenzyme NAD (thin lines) in close contact with the main chain of this loop. The motif comprises 31 amino acid residues with its amino end at position 22 and the carboxy end at Asp 52 (red), which forms a hydrogen bond to the 2'-OH of the adenosine ribose. (b) Amino acid sequences of β1-αA-β2 motifs. Sequences 1–5 are from known structures that were used to derive the fingerprint for either NAD or FAD binding. Proteins 1–3 bind NAD, 4 and 5 bind FAD. The remaining dinucleotide-binding regions were predicted using this fingerprint. (Adapted from R. Wierenga et al., *J. Mol. Biol.* 187: 101, 1986.)

(b)

1	T	C	A	V	F	G	L	G	G	V	G	L	S	V	I	M	G	C	K	A	A	G	A	A	–	–	R	I	I	G	V	D	I
2	K	I	T	V	V	G	V	G	A	V	G	M	A	C	A	I	S	I	L	M	K	D	L	A	D	–	E	V	A	L	V	D	V
3	K	I	G	I	D	G	F	G	R	I	G	R	L	V	L	R	A	A	L	S	C	G	A	Q	–	–	V	V	A	V	N	D	P
4	D	Y	L	V	I	G	G	G	S	G	G	L	A	S	A	R	R	A	A	E	L	G	A	–	–	–	R	A	A	V	V	E	S
5	Q	V	A	I	I	G	A	G	P	S	G	L	L	L	G	Q	L	L	H	K	A	G	I	–	–	–	D	N	V	I	L	E	R
6	V	I	F	V	A	G	L	G	G	I	G	L	D	T	S	K	Q	L	L	K	R	D	L	K	–	–	N	L	V	I	L	D	R
7	T	F	A	V	Q	G	F	G	N	V	G	L	H	S	M	R	Y	L	H	R	F	G	A	K	–	–	C	V	A	V	G	E	S
8	K	V	C	I	V	G	S	G	D	W	G	S	A	I	A	K	I	V	G	G	N	A	A	–	–	–	Q	L	A	Q	F	D	P
9	T	V	G	V	L	G	S	G	H	A	G	T	A	L	A	A	W	F	A	S	R	H	V	P	T	A	L	W	A	P	A	D	H
10	H	V	T	V	I	G	G	G	L	M	G	A	G	I	A	Q	V	A	A	A	T	G	H	–	–	–	T	V	V	L	V	D	Q
11	R	V	V	V	I	G	A	G	V	I	G	L	S	T	A	L	C	I	H	E	R	Y	H	S	–	–	V	L	Q	P	L	D	V

Positions for invariant glycine residues
Positions for hydrophobic residues
Invariant asparate or glutamate

1	Horse liver alcohol dehydrogenase (194–224)
2	Dogfish muscle lactate dehydrogenase (22–53)
3	Lobster glyceraldehyde-3-phosphate dehydrogenase (2–32)
4	Human erythrocyte glutathione reductase (22–51)
5	Pseudomonas fluorescens p-hydroxybenzoate hydroxylase (4–32)
6	Drosophila alcohol dehydrogenase (9–38)
7	Bovine glutamate dehydrogenase (246–275)
8	Rabbit muscle glycerol-3-phosphate dehydrogenase (4-33)
9	Agrobacterium tumefaciens nopaline synthase (34–45)
10	Pig L-3-hydroxyacyl-CoA-dehyrogenase (17–45)
11	Pig D-amine oxidase (2–31)

One region in particular, the β1-αA-β2 **motif**, has a highly conserved structure and has been used to identify NAD-binding regions in enzymes of unknown three-dimensional structure (Figure 10.12a). In this motif, comprising about 30 amino acids, there are three conserved glycine residues (yellow in the figure) with the sequence Gly-X-Gly-X-X-Gly, where X is any residue, and there are six conserved hydrophobic residues (green in the figure), which form a hydrophobic core between the helix and the β strands, and finally, there is one conserved Asp (red in the figure) at the carboxy end of β2. The glycine-rich region plays a crucial role in positioning the central part of NAD in its correct conformation close to the protein framework. This region forms the loop between β1 and α-helix A as well as the first residues of αA (Figure 10.12a). The loop region is short and very close to the strand and the helix. The amino end of α-helix A makes hydrogen bonds to the pyrophosphate group and thereby positions it as described in Chapter 2 (Figure 2.3c). Large side chains at these glycine positions would either disturb the structure of this framework or, in the case of glycines 27 and 29, push NAD out from the framework into an incorrect position.

The conserved aspartate has a special function. It is the principal means by which these enzymes discriminate between NAD and NADP as coenzyme. NAD binds to LDH, GAPDH, and LADH through hydrogen bonds between the 2'-OH of its adenosine ribose and the conserved Asp in the enzyme. But NADP has a phosphate group attached to the 2'-OH of the adenosine ribose (Figure 10.1b), and its binding to these enzymes is thus prevented by the repulsion between the negative charges of the phosphate group in the 2' position and the conserved Asp. We would expect that enzymes that bind NADP instead of NAD by this motif should have a small side chain instead of the Asp residue to make room for the phosphate group and a positively charged side chain nearby for binding the phosphate. Threonine dehydrogenase from *E. coli*, which is homologous to LADH (25% identity) and therefore has a similar structure including the β-α-β motif, uses NADP as coenzyme. As expected, this enzyme has a glycine residue instead of Asp and an Arg instead of Asn in a nearby position where the positive charge can interact with the phosphate group.

These stereochemical constraints, which are specific for enzymes that bind dinucleotides of which one-half is AMP, have been used to predict the NAD-binding regions in several dehydrogenases. Some of these sequences are given

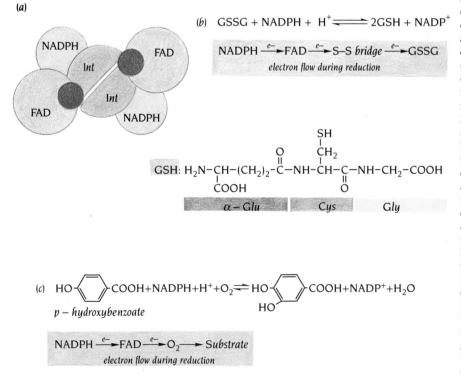

Figure 10.13 (a) The FAD requiring enzymes glutathione reductase and p-hydroxybenzoate hydroxylase are dimeric molecules where each subunit is divided into three domains—FAD-binding domain (FAD), NADPH-binding domain (NADPH), and subunit interface domain (Int). The diagram shows the domain arrangement in glutathione reductase. The red circles indicate the positions of the active sites. (Adapted from E.F. Pai and G. Schulz, *J. Biol. Chem.* 258: 1752, 1983.) (b) Glutathione reductase is a flavoenzyme that catalyzes reduction of the S-S bridge that links the two halves of the molecule in oxidized glutathione (GSSG). Each half derives from a tripeptide, NH₂-Glu-Cys-Gly-COOH, which is the reduced form of glutathione (GSH). The carboxy side chain of Glu forms the peptide link to the amino group of Cys in the tripeptide. During reduction the electrons flow from NADPH to oxidized glutathione through FAD and a redox-active S-S bridge which is formed by two Cys residues of the enzyme. (c) p-hydroxybenzoate hydroxylase is a flavoenzyme that catalyzes the addition of a hydroxyl group to p-hydroxybenzoate in a reaction that requires oxygen. During hydroxylation the electrons flow from NADPH through FAD and oxygen to the substrate, p-hydrozybenzoate. Although the reactions catalyzed by this enzyme and glutathione reductase are quite different, their FAD-binding domains have similar three-dimensional structures.

in Figure 10.12b. The structures of these proteins are not yet known, but analysis of mutants has confirmed the prediction in at least one case, *Drosophila* alcohol dehydrogenase. This enzyme has a completely different amino acid sequence and a different catalytic mechanism from that of LADH even though the two enzymes catalyze the same reaction. In one mutant the first invariant glycine was changed to Asp. The mutant enzyme is inactive, and this poor fly would not be able to thrive in Spanish wine cellars like its wild-type relatives.

FAD- and NAD-binding domains have essential similarities

The coenzyme flavin adenine dinucleotide (FAD), like nicotinamide adenine dinucleotide (NAD), contains ADP. As we have just seen in the NAD-binding domains of dehydrogenases, there is a conserved motif, β1-αA-β2, which binds the ADP moiety of the coenzyme. Furthermore, this motif is subject to strong stereochemical constraints that the amino acid sequence must respect, although other residues not at stereochemically critical positions can vary to such an extent that there is no overall sequence homology between NAD-binding domains.

In enzymes that use FAD as coenzyme, just as in those that use NAD, a β1-αA-β2 motif is responsible for binding the ADP moiety. The structures of two enzymes that use FAD and catalyze oxidation-reduction reactions, but are otherwise unrelated, both contain this motif. One of these enzymes—mammalian glutathione reductase, GRS—catalyzes the reduction of oxidized glutathione at the expense of NADPH (Figure 10.13b) and so maintains a high level of reduced glutathione in the cell. The other enzyme—p-hydroxybenzoate hydroxylase (PHBH) from soil bacteria—adds an extra -OH group to the aromatic ring of p-hydroxybenzoic acid using NADPH as the reducing agent (Figure 10.13c). PHBH plays an important role in the final steps in the biodegradation of lignin from wood. The structures of these enzymes were solved, respectively, to 2.0 Å and 2.2 Å resolution by the groups of Georg Schulz in Heidelberg, Germany, and Jan Drenth in Groningen, Holland.

Both enzymes are dimeric molecules, each with two identical subunits of about 400 amino acids that channel electrons from NADPH to a substrate (Figure 10.13). This electron flow is mediated through FAD, and in both enzymes the first reaction is a reduction of FAD by NADPH. These functional similarities are reflected in interesting structural similarities. The subunits of both enzymes are essentially divided into three domains: one each for binding NADPH and FAD and one domain that forms the subunit interface in the dimer (Figure 10.13). The substrates bind in these two enzymes in different domain interfaces. The NADPH binding and the interface domains have quite different structures in the two enzymes, but the N-terminal **FAD-binding domains** are very similar in structure, although not in amino acid sequence. Moreover, they bind FAD in a similar way from the adenine end to the flavin ring.

The structure of the FAD-binding domains is α/β type (Figure 10.14). The domain comprises from its N-terminal end a mononucleotide-binding motif with three parallel β strands and two helices on the same side of the β sheet followed by a crossover connection to a fourth β strand that is adjacent to the first β strand in the parallel β sheet. The β1-αA-β2 motif is at the beginning of this domain (Figure 10.14). The crossover connection is arranged in a separate motif comprising three antiparallel β strands (Figure 10.14), which is called a β meander. Thus, ignoring the β meander, the region of the domain described so far, including β strand 4, is formed consecutively from the amino end of the polypeptide chain and has the same structure in both GRS and PHBH. The parallel β sheet contains additional β strands, not shown in Figure 10.14, which are formed from the C-terminal regions. These have different topologies in the two enzymes and are not involved in FAD binding.

The four parallel β strands and the two α helices in the FAD-binding domains of GRS and PHBH have the same topology and structure as the corresponding part of the NAD-binding domain in dehydrogenases (compare Figures 10.7c and 10.14b). These elements form a common structural framework for

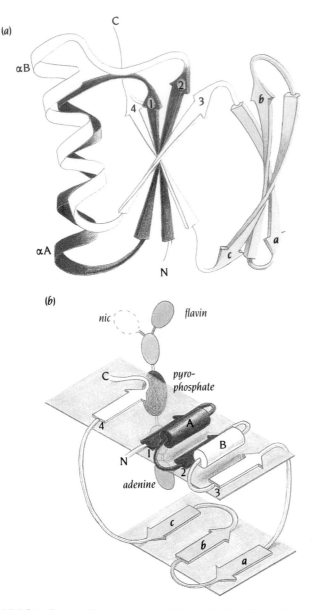

(a)

(b)

(c)

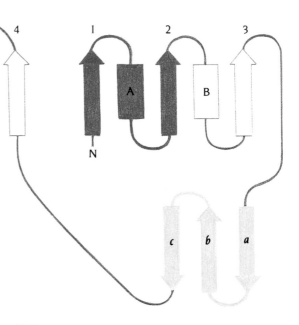

Figure 10.14 Schematic (a) and topological (b and c) diagrams of the FAD-binding domain in glutathione reductase. The β-α-β motif that is similar to the NAD-binding domains is red, and the β meander motif is yellow. The same structural framework is also present in the FAD-binding domain of p-hydroxybenzoate hydroxylase. The position of the FAD molecule is outlined in (b), which also shows the position of the nicotinamide (nic) part of NAD when bound to the similar framework in the NAD-binding domains. [(a) Adapted from J. Richardson.]

ADP binding in the two types of domain, in which the mode of binding ADP is almost identical. On the other hand, the remaining parts of these dinucleotide coenzymes, the structures of which have nothing in common, are bound quite differently in the two types of domain (Figure 10.14b).

In the FAD-binding enzymes the β1-αA-β2 motif that binds ADP is subject to the same stereochemical constraints as in the NAD-binding enzymes. Glycine residues and hydrophobic residues are conserved (Figure 10.12), and glutamate, instead of aspartate, forms the interaction with the 2′-OH group of the adenosine ribose in FAD. The amino acid sequences of other FAD-binding proteins have been searched, therefore, to find those that obey these constraints; one such example is D-amino acid oxidase (Figure 10.12). The three-dimensional structure of this enzyme is unknown, but we can guess that it contains a β-α-β motif involved in ADP binding.

Gene fusion has occurred between an FMN-binding α/β barrel and a cytochrome

Related genes are frequently used in different combinations to fulfill different biological functions. Two protein structures that participate in quite different energy-transfer processes in animal and plant cells turn up together in yeast where they couple the oxidation reaction catalyzed by one to an electron

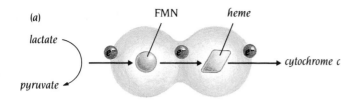

(a)

lactate

pyruvate

FMN   heme

cytochrome c

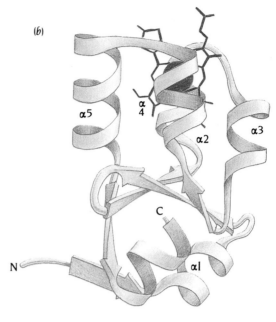

(b)

α5

α
4

α2

α3

C

N

α1

cytochrome b₅

Figure 10.15 (a) Schematic representation of the reaction catalyzed by flavocytochrome b₂. During oxidation of lactate to pyruvate an electron is transferred to the cofactor FMN in the FMN-binding domain of the enzyme. The electron is subsequently transferred to a heme group in the cytochrome domain from which it is removed by a different redox protein, cytochrome c. (b) Schematic diagram of cytochrome b₅ (and also the cytochrome domain of flavocytochrome b₂) showing the bundle of α helices (α2–α5) that binds the heme group (red). The first half of the polypeptide chain is yellow, the second is green. [(b) Adapted from J. Richardson.]

transfer function performed by the other. This biochemical process has evolved to enable yeast to metabolize on lactate, which for mammals is a waste product of anaerobic glycolysis.

The enzyme that performs this useful metabolic function for yeast cells is the mitochondrial enzyme flavocytochrome b₂, which oxidizes lactate to pyruvate and channels the released energy into the oxidative phosphorylation electron-transport chain to produce ATP (Figure 10.15a). Two separate domains of this enzyme are responsible for the oxidation and electron-transfer reactions; one binds FMN, which accepts electrons from lactate when it is oxidized, and the other, cytochrome b₂, transfers the electrons from the reduced FMN to free cytochrome c, which enters the mitochondrial electron-transport chain. This two-domain organization of flavocytochrome b₂ emerged from its crystal structure determined in 1987 to 3.0 Å resolution by the group of Scott Mathews in St. Louis, who showed that the 500-residue polypeptide chain is folded into an N-terminal FMN-binding domain and a C-terminal cytochrome with a heme group of the b₂ spectral class.

As soon as Mathews saw this structure, he realized that the N-terminal domain had a structure that was very similar to that of spinach glycolate oxidase, which had been determined to 2.2 Å resolution in 1985 by Ylva Lindqvist in the group of Carl Brändén in Uppsala, Sweden, and that the C-terminal domain looked like the structure of mammalian cytochrome b₅, which had been determined by Scott Mathews 14 years earlier.

The cytochrome-b₂ domain is homologous to mammalian cytochrome b₅

Comparisons of the amino acid sequences of flavocytochrome b₂ and mammalian microsomal cytochrome b₅ shows significant homology between these two proteins. The three-dimensional structure of the cytochrome domain contains a bundle of four α helices that form a heme-binding crevice (Figure 10.15b). The bottom of this crevice is formed by a five-stranded mixed β sheet.

Cytochrome b₅ is a transmembrane protein of the endoplasmic reticulum in animal cells. Its polypeptide chain of about 160 residues forms two domains: one is a hydrophobic membrane spanning domain of about 60 amino acids that anchors the protein in the membrane of the endoplasmic reticulum; the other is a cytochrome domain with a bound heme group. To solve the structure of this domain, Scott Mathews enzymatically cleaved it from the membrane anchor and then crystallized it.

In the membrane of the endoplasmic reticulum the intact cytochrome b_5 plays a crucial role in an electron-transport chain involving two other enzymes. Cytochrome b_5 accepts electrons from NADPH in the first enzyme and delivers them to a second enzyme where the electrons are used to desaturate fatty acids. Desaturation of fatty acids in the endoplasmic reticulum in mammalian cells and the metabolism of lactate in yeast mitochondria use essentially the same cytochrome domain.

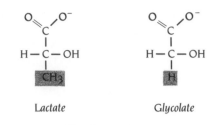

Lactate   *Glycolate*

Figure 10.16 The chemical structure of lactate and glycolate, illustrating that these two molecules differ only in a CH_3 group in lactate compared to an H atom in glycolate.

The FMN-binding domain is an α/β barrel

The **FMN-binding domain** of flavocytochrome b_2, which shows structural similarities to the FMN-dependent enzyme glycolate oxidase from spinach, is an α/β barrel with FMN bound in the ligand-binding crevice of the barrel. Glycolate oxidase is a peroxisomal enzyme that catalyzes the first step in photorespiration in green plants. The enzyme, RUBISCO, which catalyzes the initial step in the Calvin cycle of CO_2 fixation, also catalyzes an oxygenation reaction whereby glycolate is formed in large amounts. Glycolate is removed from the cell by a series of enzymatic reactions, called **photorespiration**, during which one of the carbon atoms of glycolate is oxidized to CO_2.

There are two steps in the glycolate oxidase reaction. The first step, oxidation of glycolate and reduction of FMN, is similar to the first reaction by flavocytochrome b_2. The substrates are also similar; lactate and glycolate are both α-hydroxy acids (Figure 10.16). The second step is different. Instead of channeling its redox energy into useful processes, the reduced FMN bound to glycolate oxidase is reoxidized by oxygen, hydrogen peroxide is formed, and the excess energy is dissipated as heat.

The polypeptide chain of glycolate oxidase is folded into an α/β barrel comprising about 250 residues and with the typical barrel framework discussed in Chapter 4. The remaining 130 residues form a number of additional helices scattered on the outside of one-half of the barrel (Figure 10.17). The active-site crevice of the barrel, outside the carboxy end of the parallel β strands binds FMN. This nucleotide-binding domain is thus different from those that have been

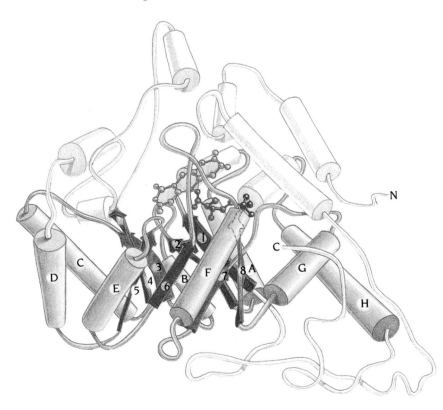

Figure 10.17 Schematic diagram of the subunit structure of the FMN requiring enzyme glycolate oxidase from spinach. The polypeptide chain is folded into an α8/β8-barrel structure, described in Chapter 4, which here is viewed from the side of the barrel. The β strands (red) are labeled 1–8 and the α helices (green) of the barrel are labeled A–H. Outside the barrel there are eight other helices (gray). The cofactor FMN (connected circles) is bound outside the carboxy ends of the β strands of the barrel. (Adapted from Y. Lindqvist and C.-I. Brändén, *Proc. Natl. Acad. Sci. (USA)* 82: 6855, 1985.)

discussed previously, which were all α/β structures with an open β sheet. A detailed comparison of the structures of glycolate oxidase and the FMN-binding domain of flavocytochrome b₂ showed that the similarities extended beyond the framework of the α/β barrels to include almost all of the extra helices and most of the loop regions. It is a general rule that when both the structure of the framework and the regions outside it are preserved, then there is significant amino acid homology among the proteins. When the sequence of glycolate oxidase was determined somewhat later it did indeed show about 35% sequence identity to the FMN-binding domain of flavocytochrome b₂.

Both of the domains in flavocytochrome b₂ thus have amino acid sequence homology and similarities of detailed three-dimensional structure to other proteins; the cytochrome-b₂ domain is similar to the soluble domain of cytochrome b₅ and the FMN-binding domain to glycolate oxidase. The two domains in flavocytochrome b₂ are arranged so that the edge of the heme ring in the cytochrome domain is close to the edge of the flavin ring in the FMN-binding domain, thus facilitating electron transfer between the two rings (Figure 10.18). The only region where there is no sequence homology between glycolate oxidase and the FMN-binding domain of flavocytochrome b₂, and where their structures differ, is a loop region in glycolate oxidase 20 residues long that is displaced in flavocytochrome b₂ to make room for the cytochrome domain. This is the only gross structural adaptation that has evolved in this domain to divert electron transfer from an external oxygen atom to a heme group bound to the same polypeptide chain.

These proteins provide one of the clearest examples of related genes being used in different combinations in different biological processes. A plausible evolutionary history is that at some stage there existed three separate genes, each coding for one function: one gene coding for a protein that oxidized α-hydroxy acids with FMN; second, a cytochrome gene encoding a protein with the cytochrome b₂/b₅ fold; and, finally, a piece of DNA specifying a polypeptide chain for anchoring proteins to membranes (Figure 10.19). Flavocytochrome b₂ evolved by fusion of the genes for the FMN-binding enzyme and the cytochrome to yield a mitochondrial protein in yeast that oxidizes lactate; cytochrome b₅ evolved by fusion of the genes for the cytochrome and the membrane anchoring polypeptide into a mammalian protein involved in desaturation of fatty acids in the endoplasmic reticulum; and, finally, glycolate oxidase evolved from the FMN-binding enzyme alone into a plant peroxisomal enzyme to oxidize glycolate and start photorespiration. It is our firm belief that as more primary and three-dimensional structures are collected and analyzed, we will find numerous examples of this type.

Hexokinase validates the theory of induced fit

Kinases comprise a large and heterogeneous family of enzymes that catalyze transfer of phosphoryl groups between trinucleotides, usually ATP, and other molecules. The three-dimensional structures of several different kinases are

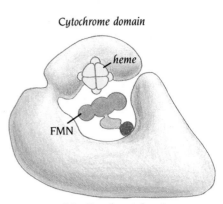

Figure 10.18 The yeast enzyme flavocytochrome b₂, which oxidizes lactate, is folded into two domains. One domain is an α8/β8-barrel domain with amino acid sequence and three-dimensional structure similar to glycolate oxidase from spinach (Figure 10.17). This domain oxidizes lactate by transferring electrons to bound FMN, thereby reducing the cofactor. The second domain binds heme and has a sequence and a fold similar to mammalian cytochrome b₅ (Figure 10.15b). Electrons are transferred from reduced FMN via the heme group to an external cytochrome-c molecule, which transiently interacts with the heme group and thereby restores the enzyme for a new round of catalysis.

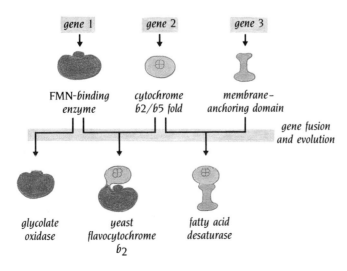

Figure 10.19 A hypothetical scheme for the evolutionary relation of glycolate oxidase, flavocytochrome b₂ and cytochrome b₅. Ancestral genes coding for three different domains were fused together in different ways to produce the genes for flavocytochrome b₂ and cytochrome b₅. The FMN-binding domain evolved by itself in plants into the enzyme glycolate oxidase.

155

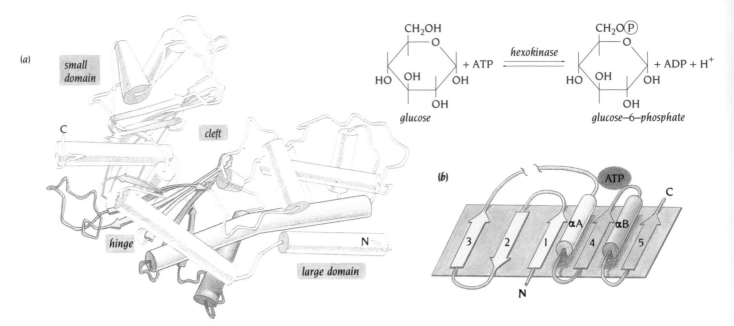

glucose

glucose–6–phosphate

(b)

known. The ATP-binding domains of these kinases are all of the α/β type with ATP bound at the predictable positions as described in Chapter 4. Most of these domains have different topologies and different ways of binding nucleotides. However, a common specific motif for ATP binding and hydrolysis has been found in three functionally unrelated proteins; hexokinase, actin, and a heat shock protein, HSC 70. Actin uses the energy released by ATP hydrolysis for muscle action. HSC 70 is a chaperone protein that is involved in disassembly of clathrin cages.

The structural similarities between these two proteins and hexokinase is quite suprising and was not anticipated when the hexokinase structure was determined in order to deduce the structural basis for the theory of induced fit, introduced by Dan Koshland, Berkeley, California, in 1960. Koshland asked the following question: Since kinases catalyze phosphoryl transfer from ATP to hydroxyl groups in metabolites, what prevents the kinases from transferring the phosphoryl group to hydroxyl groups in water? In other words, why is ATP not hydrolyzed by kinases? He postulated that in the absence of the correct substrate these kinases (as well as other enzymes) were in an inactive conformation where catalytic groups were not properly poised for catalysis. ATP could bind to this conformation of the enzyme without facing the risk of being hydrolyzed by water, since the enzyme is catalytically inactive in this form. When the correct substrate binds, however, it induces a conformational change of the enzyme that brings the catalytic groups together into the active site and phosphoryl transfer occurs from ATP to the substrate. Koshland coined the term **induced fit** for this mechanism.

In an age that was dominated by Emil Fischer's lock-and-key theory of enzyme action, this was heresy. It took 20 years before solid structural evidence confirmed Koshland's theories of induced fit through the determination of the three-dimensional structure of hexokinase in the absence and presence of its substrate glucose by the group of Tom Steitz, Yale University.

Hexokinase catalyzes the first reaction in glycolysis, the phosphorylation of glucose by ATP to form glucose-6-phosphate. The polypeptide chain of 485 residues in the yeast enzyme is folded into two domains, a smaller N-terminal domain and a larger C-terminal domain (Figure 10.20a). The large domain comprises a mixed α/β sheet and a number of additional α helices. The small domain has a predominantly antiparallel β sheet with a few α helices. Very similar β-sheet structures are present in both actin and HSC 70.

In the structure of unliganded hexokinase, which is shown in Figure 10.20a, the two domains are separated by a deep cleft. ATP binds to the large domain at the surface of this cleft in the predictable position outside the carboxy end of

Figure 10.20 The glycolytic enzyme hexokinase, which catalyzes the interconversion of glucose and glucose-6-phosphate, is folded into two domains with a hinge region between them. The large domain is built up from a mixed α/β sheet that binds ATP and from a number of additional α helices. (a) The α/β sheet is colored yellow and green. (b) Topological diagram of the α/β sheet where the red circle marks the position of bound ATP. [(a) Adapted from C. Anderson et al., *J. Mol. Biol.* 123: 1, 1978.]

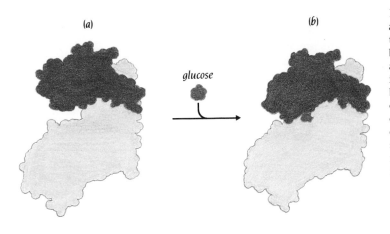

(a)   (b)

glucose

Figure 10.21 Demonstration of induced fit in an enzyme. In the open form of hexokinase (a) the two domains are separated with a deep cleft between them into which glucose can diffuse and bind. When glucose enters the binding site in the cleft, a conformational change is induced by a rotation of the domains that closes the cleft (b), thereby excluding water from the catalytic reaction. The red dot in (b) is one edge of the glucose molecule that is accessible at the surface. The remaining 90% of the glucose molecule is buried inside the protein. (Adapted from C. Anderson et al., *Science* 204: 375, 1979.)

the parallel strands of the β sheet (Figure 10.20b). No conformational change is induced by ATP binding. In contrast, when glucose binds, a large conformational change is induced. Glucose binds at the very bottom of the cleft and induces a rotation of the small domain that closes the gap between the two domains (Figure 10.21). This domain rotation has two effects, one of which is to bury the glucose molecule into the interior of the protein and thereby prevent water from entering into the active site. The glucose molecule is almost completely enclosed by the protein and the product glucose-6-phosphate cannot dissociate from it unless the conformation reverts to that of the free enzyme; precisely how this back reaction is brought about remains a matter of speculation. The second effect of the domain rotation is to rearrange the side chains in the active site in such a way that they become poised for catalytic action. In other words, water is excluded from the active site when the enzyme is ready for catalytic action, and ATP hydrolysis is thereby prevented.

The induced-fit theory was thus verified from these structural studies of hexokinase. One of the surprising features that had not been predicted was the nature of the conformational change. Movements of local small loop regions had been observed before, but movements of complete domains relative to each other were a novel feature. Such domain movements have, however, since been observed in several other enzymes, such as some of the dehydrogenases, and are probably quite common among multidomain proteins.

Conclusion

Nucleotides are used as the fundamental energy currency in cells to drive chemical reactions that require energy. Most protein domains of known three-dimensional structure that bind nucleotides for such energy transfer are of the α/β structural type.

NAD-dependent dehydrogenases comprise a family of enzymes that bind NAD through a common structural framework, the dinucleotide-binding fold. This framework binds NAD in a very similar way in these enzymes despite very different amino acid sequences. The NAD-binding domains are linked to catalytic domains, which have different structures. These enzymes could have evolved by gene fusion of a common gene for an NAD-binding protein with different genes for metabolite-binding proteins.

FAD-binding domains have the same structural motif for binding the ADP part of the dinucleotide coenzymes as the NAD-binding domains. The amino acid sequences that give rise to this motif show no overall homology, and the motif was recognized only after the three-dimensional structures of several of these enzymes had been solved. The stereochemical constraints that dictate the form of the motif can now be used to predict its presence in enzymes of unknown structure but known amino acid sequence.

One nucleotide-binding enzyme, yeast flavocytochrome b_2, has one domain for FMN binding, which is similar in amino acid sequence and structure to a plant enzyme that binds FMN, glycolate oxidase, and a second heme-binding domain, which is similar in sequence and structure to mammalian cytochrome b$_5$. These proteins are evolutionarily related and the common ancestral genes have been recruited to fulfill different functions in different organisms and, in yeast, have been fused together. Hexokinase, actin, and a chaperone protein, HSC 70, use a common specific motif for ATP binding and hydrolysis. ATP hydrolysis is prevented in hexokinase by substrate-induced conformational changes involving large domain rotations. Such domain rotations are used to poise amino acid side chains correctly for catalysis (induced fit) and also to protect the reactants from water by burying them inside the protein.

Selected readings

General

Anderson, C.M.; Zucker, F.H.; Steitz, T.A. Space-filling models of kinase clefts and conformation changes. *Science* 204: 375–380, 1979.

Branden, C.-I.; Eklund, H. Structure and mechanism of liver alcohol dehydrogenase, lactate dehydrogenase and glyceraldehyde–3-phosphate dehydrogenase. In Dehydrogenases. Ed. J. Jeffery. Cambridge, MA: Birkhäuser, 1980.

Branden, C.-I. Founding fathers and families. *Nature* 346: 607–608, 1990.

Eklund, H.; Branden, C.-I. Alcohol dehydrogenase. In Biological Macromolecules and Assemblies. Vol. 3: Active Sites of Enzymes. Eds. F.A. Jurnak; A. McPherson. New York: Wiley, 1987.

Eventoff, W.; Rossmann, M.G. The evolution of dehydrogenases and kinases. *CRC Crit. Rev. Biochem.* 3: 111–140, 1975.

Koshland, D.E., Jr. Catalysis in life and in the test tube. In Horizons in Biochemistry. Eds. M. Kasha; B. Pullman. New York: Academic Press, 1962.

Rossmann, M.G., et al. Evolutionary and structural relationships among dehydrogenases. In The Enzymes. Ed P.D. Boyer. XI: 61–102. New York: Academic Press, 1975.

Wierenga, R.K.; Terpstra, P.; Hol, W.G.J. Prediction of the occurrence of the ADP-binding βαβ-fold in proteins, using an amino acid sequence fingerprint. *J. Mol. Biol.* 187: 101–107, 1986.

Specific structures

Abad-Zapatero, C., et al. Refined crystal structure of dogfish M$_4$ apo-lactate dehydrogenase. *J. Mol. Biol.* 198: 445–467, 1987.

Adams, M. J., et al. Structure of lactate dehydrogenase at 2.8 Å resolution. *Nature* 227: 1098–1103, 1970.

Aronson, B.D., et al. The primary structure of *Escherichia coli* L-threonine dehydrogenase. *J. Biol. Chem.* 264: 5226–5232, 1989.

Buehner, M., et al. D-glyceraldehyde-3-phosphate dehydrogenase: three-dimensional structure and evolutionary significance. *Proc. Natl. Acad. Sci. USA* 70: 3052–3054, 1973.

Birktoft, J.J.; Rhodes, G.; Banaszak, L.J. Refined crystal structure of cytoplasmic malate dehydrogenase at 2.5-Å resolution. *Biochemistry* 28: 6065–6081, 1989.

Branden, C.-I., et al. Structural and evolutionary aspects of the key enzymes in photorespiration: RuBisCo and glycolate oxidase. *Cold Spring Harbor Symp. Quant. Biol.* 52: 491–498, 1987.

Eggink, G., et al. Rubredoxin reductase of *Pseudomonas oleovorans*. Structural relationship to other flavoprotein oxidoreductases based on one NAD and two FAD fingerprints. *J. Mol. Biol.* 212: 135–142, 1990.

Eklund, H., et al. Three-dimensional structure of horse liver alcohol dehydrogenase at 2.4 Å resolution. *J. Mol. Biol.* 102: 27–59, 1976.

Eklund, H., et al. Structure of a triclinic ternary complex of horse liver alcohol dehydrogenase at 2.9 Å resolution. *J. Mol. Biol.* 146: 561–587, 1981.

Eklund, H., et al. Molecular aspects of functional differences between alcohol and sorbitol dehydrogenases. *Biochemistry* 24: 8005–8012, 1985.

Eklund, H., et al. Comparison of three classes of human liver alcohol dehydrogenase. *Eur. J. Biochem.* 193: 303–310, 1990.

Flaherty, K.M.; de Luca-Flaherty, C.; McKay, D.B. Three dimensional structure of the ATP ase fragment of the 70K heat shock cognate protein. *Nature* 346: 623–628, 1990.

Hill, E., et al. Polypeptide conformation of cytoplasmic malate dehydrogenase from an electron density map at 3.0 Å resolution. *J. Mol. Biol.* 72: 577–591, 1972.

Kabsch, W., et al. Atomic structure of the actin: DNase I complex. *Nature* 347: 37–44, 1990.

Karplus, P.A.; Schulz, G.E. Refined structure of glutathione reductase at 1.54 Å resolution. *J. Mol. Biol.* 195: 701–729, 1987.

Lindqvist, Y. Refined structure of spinach glycolate oxidase at 2 Å resolution. *J. Mol. Biol.* 209: 151–166, 1989.

Mathews, F.S.; Levine, M.; Argos, P. The structure of calf liver cytochrome b₅ at 2.8 Å resolution. *Nature New Biol.* 233: 15–16, 1971.

Ohlson, I.; Nordström, B.; Branden, C.-I. Structural and functional similarities within the coenzyme-binding domains of dehydrogenases. *J. Mol. Biol.* 89: 339–354, 1974.

Rossmann, M.G.; Moras, D.; Olsen, K.W. Chemical and biological evolution of a nucleotide-binding protein. *Nature* 250: 194–199, 1974.

Schreuder, H.A., et al. Crystal structure of p-hydroxybenzoate hydroxylase complexed with its reaction product 3,4-dihydroxybenzoate. *J. Mol. Biol.* 199: 637–648, 1988.

Skarzynski, T.; Wonacott, A.J. Coenzyme-induced conformational changes in glyceraldehyde-3-phosphate dehydrogenase from *Bacillus stearothermophilus. J. Mol. Biol.* 203: 1097–1118, 1988.

Wierenga, R.K.; Drenth, J.; Schulz, G.E. Comparison of the three-dimensional protein and nucleotide structure of the FAD-binding domain of p-hydroxy-benzoate hydroxylase with the FAD—as well as NADPH-binding domains of glutathione reductase. *J. Mol. Biol.* 167: 725–739, 1983.

Xia, Z.-X.; Mathews, F.S. Molecular structure of flavocytochrome b₂ at 2.4 Å resolution. *J. Mol. Biol.* 212: 837–863, 1990.

The Structure of Spherical Viruses

11

All viruses for their existence depend on their ability to infect cells and cause them to make more virus particles. If the virus is successful, the cells almost invariably die in the process. Even though the infection cycles of different species of viruses vary in detail, they all follow the same basic pattern. First, the viruses deliver their nucleic acid into the host cell; the mechanism of delivery varies depending on the virus, but the end result is the same. Second, the host cell's biosynthetic machinery is subverted for the replication, transcription, and translation of the viral genes at the expense of cellular gene expression. Finally, progeny virus particles assemble in the infected cell and by one route or another leave it to infect a fresh host. Fortunately, individual viruses can only infect a restricted range of hosts, added to which the host organisms usually have defense mechanisms, ranging from restriction enzymes in bacteria to the immune system of vertebrates. The combination of restricted host ranges and host defense mechanisms keeps in check the extent of viral infections.

Viruses come in many different sizes, shapes, and compositions. They all have a genomic nucleic acid molecule, but this is DNA in some viruses and RNA in others, double stranded in some but single stranded in others, a single molecule in some but segmented into several different molecules in others. The nucleic acid genome is always surrounded by a protein shell, a **capsid**, and some viruses also have a lipid bilayer membrane, an **envelope**, enclosing the capsid. One example is the influenza viruses discussed in Chapter 5. Other viruses have an even more complex structure. For example, the structure of bacteriophage T4, which must penetrate the bacterial cell wall at specific sites and inject its DNA into the cell, is seen in electron micrographs to resemble a hypodermic syringe.

A nucleic acid can never code for a single protein molecule that is big enough to enclose and protect it. Therefore, the protein shell of viruses is built up from many copies of one or a few polypeptide chains. The simplest viruses have just one type of polypeptide chain, which forms either a rod-shaped or a roughly spherical shell around the nucleic acid. All such simple viruses whose three-dimensional structures are known are plant viruses: the rod-shaped tobacco mosaic virus, the spherical satellite tobacco necrosis virus, tomato bushy stunt virus, and southern bean mosaic virus. Certain mammalian viruses, for example, the picorna viruses, are almost as simple in their construction, having a spherical shell built up of four different polypeptide chains. The structures of four picorna viruses are known: common cold virus, Mengo virus, poliovirus, and foot-and-mouth disease virus. Schematic diagrams illustrating the size and shape of some viruses are given in Figure 11.1.

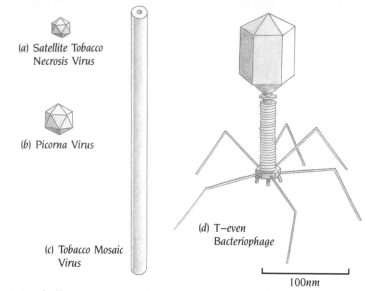

(a) Satellite Tobacco Necrosis Virus

(b) Picorna Virus

(c) Tobacco Mosaic Virus

(d) T–even Bacteriophage

100nm

Figure 11.1 Viruses vary in size and shape from the simplest satellite viruses (a) that need another virus for their replication to the T-even bacteriophages (d) that have developed sophisticated mechanisms for injecting DNA into bacteria. Four different virus particles are shown in correct relative scale.

The protein shells, or capsids, of the simple spherical animal and plant viruses are made up of symmetrically disposed protein subunits. This symmetry of subunits has been exploited by crystallographers to facilitate the determination of the structure of the capsids at atomic resolution by x-ray methods. Consequently, these viruses are the largest aggregates of biological macromolecules whose structure has been determined at high resolution. Unfortunately, however, the nucleic acid genome, which is inside the protein shell of the virus, lacks symmetry and is not seen in virus x-ray structures. We therefore have detailed knowledge of the structures of the protein shells of some of the simplest viruses but do not know the structure of the nucleic acid inside the shell or how it interacts with the protein.

This means our understanding of viral structure is necessarily incomplete. However, detailed knowledge of the structure of viral capsids alone is valuable for both practical and theoretical reasons. In principle, it should be possible to exploit this information for the rational design of antiviral drugs. For instance, the entry of mammalian viruses into cells depends upon specific recognition by the viral capsid of proteins or carbohydrates that occur on the surface of mammalian cells. This interaction thus plays a crucial part in determining the viral host range, and compounds that interfere with it should block infection. Understanding the capsid structure should help in designing such compounds. Because viral binding sites for surface molecules of host cells are also a target for protective antibodies, the immune system of the host has in all probability played a major role in the evolution of animal viruses. Detailed knowledge of the surface regions of these viruses should therefore teach us more about the structural adaptations that have evolved as the virus's evasive response to host defences.

In this chapter we will examine the construction principles of spherical viruses , the structures of individual subunits and the binding properties of the surface of one of the picorna viruses, the common cold virus.

The protein shells of spherical viruses have icosahedral symmetry

Two basic principles govern the arrangement of protein subunits within the shells of spherical viruses. The first is specificity; subunits must recognize each other with precision to form an exact interface of non-covalent interactions since virus particles assemble spontaneously from their individual components. The second principle is genetic economy; the shell is built up from many copies of a few kinds of subunit. These principles together imply symmetry; specific, repeated bonding patterns of identical building blocks lead to a symmetrical final structure.

The first question then is, what types of symmetry can form roughly spherical objects that are built up from identical units, or, in other words, how can one make a sphere by arranging identical objects symmetrically on its surface?—a problem also faced by producers of soccer balls. This question was answered more than 2000 years ago by the classical Greek mathematicians, who showed that there is only a very limited number of ways to build such objects. Among these, the icosahedron (Figure 11.2) and dodecahedron have the highest possible symmetry, generally called **icosahedral symmetry**, and therefore allow a maximum number of identical objects to form a closed symmetrical shell. Do these geometric principles apply to viral structures?

Fortunately, virus particles are large enough to be studied by electron microscopy using negative staining. In the 1950s and 1960s such studies, mainly by Aaron Klug and his collaborators in Cambridge, England, showed that indeed all the spherical viruses examined exhibited icosahedral symmetry. They could also demonstrate from the x-ray diffraction patterns of virus crystals that not only does the external surface obey this symmetry, but so also does the whole arrangement of subunits in the virus's shell. Since icosahedral symmetry is central to an understanding of the architecture of spherical viruses, it will be examined in detail.

The icosahedron has high symmetry

The icosahedron is a roughly spherical object that is built up from 20 identical equilateral triangles. These triangular tiles are arranged side by side in such a way that they enclose the volume inside the icosahedron. Figure 11.3 shows a view of an icosahedron where there are 5 tiles at the top, 5 at the bottom, and 10 in a band around the middle region. These faces are identical: in other words, identical icosahedra are obtained irrespective of which of the vertices is at the top in Figure 11.3a.

The symmetry of the icosahedron is described by the different types of rotations (Figure 11.3) that bring it into self-coincidence. There are 12 vertices, each of which has a fivefold rotation axis that passes through it and through the center. Rotation by a fifth of a revolution about one of these axes brings the icosahedron to an orientation indistinguishable from the starting orientation.

Figure 11.2 The icosahedron (*top*) and dodecahedron (*bottom*) have identical symmetries but different shapes. Protein subunits of spherical viruses form a coat around the nucleic acid with the same symmetry arrangement as these geometrical objects. Electron micrographs of these viruses have shown that their shapes are best represented by icosahedra. We, therefore, in this chapter will use the icosahedral shape and symmetry when discussing the structure of spherical viruses. One each of the twofold, threefold, and fivefold symmetry axes is indicated by a square, triangle, and pentagon, respectively.

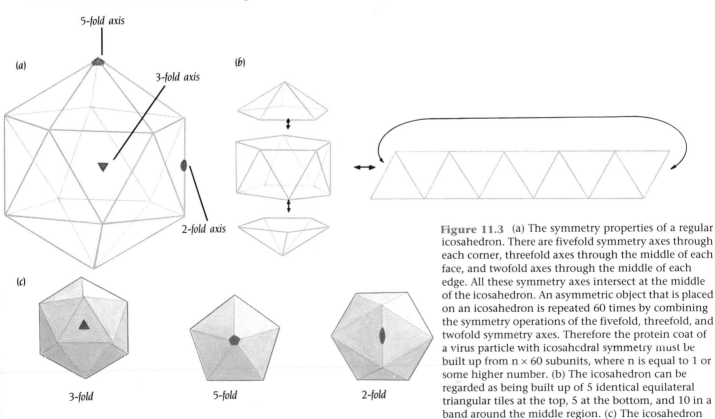

Figure 11.3 (a) The symmetry properties of a regular icosahedron. There are fivefold symmetry axes through each corner, threefold axes through the middle of each face, and twofold axes through the middle of each edge. All these symmetry axes intersect at the middle of the icosahedron. An asymmetric object that is placed on an icosahedron is repeated 60 times by combining the symmetry operations of the fivefold, threefold, and twofold symmetry axes. Therefore the protein coat of a virus particle with icosahedral symmetry must be built up from n × 60 subunits, where n is equal to 1 or some higher number. (b) The icosahedron can be regarded as being built up of 5 identical equilateral triangular tiles at the top, 5 at the bottom, and 10 in a band around the middle region. (c) The icosahedron viewed along each of its different symmetry axes.

163

(a)

(b)

Figure 11.4 The division of the surface of an icosahedron into asymmetric units. (a) One triangular face is divided into three asymmetric units into which an object is placed. These are related by the threefold symmetry axis. (b) Triangular tiles with the three symmetry related objects have been positioned on the surface of the icosahedron in such orientations that the objects are related by the fivefold and twofold symmetry axes.

In addition, there are threefold symmetry axes through the center of the icosahedron and the middle of each of the 20 faces and twofold axes through the middle of each of the 30 edges. All this symmetry has important consequences for the way in which the shell of spherical viruses are constructed.

Any symmetrical object is built up from smaller pieces that are identical and that are related to each other by symmetry. An icosahedron can therefore be divided into a number of smaller identical pieces called symmetry-related units. Protein subunits are asymmetric objects; hence, a symmetry axis cannot pass through them. The minimum number of protein subunits that can form a virus shell with icosahedral symmetry is therefore equal to the number of symmetry-related units in an icosahedron. This number can easily be deduced by regarding the 20 triangular tiles on the surface of the icosahedron. These tiles have threefold symmetry, and therefore, by definition, three identical objects are required to form one tile. One obvious way to regard such tiles is shown in Figure 11.4a, which illustrates the division of one such triangle into three equal parts, each of which contains an object. Since there are twenty tiles the total number of units in the icosahedron is $20 \times 3 = 60$. We could equally well have regarded the 12 different corners with fivefold symmetry or the 30 edges with twofold symmetry. Such units are called asymmetric units, thus the icosahedron and by inference spherical viruses have 60 asymmetric units (Figure 11.4b). Hence at least 60 protein subunits are required to form a virus shell with icosahedral symmetry.

The symmetry properties of an icosahedron are not restricted to the surface but extend through the whole volume. An asymmetric unit is therefore a part of this volume; it is a wedge from the surface to the center of the icosahedron. Sixty such wedges completely fill the volume of the icosahedron.

The simplest virus has a shell of 60 protein subunits

The asymmetric unit of an icosahedron can contain one or several polypeptide chains. The protein shell of a spherical virus with icosahedral symmetry must therefore have a minimum number of 60 identical protein subunits. Is there any virus known with only 60 protein subunits, or is the volume inside such a shell too small to enclose the viral genome?

No known self-sufficient virus has as few as 60 identical subunits in its shell. The smallest known viruses, however, are the satellite viruses, which do not themselves encode all of the functions required for their replication and are therefore not self-sufficient. The first satellite virus to be discovered, **satellite tobacco necrosis virus**, which is also one of the smallest known with a diameter of 180 Å, has a protein shell of 60 subunits. This virus cannot replicate on its own inside a tobacco cell but needs a helper virus, tobacco necrosis virus, to supply the functions it does not encode. The RNA genome of the satellite virus is very small, 1120 nucleotides, which code for the viral coat protein of 195 amino acids but no other protein. With this minimal genome the satellite viruses are obligate parasites of the viruses that parasitize cells.

The structure of satellite tobacco necrosis virus has been determined to 2.5 Å resolution by the group of Bror Strandberg in Uppsala, Sweden. As expected, the viral shell, or capsid, has icosahedral symmetry with one polypeptide chain in the asymmetric unit. This is schematically illustrated in Figure 11.5a. The

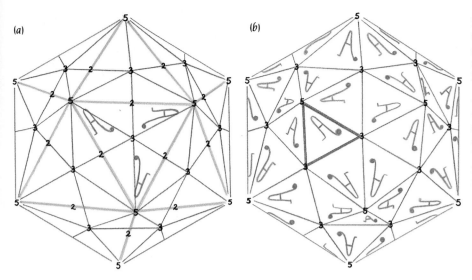

Figure 11.5 Schematic illustration of the way the 60 protein subunits are arranged around the shell of satellite tobacco necrosis virus. Each subunit is shown as an asymmetric A. The view is along one of the threefold axes, as in Figure 11.3a. (a) Three subunits are positioned on one triangular tile of an icosahedron, in a similar way to that shown in 11.4a. The red lines represent a different way to divide the surface of the icosahedron into 60 asymmetric units. This representation will be used in the following diagrams because it is easier to see the symmetry relations when there are more than 60 subunits in the shells. (b) All subunits are shown on the surface of the virus, seen in the same orientation as 11.4a. The shell has been subdivided into 60 asymmetric units by the red lines. When the corners are joined to the center of the virus, the particle is divided into 60 triangular wedges, each comprising an asymmetric unit of the virus. In satellite tobacco necrosis virus each such unit contains one polypeptide chain.

subunits are arranged so that they are quite close together around the fivefold axes forming tight contacts at the surface of the particle (Figure 11.5b). On the inside of the shell, on the other hand, the polypeptide chains are clustered around the threefold axes, one α helix from each subunit forming a bundle of three α helices. These tight interactions around the fivefold and threefold axes form a scaffold that links all subunits together to complete the shell.

The outer diameter of the shell is approximately 180 Å and the inner diameter about 125 Å except around the threefold axes, where the N-terminal α helices project around 20 Å into the core. The RNA molecule that is present in the core is not visible in the electron density map.

Complex spherical viruses have more than one polypeptide chain in the asymmetric unit

We have seen in the structure of this simple satellite virus that 60 subunits are sufficient to form a shell around an RNA molecule that codes for the subunit protein, but there is little room for additional genetic information. Self-sufficient viruses have longer genomes, which code for enzymes essential for the replication of the viral nucleic acid in addition to the structural proteins of the capsid. These genomes need a larger volume inside the capsid and therefore need a larger capsid. How are such larger shells constructed within the constraints of icosahedral symmetry? Very little is gained by increasing the size of each subunit because the shell becomes thicker as it is enlarged. The only way out is to increase the number of subunits. To preserve icosahedral symmetry, the asymmetric unit itself must then contain more than one subunit. Since there are 60 asymmetric units, the total number of subunits in the shell must therefore be a multiple of 60. These subunits can be either identical or different. From the point of view of genetic economy, it is advantageous for the virus to use many identical subunits because then only one gene is needed to code for the entire protein shell. There are many examples of viruses of this type, and we will examine one example before describing viruses that contain different types of subunits in their capsids.

Can any number of identical subunits be accommodated in the asymmetric unit while preserving specificity of interactions within an icosahedral arrangement? This question was answered by Don Caspar at Brandeis University, Massachusetts, and Aaron Klug in Cambridge, England, who showed in a classical paper in 1962 that only certain multiples (1, 3, 4, 7 . . .) of 60 subunits are likely to occur. They called these multiples **triangulation numbers, T**. Icosahedral virus structures are frequently referred to in terms of their triangulation numbers; a T = 3 virus structure therefore implies that the number of subunits in the icosahedral shell is $3 \times 60 = 180$.

Caspar and Klug based their arguments on the principle of specificity, or that the protein-protein contacts should be approximately similar. When T is larger than one, it is no longer possible to pack the protein subunits into an icosahedral shell in a strictly equivalent way as it is for a T = 1 structure like satellite tobacco necrosis virus where all subunits have the same environment and hence the same packing interactions. What Caspar and Klug showed is that for certain specific values of T (1, 3, 4, 7 . . .) it is possible to pack the subunits with only slightly different bonding patterns, in a **quasi-equivalent** way. These triangulation numbers were derived from the relation $T = h^2 + hk + k^2$, where h and k are any integers.

As examples of such quasi-equivalent arrangement of subunits, we will examine the T = 3 and T = 4 packing modes. In the T = 3 structure, which has 180 subunits (3 × 60), each asymmetric unit contains three protein subunits with different environments, which we will call A, B, and C. These are arranged so that the A subunits interact around the fivefold axes, and the B and C subunits alternate around the threefold axes (Figure 11.6). There are thus six subunits, three each of B and C, arranged in a pseudosymmetrical way around the threefold axes, which therefore are also pseudosixfold axes. Caspar and Klug argued that the subunits could pack around a fivefold and a pseudosixfold axis in a rather similar fashion with only minor alterations of the packing mode. We will see in the structure of **tomato bushy stunt virus** that they were right. This viral capsid is made from chemically identical proteins that are able to accommodate the necessary differences in packing mode by changing their conformation according to their different environments.

In the T = 4 structure there are 240 subunits (4 × 60) in four different environments, A, B, C, and D, in the asymmetric unit. The A subunits interact around the fivefold axes, and the D subunits around the threefold axes (Figure 11.7). The B and C subunits are arranged so that two copies of each interact around the twofold axes in addition to two D subunits (Figure 11.7). For a T = 4 structure the twofold axes thus form pseudosixfold axes. The A, B, and C subunits interact around pseudothreefold axes clustered around the fivefold axes. There are 60 such pseudothreefold axes. The T = 4 structure therefore has a total of 80 threefold axes: 20 with strict icosahedral symmetry and 60 with pseudo symmetry. These aspects of a T = 4 structure, which were theoretically derived by Caspar and Klug, proved important in deducing the structural organization of Sindbis virus, as we will see.

Structural versatility gives quasi-equivalent packing in T = 3 plant viruses

The molecular basis for quasi-equivalent packing was revealed by the very first structure determination to high resolution of a spherical virus, tomato bushy stunt virus. The structure of this T = 3 virus was determined to 2.9 Å resolution in 1978 by Stephen Harrison and co-workers at Harvard University. The virus shell contains 180 chemically identical polypeptide chains, each of 386 amino acid residues. Each polypeptide chain folds into distinct modules (Figure 11.8): an internal domain R that is disordered in the structure, a region (a) that connects R with the S domain that forms the viral shell, and, finally, a domain P that projects out from the surface. The S and P domains are joined by a hinge region.

When they form the three subunits A, B, and C of the asymmetric unit, the identical polypeptides adopt different three-dimensional structures. The C subunit in particular is distinct from the A and B structures, its hinge region assuming a different conformation so that the S and P domains are quite differently oriented. In addition, the arm region (a) is ordered in C but disordered in A and B. There are no gross conformational differences between A and B, but there are significant local structural differences, especially in the interaction areas. The subunits thus accommodate to the three different environments and preserve quasi-equivalence in most contact regions by conformational differences of the protein chain; structural diversity has been achieved without sacrificing genetic economy.

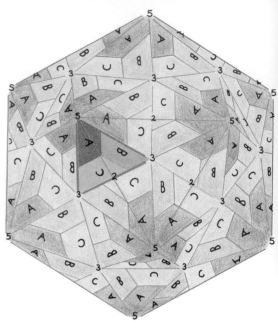

Figure 11.6 A T = 3 icosahedral virus structure contains 180 subunits in its protein shell. Each asymmetric unit (one such unit is shown in thick lines) contains three protein subunits A, B, and C. The icosahedral structure is viewed along a threefold axis, the same view as in Figure 11.5. One asymmetric unit is shown in dark colors.

Figure 11.7 A T = 4 icosahedral virus structure contains 240 subunits in its protein shell. Each asymmetric unit (one such unit is shown in dark colors) contains four protein subunits, A, B, C, and D. We have now twisted the icosahedron somewhat, compared to Figure 11.6, so that we view the structure along one of the twofold axes, the same view as in Figure 11.3c.

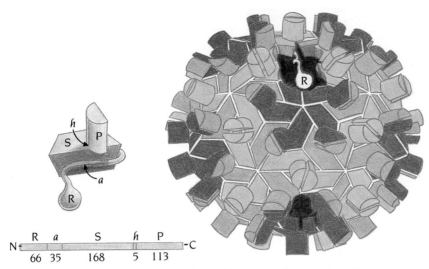

Figure 11.8 Architecture of the tomato bushy stunt virus particle. The polypeptide chain of each subunit folds into three domains (R, S, P) with a 35-residue connecting arm (a) between R and S and a hinge (h) between S and P. The number of amino acid residues in each structural module is indicated. The subunits pack into the virus particle in one of three conformations, colored red (A), blue (B), and green (C). The S domains of the red subunits pack around fivefold axes; the S domains of the blue and green subunits alternate around the threefold axes. The P domains project out from the surface of the particle. The cutaway region in the large diagram shows the connecting arm (a) between domains S and R and the region that the disordered domain R might occupy. (Adapted from S.C. Harrison et al., *Nature* 276: 370, 1978.)

The S domains form the viral shell by tight interactions in a manner predicted by the Caspar and Klug theory and shown in Figure 11.8. The P domains interact pairwise across the twofold axes and form protrusions on the surface. There are 30 twofold axes with icosahedral symmetry that relate the P domains of C subunits (green) and in addition 60 pseudotwofold axes relating the A (red) and B (blue) subunits (Figure 11.9). By this arrangement the 180 P domains form 90 dimeric protrusions.

One of the remarkable features of this structure as well as other T = 3 plant virus structures is the way in which the ordered connecting arms of C subunits interdigitate to form an internal framework (Figure 11.10). These arms (a in Figure 11.8) extend along the inner face of the S domain and loop around icosahedral threefold axes. Three such C subunit arms contact each other in this way, and all 60 C subunits form a coherent network (Figure 11.10). The function of this framework is to determine the size of the particle—that is, to ensure that the viral shell closes round on itself correctly during assembly.

The size of this viral particle is of course larger than that of a virus with only 60 subunits. The diameter of tomato bushy stunt virus is 330 Å compared with 180 Å for satellite tobacco necrosis virus. The increase in volume of the capsid means that a roughly four times larger RNA molecule can be accommodated.

The protein capsid of picorna viruses contains four polypeptide chains

A protein capsid built up from identical subunits is an economical way for the virus to protect its RNA. However, more elaborate schemes have also evolved, such as that found in **picorna viruses**. These viruses comprise a large family of animal viruses with single-stranded RNA genomes. They have been classified into four genera: (1) cardioviruses, such as encephalomyocarditis virus and **Mengo virus**; (2) enteroviruses such as **poliovirus** and hepatitis A virus; (3) **rhinoviruses**, which cause common colds and of which there are about 100 serotypes; and (4) aphtoviruses, which include **foot-and-mouth disease virus**. The three-dimensional structures of representative examples from each of these genera are known. The group of Michael Rossmann at Purdue University, Indiana, has determined the structure of Mengo virus to 3.0 Å resolution and that of human rhinovirus strain 14 to 2.6 Å resolution. The structure of poliovirus was determined by the group of James Hogle at Scripps Clinic, La Jolla, California, to 2.9 Å resolution, and the structure of foot-and-mouth disease virus, by the group of David Stuart at Oxford University to 2.5 Å resolution.

All these viruses have a molecular mass of around 8.5 million daltons, one long RNA molecule of around 8000 nucleotides accounts for 30% of this molecular mass. The RNA chain occupies the interior of the spherical virion and

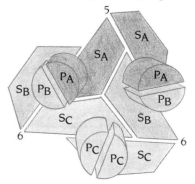

Figure 11.9 Contacts between P domains in tomato bushy stunt virus. S_A, S_B, and S_C are the shell domains of subunits A, B, and C, respectively. P_A, P_B, and P_C are the protruding domains of subunits A, B, and C, respectively.

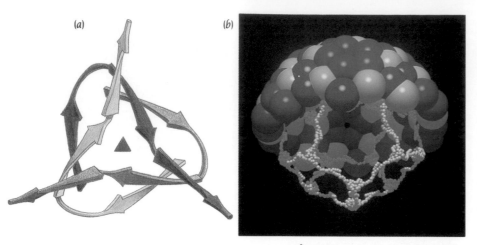

(a)   *(b)*

Figure 11.10 The arms of all 60 C subunits in tomato bushy stunt virus form an internal framework. (a) Configuration of interdigitated arms from the three C subunits, viewed down a threefold axis. The β strands are shown as arrows. (b) Cutaway view of the virus particle, emphasizing the framework function of the C-subunit arms. These arms are shown as chains of small balls, one per residue. The region where three arms meet and interdigitate is shown schematically in (a). The main part of each subunit is represented by large balls. Only about one hemisphere of these is drawn, but all the C-subunit arms are included. [(a) Adapted from R.J. Olson et al., *J. Mol. Biol.* 171: 89, 1983. (b) Courtesy of A. Olson.]

extends from its center to a radius of about 110 Å (Figure 11.11). A continuous protective shell of protein subunits surrounds the RNA chain. This continuous shell has an average thickness of about 30 Å, giving the virus particle an approximate total diameter of 300 Å. The outer surface is not smooth, however, but is covered by parts of the polypeptide chains that project out from the continuous protein shell. The protruding loops are recognized by the immune defense mechanism of the host and contain most of the antigenic-binding sites for host antibodies. The surface cavities bounded by these protrusions are believed in some cases to be specific binding sites for receptors on the surface of the host cell. The structural properties of the protein molecules that form the shell must thus include not only the usual requirements for assembly into an icosahedron but also the ability to preserve the receptor sites that render the virus infectious, while accommodating mutations in neighboring antigenic sites in order to escape the immune defenses of the host.

There are four different structural proteins in picorna viruses

The shell of all picorna viruses is built up from 60 copies each of four polypeptide chains, called VP1 to VP4. These are translated from the viral RNA into one single polypeptide, which is posttranslationally processed by stepwise proteolysis involving virally coded enzymes. First, the polypeptide chain is cleaved into three proteins VP0 (which is the precursor for VP2 and VP4) and VP1 and VP3. These proteins begin the assembly process. The last step of the processing cascade occurs during completion of the virion assembly; the precursor protein VP0 is cleaved into VP2 and VP4 by a mechanism that is probably autocatalytic but may also involve catalytic action by the viral RNA. VP1, VP2, and VP3 have molecular masses around 30.000 daltons, whereas VP4 is small, 7.000 daltons, and is completely buried inside the virion.

The arrangement of subunits in the shell of picorna viruses is similar to that of T = 3 plant viruses

Figure 11.11 Schematic diagram of a picorna virus particle, illustrating the volume occupied by RNA and protein. The surface of the particle contains protrusions and depressions.

The capsids of many plant viruses, like tomato bushy stunt virus, have 180 polypeptide subunits (T = 3) that are chemically identical. Picorna viruses also have 180 polypeptide subunits in their capsids; these, however, are 60 copies each of three different types, VP1, VP2 and VP3, with no significant amino acid sequence homology between them. The 60 copies of the fourth small subunit, VP4, we consider as a continuation of VP2 that has become detached. Given this diversity of subunits, the picorna viruses could, in principle, form an icosahedral shell with a completely different subunit arrangement and different subunit structures from those of the plant viruses. This is not the case, however. The 180 large subunits of the picorna viruses are arranged in an extraordinarily similar way to the 180 subunits of the T = 3 plant viruses. Quasi-equivalent packing is achieved not by conformational changes, as is the case with the single polypeptide of the plant viruses, but

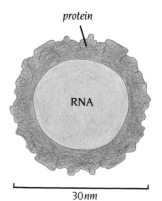

protein

RNA

30 *nm*

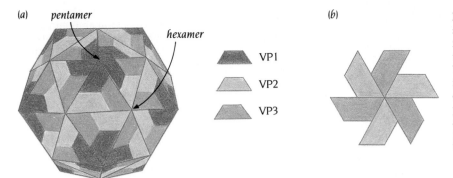

(a)

pentamer

hexamer

VP1
VP2
VP3

(b)

Figure 11.12 (a) Schematic diagram of the surface of a picorna virus. Each triangle represents an asymmetric unit that contains three protein subunits on its surface: VP1 is red, VP2 is green, and VP3 is blue. The diagram illustrates that, at the surface, VP1 subunits are clustered around the fivefold axes, forming pentamers, whereas VP2 and VP3 alternate around the threefold axes, forming hexamers. (b) The hexameric arrangement of VP2 and VP3.

by having within the shell three chemically different polypeptide chains with local structural differences.

The asymmetric unit contains one copy each of the subunits VP1, VP2, VP3, and VP4. VP4 is buried inside the shell and does not reach the surface. The arrangement of VP1, VP2, and VP3 on the surface of the capsid is shown in Figure 11.12a. These three different polypeptide chains build up the virus shell in a way that is analogous to that of the three different conformations A, B, and C of the same polypeptide chain in tomato bushy stunt virus. The viral coat assembles from 12 compact aggregates, or pentamers, which contain five of each of the coat proteins. Each pentamer is shaped like a molecular mountain: the VP1 subunits, which correspond to the A subunits in T = 3 plant viruses, cluster at the peak of the mountain; VP2 and VP3 alternate around the foot; and VP4 provides the foundation. The amino termini of the five VP3 subunits of the pentamer intertwine around the fivefold axis in the interior of the virion to form a β structure that stabilizes the pentamer and in addition interacts with VP4.

Subunits VP2 and VP3 from different pentamers alternate around the threefold symmetry axes like subunits B and C in the plant viruses (Figure 11.12). Since VP2 and VP3 are quite different polypeptide chains, they cannot be related to each other by strict symmetry. To a first approximation, however, they are related by quasi-symmetry since they both participate in the hexamer formation. The strict threefold axis is thus also a quasi-sixfold symmetry axis. This quasi-symmetry is, of course, more regular for the T = 3 plant viruses where the six subunits are chemically identical.

The fact that spherical plant viruses and animal viruses build their icosahedral shells using essentially similar asymmetric units raises the possibility that they have a common evolutionary ancestor. The folding of the main chain in the protein subunits of these viruses supports this notion.

The coat proteins of spherical plant and animal viruses have similar structure, the jelly roll barrel structure, indicating an evolutionary relationship

One of the most striking results that has emerged from the high-resolution crystallographic studies of these icosahedral viruses is that their coat proteins have the same basic core structure, that of a jelly roll barrel, which was discussed in Chapter 5. This is true of plant, insect, and mammalian viruses. In the case of the picorna viruses, VP1, VP2, and VP3 all have the same jelly roll structure as the subunits of satellite tobacco necrosis virus, tomato bushy stunt virus, and the other T = 3 plant viruses. Not every spherical virus has, however, subunit structures of the jelly roll type. As we will see, the subunits of a small bacteriophage, MS2, have a very different structure. Nevertheless, they form an icosahedral shell similar to those of other spherical viruses.

The canonical jelly roll barrel is schematically illustrated in Figure 11.13. Superposition of the structures of coat proteins from different viruses show that the eight β strands of the jelly roll barrel form a conserved core. This is illustrated in Figure 11.14, which shows structural diagrams of three different coat proteins. These diagrams also clearly show that the β strands are arranged in two sheets of four strands each: β strands 1, 8, 3, and 6 form one sheet and strands

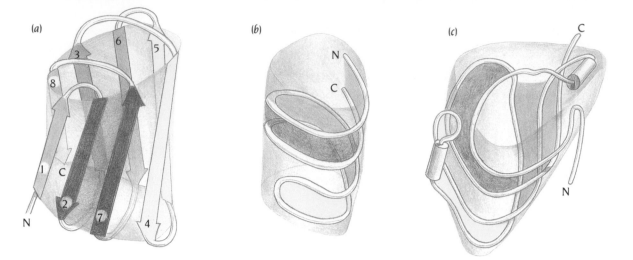

(a) *(b)* *(c)*

Figure 11.13 (*above*) The known subunit structures of plant, insect, and animal viruses are of the jelly roll antiparallel β barrel type, described in Chapter 5. This fold, which is schematically illustrated in two different ways, (a) and (b), forms the core of the S domain of the subunit of tomato bushy stunt virus (c). [(b) and (c) Adapted from A. Olson et al., *J. Mol. Biol.* 171: 79, 1983.]

Figure 11.14 (*below*) Schematic diagrams of three different viral coat proteins, viewed in approximately the same direction. β strands 1 through 8 form the common jelly roll barrel core. (a) Satellite tobacco necrosis virus coat protein. (Adapted from L. Liljas et al., *J. Mol. Biol.* 159: 93, 1982.) (b) Subunit VP1 from poliovirus. (Adapted from J. Hogle et al., *Science* 229: 1358, 1985.) (c) Subunit VP2 from human-rhinovirus. (Adapted from M.G. Rossmann et al., *Nature* 317: 145, 1985.) (d) Canonical jelly roll barrel, illustrating the wedge-shaped arrangement of the β strands in the subunit structures of viruses.

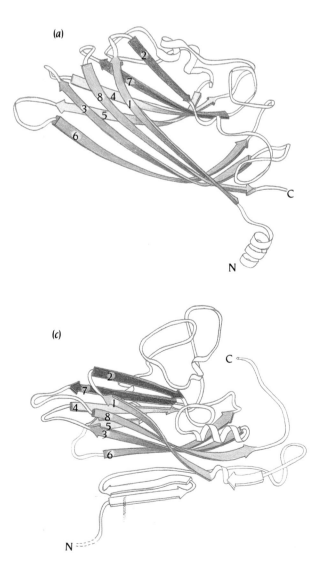

(a)

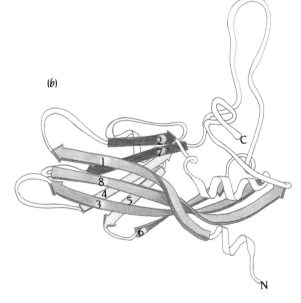

(b)

(c)

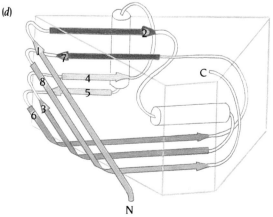

(d)

2, 7, 4, and 5 form the second sheet. Hydrophobic residues from these sheets pack inside the barrel.

In all jelly roll barrels the polypeptide chain enters and leaves the barrel at the same end, the base of the barrel. In the viral coat proteins a fairly large number of amino acids at the termini of the polypeptide chain is usually left outside the actual barrel structure. These regions vary considerably both in size and conformation between different coat proteins. In addition, there are three loop regions at this end of the barrel that usually are quite long and that also show considerable variation in size in the plant viruses as well as the picorna viruses. In contrast, the four loop regions at the other end of the barrel, the top, are short and exhibit only minor variations in size. As a consequence all coat proteins are shaped like a wedge or flattened cone with a broad base formed by the long loops and a narrow tip where the short loops keep the β strands together (Figure 11.14d).

The finding that the coat proteins of these spherical viruses have the same general topology and that the subunit arrangements of picorna viruses and T = 3 plant viruses are very similar means that we have to take very seriously the suggestion that all these viruses are evolutionarily related. Clearly, not all viruses can be evolutionarily related. Tobacco mosaic virus, which is a rod-shaped plant virus, has a completely different subunit topology (a four-helix bundle as discussed in Chapter 3) and a very different subunit arrangement from the spherical viruses. The discussion of common ancestry is thus restricted to the small spherical viruses whose subunits have the same topology.

Amino acid sequence comparisons reveal relationships within only the same class of viruses. The three major protein chains of a picorna virus, VP1, VP2, and VP3, which range in size from 230 to 280 amino acids, have different sequences and show no obvious homology with each other, either at the gene or protein level. Furthermore, there is no sequence homology to the coat proteins of spherical viruses of other families such as the plant viruses. In contrast, the same protein from different picorna viruses show significant sequence homology. For example, VP2 from Mengo virus and rhinovirus have about 30% sequence identity. Similar sequence homology is found in other proteins that are coded by the picorna viral genome, such as RNA polymerase. Picorna viruses obviously form an evolutionarily related family of viruses.

From sequence data alone no one would suggest that picorna viruses and T = 3 plant viruses are related. However, in order to obtain both the same topology of the subunits and similar packing arrangement, there must be constraints on a large number of amino acids. Even though they are not the same in the different proteins, they must fulfill the same functional role. Despite their different sequences the common topology and packing properties suggest that they all derive from the same ancestor. During evolution, mutations that could be accommodated by a number of successive small local structural changes in the protein have accumulated in genes. As we have already discussed in Chapter 3, the hemoglobins provide a precedent for this. The alternative, namely, that both the same topology and packing arrangements of the coat proteins have arisen independently many times during evolution, seems much less likely.

Drugs against common cold may be designed from the structure of rhinovirus

Most drugs are effective because they bind to a specific receptor site and block the physiological function of a protein. Classical drug design has been based on this concept for several decades. Compounds that are structural variants of substrates for a suitable target protein are synthesized and tested in binding studies. It is usually possible to obtain in this way inhibitors that bind better than the physiological substrates, although several thousands of compounds may have to be synthesized and tested before a suitable drug is found. It has, therefore, been the dream of many pharmaceutical chemists to be able to design new drugs in a more rational way based on a knowledge of the three-dimen-

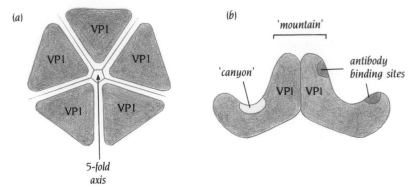

Figure 11.15 Subunits VP1 in picorna viruses cluster around the fivefold axis (a) so that their tips form a protrusion, or "mountain," on the viral surface. In rhinoviruses a depression, or "canyon," is formed outside this mountain, which probably contains a binding site for initial attachment of the virus to its receptor, the adhesion molecule ICAM-1. (b) Antiviral agents bind below the floor of this canyon in a cavity within VP1. Epitope mapping has shown that antibodies can bind to the rim of the canyon but cannot interact with residues at the bottom. [(b) Adapted from M. Lou et al., *Science* 235: 182, 1987.]

sional structure of the binding site of the relevant receptor protein. Understandably, Michael Rossmann attracted considerable attention in 1986 by reporting a detailed crystallographic analysis on binding studies of an antiviral drug called WIN 51711 to human rhinovirus. This work raises hopes that it might eventually be possible to design efficient drugs against the common cold, as well as many other viral diseases.

The cleft where this drug binds is inside the jelly roll barrel of subunit VP1. All spherical viruses of known structure have the tip of one type of subunit close to the fivefold symmetry axes (Figure 11.15a). In all the picorna viruses this position is, as we have described, occupied by the VP1 subunit. Two of the four loop regions at the tip are considerably longer in VP1 than in the other viral coat proteins. These long loops at the tips of VP1 subunits protrude from the surface of the virus shell and form mountains around its 12 fivefold axes (Figure 11.15b).

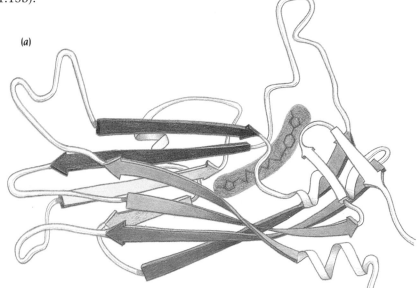

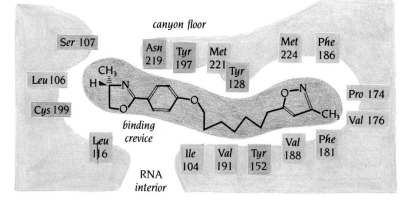

Figure 11.16 Schematic diagrams illustrating the binding of an antiviral agent to human rhinovirus strain 14. (a) The drug binds in a hydrophobic pocket of VP1 below the floor of the canyon. (b) Schematic diagram of VP1 illustrating the pocket in the jelly roll barrel where the drug binds. (Adapted from T.J. Smith et al., *Science* 233: 1233, 1986.)

In rhinoviruses there are depressions, or "canyons," which are 25 Å deep and 12 to 30 Å wide and which encircle the mountains (Figure 11.15b). One wall of the canyons is lined by residues from the base of VP1. The structure of VP1 is such that the barrel is open at the base and permits access to the hydrophobic interior of the barrel, as in the up-and-down barrel structure of the retinol-binding protein described in Chapter 5.

The antiviral compound that Rossmann studied binds inside the barrel of VP1 (Figure 11.16a). It appears to gain access to this site through an opening of the floor of the canyon. One end of the compound is a 3-methylisoxazole group that is inserted into the hydrophobic interior of the VP1 barrel (Figure 11.16b). This end is connected by an aliphatic chain to a 4-oxazolinyl phenoxy group OP. Both the aliphatic chain and the OP group are bound by side chains of the amino acids that line the bottom of the canyon. It is quite apparent by looking at the geometries of the cleft and the antiviral compound that the fit is not optimal and that it should be possible to design compounds with stronger binding to this binding site. In fact, such compounds have been designed and synthesized and are now subject to clinical trials.

Rossmann has suggested, but not proved, that the canyons form the binding site for the rhinovirus receptor on the surface of the host cells (Figure 11.15b). The receptor for the major group of rhinoviruses is an adhesion protein known as ICAM-1. Residues that line the bottom of this canyon are invariant between different strains of both poliovirus and rhinovirus, which suggests that they have an important conserved function. Residues at the rim of the canyon, on the other hand, are frequently mutated. They form antigenic epitopes: in other words, they are recognized by antibodies produced by the host's immune response (Figure 11.15b). Viruses can escape this response by mutation, provided that the change does not affect an important function. Mutations in residues at the rim do not impair function, whereas those at the bottom of the canyons would do so if the residues there formed the receptor binding site. Mutations at the rim of the canyon by which rhinoviruses escape the defense system of the host would therefore neither affect receptor binding nor drug binding.

Bacteriophage MS2 has a different subunit structure

About a dozen different plant, insect, and animal spherical virus structures have so far been determined to high resolution, and in all of them the subunit structures have the same jelly roll topology. It was therefore a major surprise to find a very different fold of the subunit in **bacteriophage MS2** whose structure was determined to 3 Å resolution by Karin Valegård in the laboratory of Lars Liljas, Uppsala.

MS2 belongs to a family of small single-stranded RNA phages—MS2, R17, f2, and Qβ—that infect *E. coli*. These phages are about 250 Å in diameter and contain only four genes in their 4000-nucleotide-long chromosome. One of these genes codes for the 129 amino acid coat protein that is present in 180 copies in the T = 3 virus particle. A second gene encodes a protein 393 amino acids long that is present in a single copy in each virus particle and is needed both for attachment of the phage to its host and penetration of the viral RNA into the bacterium. A third gene codes the phage replicase subunit that contains the polymerase active site.

The x-ray structure of the MS2 particle was determined by similar methods to those applied to other viral structures; the icosahedral symmetry of the particle was used to compute an average electron density for the subunits. The electron densities for both the attachment protein and the RNA, which are present in only one copy each, are smeared out in this process, and only the structure of the coat protein is obtained. The subunit (Figure 11.17) folds into a five-stranded up-and-down antiparallel β sheet with an additional short hairpin at the amino terminus and two α helices at the other end of the chain. The α helices are responsible for interactions with a second subunit to form a tight dimer. The two β sheets of each subunit in the dimer are aligned at their edges

Figure 11.17 The subunit structure of the bacteriophage MS2 coat protein is different from those of other spherical viruses. The 129 amino acid polypeptide chain is folded into an up-and-down antiparallel β sheet of five strands, β 3–β 7, with a hairpin at the amino end and two C-terminal α helices. (Adapted from a diagram provided by L. Liljas.)

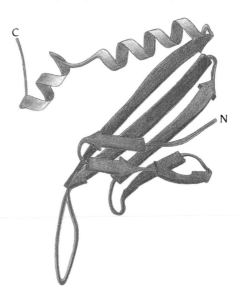

so that a large continuous β sheet of 10 adjacent antiparallel β strands is formed (Figure 11.18), with α helices from one subunit packing against the β sheets of the second subunit.

This subunit structure is quite different from the jelly roll structure found in all other spherical viruses so far. Since all members of this family of RNA phages have homologous coat proteins, their subunits are expected to have the same three-dimensional structure. It remains to be seen if the MS2 fold is also present in any other unrelated viruses. The fold is so far unique for the MS2 subunit, but similar structures have been observed in other proteins such as the major histocompatibility antigen, HLA, which will be discussed in Chapter 12.

Both the core and the spikes of enveloped viruses have icosahedral symmetry

No detailed x-ray structure is available for an enveloped virus. However, using cryoelectron microscopy, Stephen Fuller at EMBL, Heidelberg, has obtained a three-dimensional, low-resolution (35 Å) image of **Sindbis virus**, where not only the surface but also the interior of the virus particle is visualized.

Sindbis virus has a nucleocapsid core containing RNA of about 400 Å diameter, which is surrounded by a membrane about 40 Å thick. Anchored in this membrane and projecting out from the surface of the virus are spike proteins (Figure 11.19). Biochemical studies had shown that the virus contains about 200 copies each of three different proteins, one that forms the nucleocapsid and two that form the spikes.

It was no surprise when the structure determination showed that the nucleocapsid of the Sindbis virus has icosahedral symmetry with T = 3 structure similar to the plant viruses described earlier and hence contains exactly 180 subunits. It was more surprising to find that the membrane does not form a spherical envelope around this core but has a polyhedral arrangement. Finally, the spikes are also arranged according to icosahedral symmetry. These spikes are large (Figure 11.19), and it can clearly be seen in the reconstructed electron microscopy image that there are 80 of them arranged so that they form pseudosixfold axes along the twofold axes but not along the threefold axes. As previously discussed in this chapter, this arrangement demonstrates that the spikes are arranged in a T = 4 lattice. The volume of each spike, as measured from the electron-density maps, corresponds to three copies each of the two different spike polypeptide chains. Hence, Sindbis virus contains a total of 240 copies of each spike polypeptide arranged as 80 spikes with local threefold symmetry.

The result that the spikes of the virus and the nucleocapsid have different symmetries raises the problem of interaction between these two layers. Fuller has shown that this interaction is complementary; holes between the subunits of the T = 3 nucleocapsid lie directly beneath the centers of the spikes. Thus the cytoplasmic part of the spike proteins can be accommodated in these holes and interact with the nucleocapsid proteins. Presumably these tight complementary interactions forces the envelope with its spike proteins to adopt icosahedral symmetry.

The subunits in polyoma virus have nonequivalent environments

Members of the papovavirus family, polyoma virus, simian virus 40 (SV40), and the papilloma viruses were for a long time thought to have a simple T = 7 design. Electron micrographs by Klug and co-workers of negatively stained particles showed strongly contrasted projections, or capsomers, that would correspond to clusters of five and six subunits around the 12 icosahedral fivefold positions and the 60 pseudosixfold positions of a T = 7 structure. This arrangement would give 72 capsomers and 420 subunits (7 × 60 = 12 × 5 + 60 × 6), per virus particle.

X-ray studies at 22.5 Å resolution of **polyoma virus** by I. Rayment and co-workers at Brandeis University confirmed the presence of these 72 capsomers at

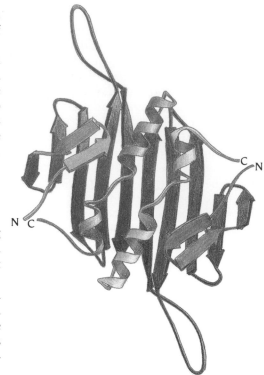

Figure 11.18 A dimer is the basic unit that builds up the capsid of bacteriophage MS2. The two subunits (red and blue) are arranged so that the dimer has a β sheet of 10 antiparallel strands on one side and the hairpins and α helices on the other side. The helices from one subunit pack against β strands from the other subunit and vice versa. (Adapted from a diagram provided by L. Liljas.)

Figure 11.19 Sindbis virus is a large enveloped virus with a nucleocapsid core surrounded by a membrane. Spike proteins, which are membrane bound, project out from the surface. (Courtesy of S. Fuller.)

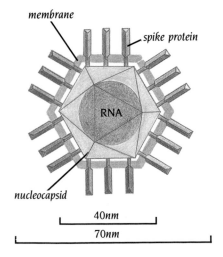

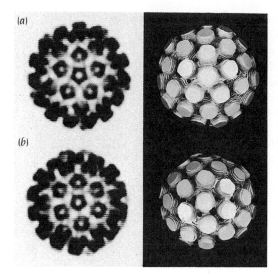

Figure 11.20 Polyoma virus was believed from electron micrographs to have a simple T = 7 design until x-ray studies showed otherwise. A view down the fivefold symmetry axis of the icosahedral structure (a) shows that the central capsomer is pentameric in shape and surrounded by five other capsomers as expected. The view down the pseudosixfold axis (b) shows, however, that the central capsomer is pentameric in shape and not hexameric as required for a T = 7 structure. (Adapted from I. Rayment et al., *Nature* 295: 110, 1982.)

the expected positions and hence the icosahedral arrangement. However, even at this low resolution it could clearly be seen that the shapes of all 72 capsomers were the same. They are all pentameric assemblies (Figure 11.20). Thus when we look down the fivefold axis in Figure 11.20a, the central capsomer is pentameric in shape and surrounded by five capsomers as expected. However, when we look down the presumed pseudosixfold axis, the central capsomer is surrounded by six other capsomers, as expected, but each individual capsomer is pentameric in shape, not hexameric. This is a very unexpected result, which implies that the subunits have several quite different bonding arrangements among them. Each asymmetric unit thus has six subunits and not the expected seven for a T = 7 structure, and therefore the complete capsid contains only 360 subunits. All six asymmetric subunits have nonequivalent environments with different bonding patterns. These results have recently been confirmed by Stephen Harrison from a structural study of SV40 to 3.8 Å resolution, which also shows how the different bonding arrangements are achieved.

Quasi-equivalence was conceived to explain why icosahedral symmetry should be selected for the design of closed containers built of a large number of identical structural units with conserved binding specificity. However, as the subunit interaction areas are different in these papovavirus capsids, the reason for the icosahedral symmetry is no longer obvious. Why has an icosahedral capsid design of 72 pentamers been selected for this family of viruses? As yet, we cannot answer that question.

Conclusion

Small spherical viruses have a protein shell around their nucleic acid that is constructed according to icosahedral symmetry. Objects with icosahedral symmetry have 60 identical units related by fivefold, threefold, and twofold symmetry axes. Each such unit can accommodate one or several polypeptide chains. Hence, virus shells are built up from multiples of 60 polypeptide chains. To preserve quasi-equivalent symmetry when packing subunits into the shell, only certain multiples (T = 1, 3, 4, 7 . . .) are allowed.

Satellite tobacco necrosis virus is an example of a T = 1 virus structure. The 60 identical subunits interact tightly around the fivefold axes on the surface of the shell and around the threefold axes on the inside. These interactions form a scaffold that links all subunits together to complete the shell.

Tomato bushy stunt virus is a T = 3 plant virus with 180 chemically identical subunits. Each polypeptide chain is divided into several domains. The subunits preserve quasi-equivalent packing in most contact regions by conformational differences of the protein chains, especially a large change in the orientation of the domains. A 35-residue-long region in 60 of the subunits forms a connected network of interactions around the icosahedral threefold axis that determines the size and shape of the particle.

Picorna viruses construct their shells from 60 copies each of three different polypeptide chains. These 180 subunits are arranged within the shell in a manner very similar to the 180 identical subunits of bushy stunt virus. In some picorna viruses there are protrusions around the fivefold axes, which are surrounded by deep "canyons." In rhinoviruses, the canyons probably form the virus's attachment site for protein receptors on the surface of the host cells, and they are adjacent to cavities that bind antiviral drugs.

All subunits of plant and animal spherical viruses whose structure has so far been determined have the same topology in their core, that of a jelly roll antiparallel β barrel structure. This implies evolutionary relationships even though there is no significant sequence homology. The barrel is flattened with short loop regions at the top and long at the base, giving the barrel a wedge shape. The subunit structure of the bacteriophage MS2 is constructed by a different arrangement of antiparallel β strands. The capsids of polyoma virus and the related SV40 have been shown to have icosahedral symmetry but nonequivalent bonding patterns of the subunits. There are six subunits per asymmetric unit, all with different environments.

A three-dimensional reconstruction of cryoelectron microscopy images of the enveloped Sindbis virus showed that the nucleopcapsid core has a T = 3 icosahedral structure. The membrane is polyhedral and not spherical. The spikes on the surface are arranged according to T = 4 icosahedral symmetry and are positioned above holes in the core structure.

Selected readings

General

Baltimore, D. Picornaviruses are no longer black boxes. *Science* 229: 1366–1367.

Caspar, D.L.D.; Klug, A. Physical principles in the construction of regular viruses. *Cold Spring Harbor Symp. Quant. Biol.* 27: 1–24, 1962.

Harrison, S.C. Structure of simple viruses: specificity and flexibility in protein assemblies. *Trends Biochem. Sci.* 3: 3–7, 1978.

Harrison, S.C. Virus structure: high-resolution perspective. *Adv. Virus Res.* 28: 175–240, 1983.

Harrison, S.C. Multiple modes of subunit association in the structures of simple spherical viruses. *Trends Biochem. Sci.* 9: 345–351, 1984.

Harrison, S.C. Finding the receptors. *Nature* 338: 205–206, 1989.

Hogle, J.M.; Chow, M.; Filman, D.J. The structure of poliovirus. *Sci. Am.* 256(3): 28–35, 1987.

Hurst, C.J.; Benton, W.H.; Enneking, J.M. Three-dimensional model of human rhinovirus type 14. *Trends Biochem. Sci.* 12: 460, 1987.

Liljas, L. The structure of spherical viruses. *Prog. Biophys. Mol. Biol.* 48: 1–36, 1986.

McKinlay, M.A.; Rossmann, M.G. Rational design of antiviral agents. *Annu. Rev. Pharmacol. Toxicol.* 29: 111–122, 1989.

Rossmann, M.G. The canyon hypothesis. *J. Biol. Chem.* 264: 14587–14590, 1989.

Rossmann, M.G. Virus structure, function, and evolution. *Harvey Lectures*, Series 83: 107–120, 1989.

Rossmann , M.G.; Johnson, J.E. Icosahedral RNA virus structure. *Annu. Rev. Biochem.* 58: 533–573, 1989.

Specific structures

Abad-Zapatero, C., et al. Structure of southern bean mosaic virus at 2.8 Å resolution. *Nature* 286: 33–39, 1980.

Acharya, R., et al. The three-dimensional structure of foot-and-mouth disease virus at 2.9 Å resolution. *Nature* 337: 709–716, 1989.

Arnold, E.; Rossmann, M.G. Analysis of the structure of a common cold virus, human rhinovirus 14, refined at a resolution of 3.0 Å. *J. Mol. Biol.* 211: 763–801, 1990.

Arnold, E., et al. Implications of the picornavirus capsid structure for polyprotein processing. *Proc. Natl. Acad. Sci. USA.* 84: 21–25, 1987.

Badger, J., et al. Structural analysis of a series of antiviral agents complexed with rhinovirus 14. *Proc. Natl. Acad. Sci. USA.* 35: 3304–3308, 1988.

Badger, J., et al. Three-dimensional structures of drug-resistant mutants of human rhinovirus 14. *J. Mol. Biol.* 207: 163–174, 1989.

Chen, Z., et al. Protein-nucleic acid interactions in a spherical virus: the structure of beanpod mottle virus at 3.0 Å resolution. *Science* 245: 154–159, 1989.

Filman, D.J., et al. Structural factors that control conformational transitions and serotype specificity in type 3 poliovirus. *EMBO J.* 8: 1567–1579, 1989.

Fuller, S.D. The T = 4 envelope of sindbis virus is organized by interactions with a complementary T = 3 capsid. *Cell* 48: 923–934, 1987.

Harrison, S.C., et al. Tomato bushy stunt virus at 2.9 Å resolution. *Nature* 276: 368–373, 1978.

Hogle, J.M.; Chow, M.; Filman, D.J. Three-dimensional structure of poliovirus at 2.9 Å resolution. *Science* 229: 1358–1365, 1985.

Hogle, J.M.; Maeda, A.; Harrison, S.C. Structure and assembly of turnip crinkle virus. I. X-ray crystallographic structure analysis at 3.2 Å resolution. *J. Mol. Biol.* 191: 625–638, 1986.

Hosur, M.V., et al. Structure of an insect virus at 3.0 Å resolution. *Prot.: Struct. Funct. Gen. 2:* 167–176, 1987.

Jones, T.A.; Liljas, L. Structure of satellite tobacco necrosis virus after crystallographic refinement at 2.5 Å resolution. *J. Mol. Biol.* 177: 735–768, 1984.

Kim, S., et al. Crystal structure of human rhinovirus serotype 1A (HRV1A). *J. Mol. Biol.* 210: 91–111, 1989.

Krishnaswamy, S.; Rossmann, M.G. Structural refinement and analysis of mengo virus. *J. Mol. Biol.* 211: 803–844, 1990.

Luo, M., et al. The atomic structure of mengo virus at 3.0 Å resolution. *Science* 235: 182–191, 1987.

Olson, A.J.; Bricogne, G.; Harrison, S.C. Structure of tomato bushy stunt virus IV. *J. Mol. Biol.* 171: 61–93, 1983.

Rayment, I., et al. Polyoma virus capsid structure at 22.5 Å resolution. *Nature* 295: 110–115, 1982.

Robinson, I.K.; Harrison, S.C. Structure of the expanded state of tomato bushy stunt virus. *Nature* 297: 563–568, 1982.

Rossmann, M.G. Antiviral agents targeted to interact with viral capsid proteins and a possible application to human immunodeficiency virus. *Proc. Natl. Acad. Sci. USA*, 85: 4625–4627, 1988.

Rossmann, M.G., et al. Subunit interactions in southern bean mosaic virus. *J. Mol. Biol.* 166: 37–72, 1983.

Rossmann, M.G., et al. Structure of a human common cold virus and functional relationship to other picornaviruses. *Nature* 317: 145–153, 1985.

Smith, T.J., et al. The site of attachment in human rhinovirus 14 for antiviral agents that inhibit uncoating. *Science* 233: 1286–1293, 1986.

Valegård, K., et al. The three-dimensional structure of the bacterial virus MS2. *Nature* 344: 36–41, 1990.

Recognition of Foreign Molecules by the Immune System

12

The immune system in vertebrates provides a defense mechanism against foreign parasites such as viruses and bacteria. Three main properties are essential to its operation: specific recognition of foreign molecules, which also involves discrimination between self and nonself; the ability to destroy the foreign parasite; and a memory mechanism that results in a more rapid response to a second infection by the same microorganism.

Foreign invaders are recognized through specific and tight binding of the proteins of the immune system to molecules specific to the foreign organisms. The sites on foreign molecules that are recognized by the immune system are called **antigenic determinants**, and they interact with two different classes of antigen receptors on the surface of the two major cell types of the immune system. **Antibodies**, which are also known as **immunoglobulins**, act as antigen receptors on the surface of B cells, which are stimulated by antigen binding to secrete antibodies into the bloodstream (Figure 12.1a). A related, though quite distinct molecule, the **T-cell receptor**, acts as the antigen re-

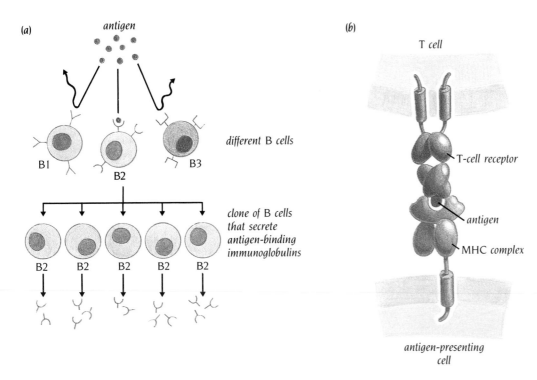

(a)

antigen

different B cells

B1

B2

B3

clone of B cells that secrete antigen-binding immunoglobulins

B2 B2 B2 B2 B2

(b)

T cell

T-cell receptor

antigen

MHC complex

antigen-presenting cell

Figure 12.1 (a) The clonal selection theory. An antigen (red) activates those B cells that have immunoglobulin molecules on their surfaces that can recognize and bind the antigen. This binding triggers production of a clone of identical B cells that secrete soluble antigen-binding immunoglobulins into the blood stream. (b) T cells recognize foreign viral antigens (red), through a T-cell receptor (blue) that can bind degraded fragments of the antigen when they are associated with an MHC molecule (green). The polypeptide chains of both T-cell receptors and MHC molecules are folded into immunoglobulin-like domains, represented as ellipsoids in the diagram.

ceptor for T cells one of whose principal functions is to recognize and destroy virus-infected cells (Figure 12.1b).

Antibody molecules at the surface of B cells recognize and bind intact soluble molecules or molecules on the surface of invading microorganisms. This stimulates the B cells first to multiply and to differentiate and then to synthesize and to secrete into the blood more copies of the antibody. The soluble antibody molecules bind to the cognate antigen and the complexes are recognized and removed by macrophages.

T cells, by contrast, directly destroy infected cells, and their receptors recognize antigenic determinants only when presented as part of a complex with a third important class of protein molecules, the molecules encoded by the major histocompatibility complex, or MHC, on the surface of cells. MHC molecules are so called because of the part they play in transplant rejection: each individual carries a selected set of MHC molecules on the surface of his or her cells, and the difference in the precise set of MHC molecules carried by the cells of different individuals forms the basis of discrimination between self and nonself. The physiological function of the MHC molecules, however, is to bind to degraded fragments of antigen generated inside infected cells and display them for recognition by T cells.

Thus T-cell receptors exist only on the surface of T cells, and the antigenic determinants they recognize are peptides derived from foreign proteins and complexed with MHC molecules on the surfaces of cells. Antibodies, by contrast, exist both as cell surface and as secreted soluble molecules and interact directly with intact antigens by forming very specific antigen-antibody complexes. Antibodies are the most structurally diverse of all known proteins, and any antigen presented to the immune system triggers the production of many different and specific antibodies.

In this chapter we will see that all three of the molecules directly involved in the specific recognition of antigen—immunoglobulin, the T-cell receptor, and the molecules of the MHC—belong to a family of proteins that seems to have evolved by duplication and diversification from a single ancestral domain. Because antibodies are naturally produced in very large quantities as soluble molecules, the crystal structure of immunoglobulin was solved long before either of the other two molecules could be approached, and hence the presumed ancestral domain structure is known as an immunoglobulin or immunoglobu-

Figure 12.2 (a) The immunoglobulin molecule, IgG, is built up from two copies each of two different polypeptide chains, heavy (H) and light (L). The L chain folds into two domains: V_L with variable sequence between different IgG molecules and C_L with constant sequence. The H chain folds into four domains: one variable V_H and three constant domains, C_{H1}, C_{H2}, and C_{H3}. Disulfide bridges connect the four chains. The antigen binding sites are at the ends of the variable domains. (b) Schematic polypeptide chain structure of different immunoglobulin molecules. IgM is a pentamer where the heavy chains have one variable domain and four constant domains. An additional polypeptide chain, the J chain, (black bar) is associated with the pentameric molecule. The J chain also links two units to form the dimeric IgA. IgG, IgD and IgE are monomeric immunoglobulin molecules.

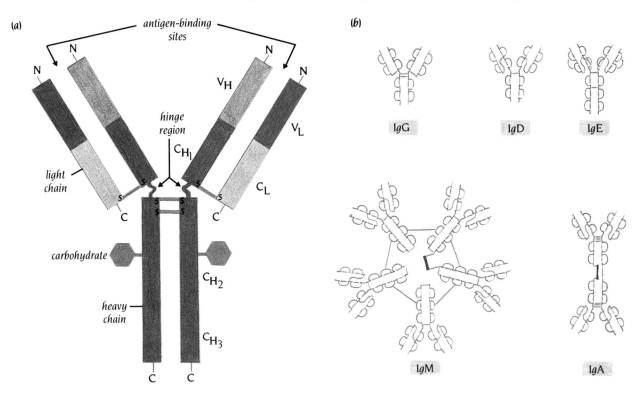

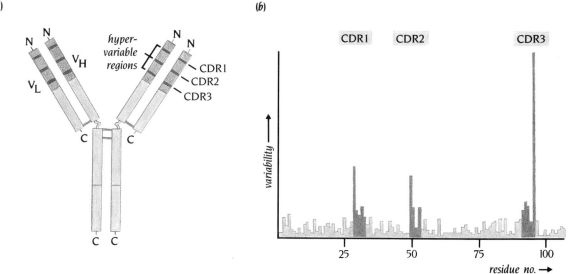

(a)

(b)

Figure 12.3 Certain regions within the 110 amino acid variable domains show a high degree of sequence variations. These regions determine the antigen specificity and are called hypervariable regions or complementarity determining regions, CDR (a). There are three such regions, CDR1–CDR3, in each variable domain. The sequences of a large number of variable domains in L chains have been compared, and the sequence variability is plotted as a function of residue number along the polypeptide chain in (b). The three large peaks in this diagram correspond to the three hypervariable regions in (a): CDR1–CDR3. [(b) Adapted from T.T. Wu and E.A. Kabat, *J. Exp. Med.* 132: 211, 1970.]

lin-like domain. The T-cell receptor has only just become available in soluble form for crystallization, but the crystal structure of an MHC molecule is now known. We examine the three-dimensional structures of antibodies and their bound antigens and discuss the relationship between the genetic basis for the diversification of antibodies and the structural basis for specific recognition. We also describe strategies for the design of novel antibodies including antibodies with enzymatic function. Finally, we examine the structure of an MHC molecule and show that the structural principles for recognition and binding of antigens to MHC are quite different from those of antibodies and that they reflect the special requirement for co-recognition of MHC and antigen by the T-cell receptor.

The polypeptide chains of antibodies are divided into domains

The basic structure of all immunoglobulin (Ig) molecules comprises two identical light chains and two identical heavy chains linked together by disulfide bonds (Figure 12.2a). There are two different classes, or isotypes, of light chains, λ and κ; there is no known functional distinction between light-chain isotypes. Heavy chains, by contrast, have five different isotypes that divide the immunoglobulins into different functional classes, IgG, IgM, IgA, IgD, and IgE (Figure 12.2b), each with different effector properties in the elimination of antigen. Each class of heavy chains can combine with either of the two different classes of light chains.

Immunoglobulins of class G, IgG, which is the major type of immunoglobulin in normal human serum and which we will discuss in this chapter, are monomers of the basic unit. Each chain of an IgG molecule is divided into domains of about 110 amino acid residues. The **light chains** have two such domains, and the **heavy chains** have four.

The most remarkable feature of the antibody molecule is revealed by comparison of the amino acid sequences from many different immunoglobulin IgG molecules. This comparison shows that the amino-terminal domain of each polypeptide chain is highly variable whereas the remaining domains have constant sequences for each class. A light chain is thus built up from one amino-terminal variable domain (V_L) and one carboxy-terminal constant domain (C_L) and a heavy chain from one amino-terminal **variable domain** (V_H), followed by three **constant domains** (C_{H1}, C_{H2}, and C_{H3}).

The variable domains are not uniformly variable throughout their lengths: in particular, three small regions show much more variability than the rest of the chain; they are called hypervariable regions or **complementarity determining regions**, CDR1–CDR3 (Figure 12.3). They vary both in size and in

germ-line DNA

sequence between different immunoglobulins. These are the regions that determine the specificity of the antigen-antibody interactions. The remaining parts of the variable domains (except a fourth hypervariable region of unknown function in the heavy chain) have quite similar amino acid sequences. In fact, all variable domains show significant overall sequence homology and hence evolutionary relationships among each other. Similarly, the different constant domains—C_L, C_{H1}, C_{H2}, and C_{H3}—show significant (30–40%) sequence identity to each other and to constant domains from immunoglobulin chains of different classes. This strongly suggests that the genes that code for current constant domains arose by successive gene duplication of an ancestral antibody gene. In contrast, there is no significant sequence homology between constant and variable domains, even though, as we will see, there are striking structural similarities. Before describing the structures of the immunoglobulin domains, however, we must pause to outline the genetic mechanisms underlying the variability of immunoglobulin molecules.

Antibody diversity is generated by several different mechanisms

It has been estimated that a human body generates 50 million new antibody-producing B cells every day. The vast majority of these cells produce an antibody molecule that is unique to that cell and its progeny. The genetic information for this diversity is contained in about 1,000 small segments of DNA. The elucidation of the mechanisms for the generation of this **antibody diversity** is one of the truly outstanding achievements of modern biology.

The gene segments encoding the variable and constant regions of antibodies are clustered in three gene pools, one for the class G heavy chain and one for each of the two light-chain isotypes, on separate chromosomes. In the heavy-chain gene pool, the variable domain is encoded by three types of segments, V, D, and J (Figure 12.4). V codes for approximately the first 90 residues, D for the hypervariable region CDR3, and J for the remaining 15 residues of the variable domain.

There are about 1,000 different V segments, about 10 different D segments of variable lengths, and about 4 different J segments in the heavy-chain gene pool. The DNA for the variable domain of a new B cell is assembled by random joining of one of each of these segments into a single continuous exon (Figure 12.5). This process is called **combinatorial joining**, and there are about 40,000

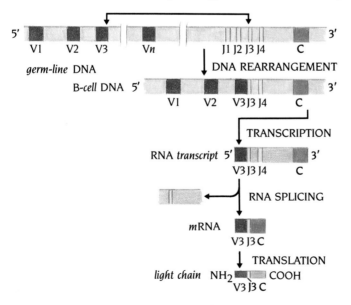

Figure 12.4 Variable domains in immunoglobulins (and T-cell receptors) are made by combinatorial joining of gene segments. Three such segments, V, D, and J, are joined to make the variable domain of a heavy chain. In the mouse the gene-segment pool for an H chain contains about 1,000 V segments, 12 D segments, 4 J segments, and an ordered cluster of C segments, each encoding a different class of H chains.

Figure 12.5 The V–J joining process involved in making a κ light chain in the mouse. In the germ-line DNA the cluster of 4 J gene segments is separated from the C gene segment by a short intron and from the about 250 V gene segments by long regions of DNA. During B-cell development the chosen V gene segment (V3) is moved to lie precisely next to the chosen J segment (J3). The "extra" J gene (J4) and the intron sequence are transcribed along with the V3, J3, and C segments and then removed by RNA splicing to generate mRNA molecules.

different ways it can occur. Furthermore, the joining of these segments is not precise. This creates additional diversity, called **junctional diversity**, where D recombines with V and J. In addition, extra nucleotides can be added during the joining procedure, giving still further diversity. The D segment encodes CD3, which is thus a focus of particular diversity in the heavy chain.

The newly created V-D-J exon becomes joined to one of the eight C segments that encode the constant region domains of the heavy chain as separate exons. Usually, a variable region is first expressed in an IgM molecule. Later, further genetic rearrangements will delete the mu (μ) exons encoding the IgM constant region, and the variable region will become joined to a second set of constant-region exons, most commonly those encoding IgG.

The genetic mechanisms underlying light-chain diversity are similar except that there are no D segments, and the diversity of CDR3 therefore depends largely upon junctional diversity generated during V–J joining.

By these genetic mechanisms it has been estimated that at least 90,000 different heavy chains and 3,000 different light chains can be produced. We will see that from structural considerations it is reasonable to assume that almost any heavy chain can combine with almost any light chain to produce a functional antibody. Since the antigen binding sites are built up from the CDR regions of both chains, this ability of light and heavy chains to combine and form functionally viable pairs generates considerable additional diversity. By this combinatorial association the number of different antibody molecules that a new B cell can choose from is increased to $3,000 \times 90,000 = 270$ million. The actual number of antibody molecules with different antigen binding sites is, however, several orders of magnitude higher because of yet one more mechanism for increasing diversity in mature B cells. In these cells point mutations can be introduced into the exons for the variable domains by somatic hypermutation.

All immunoglobulin domains have similar three-dimensional structure

Complete IgG molecules are difficult to crystallize, presumably because of the conformational flexibility in the hinge region between C_{H1} and C_{H2} of the heavy chains (Figure 12.2). When an IgG molecule is treated mildly with the proteolytic enzymes papain or pepsin, however, the heavy chains are cleaved in the hinge region to give two identical Fab fragments and one Fc fragment (Figure 12.6) that can be crystallized relatively easily. The Fab molecule comprises one complete light chain that is linked by a disulfide bridge to a fragment of the heavy chain consisting of V_H and C_{H1}. The Fc molecule comprises C_{H2} and C_{H3} from both heavy chains, which are also linked by disulfide bridges. All detailed high-resolution x-ray structural information on immunoglobulins have been obtained from structure determinations of Fab, Fc, or similar fragments including those of complexes with antigen. Until recently, because of the extreme

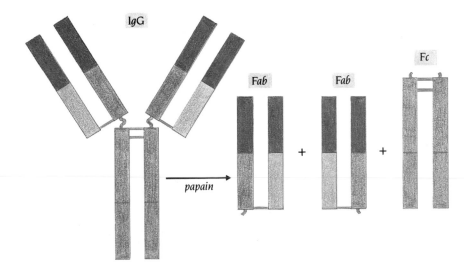

Figure 12.6 Enzymatic cleavage of immunoglobulin IgG. The enzyme papain splits the molecule in the hinge region yielding two Fab fragments and one Fc fragment.

diversity of serum antibodies, the only source of homogeneous immunoglobulins in quantities large enough for sequence and x-ray studies was a naturally occurring tumor, myeloma, which produces large quantities of a single immunoglobulin. Today monoclonal antibodies directed against almost any antigen can be routinely produced in large amounts using hybridoma techniques.

The first detailed description of an immunoglobulin domain was obtained before the advent of monoclonal antibody technology by Allen Edmundson and collaborators at Argonne National Laboratories, U.S.A. The high-resolution structure they determined in 1973 was that of a Bence-Jones protein, a tumor-derived molecule that is a dimer of two light chains. Roberto Poljak and collaborators, at Johns Hopkins Medical School, Baltimore, determined also in 1973 the first high-resolution structure of an intact immunoglobulin Fab fragment, and the same year David Davies and collaborators at the U.S. National Institutes of Health (NIH) demonstrated the binding of an antigenic determinant to the antigen binding site in a Fab fragment. Since then the structures of various other proteolytic fragments of immunoglobulin molecules have been solved to high resolution, and the structures of a few intact IgG molecules have been solved to low or medium resolution. All of the domains, both constant and variable, from both heavy and light chains, have proved to have a similar structure, which has become known as the immunoglobulin fold.

The immunoglobulin fold is best described as two antiparallel β sheets packed tightly against each other

All constant domains are built up from seven β strands arranged so that four strands form one β sheet and three strands form a second sheet (Figure 12.7). The sheets are closely packed against each other and joined together by a disulfide bridge from β strand 6 in the three-strand sheet to β strand 2 in the four-strand sheet. The topology of the sheets is quite simple, as can be seen from the topological diagrams in Figure 12.7. Even though the actual structure consists of two separate β sheets, it is convenient to regard the topology in the form of a Greek key barrel.

The loops are short, and as a result, a majority of the residues of the domain are in the two β sheets. These framework residues have the same structure in all constant domains of all immunoglobulins that have been studied. In other words, the structure is the same in both heavy and light chains.

Most of the invariant residues of the constant domain, including the disulfide bridge, are found in the sheets of the framework region. These invariant residues have two important functions. One is to form and stabilize the framework by packing the β sheets through hydrophobic interactions to give a hydrophobic core between the sheets. Their second function involves interactions between constant domains of different chains to form a complete immunoglobulin molecule.

The remaining residues that are not part of the framework region form the loops between the β strands. These loops may vary in length and sequence between immunoglobulin chains of different classes but are constant within each class; the sequence of the loops is invariant. The functions of these loops are not known, but they are probably involved in the effector functions of antibodies. When an antibody-antigen complex has been formed, signals are mediated through the Fc region to different systems, such as phagocytic cells or the complement system, which will destroy the complex. These activities are triggered by specific interactions of ligands with the constant domains of immunoglobulins.

The overall structure of the variable domain is very similar to that of the constant domain but there are nine β strands instead of seven. The two additional β strands are inserted into the loop region that connects β strands 3 and 4 (red in Figure 12.8). Functionally, this part of the polypeptide chain is important since it contains the hypervariable region CDR2. The two extra β strands provide the framework that positions CDR2 close to the other two hypervariable regions in the domain structure (Figure 12.8).

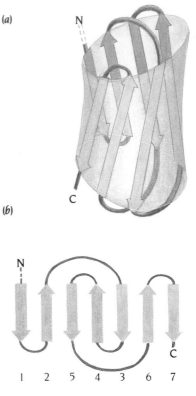

Figure 12.7 The constant domains of immunoglobulins are folded into a compressed antiparallel β barrel built up from one three-stranded β sheet packed against a four-stranded sheet (a). The topological diagrams (b) show the connected Greek key motifs of this fold.

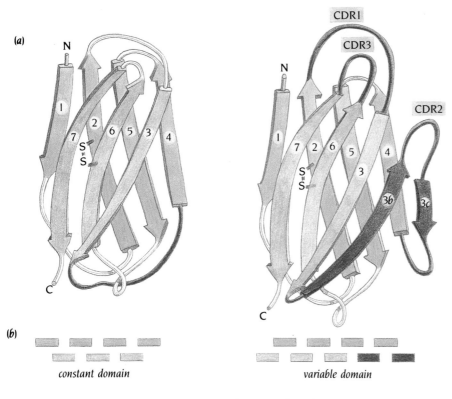

The hypervariable regions are clustered in loop regions at one end of the variable domain

The specificity of immunoglobulins is determined by the sequence and size of the hypervariable regions in the variable domains. From amino acid sequence comparisons of several hundred different variable domains these hypervariable regions have been defined as amino acids 24–34, 50–56, and 89–97 in the light chains and 31–35, 50–65, and 95–102 in the heavy chains. In the three-dimensional structures these residues occur in the loop regions that connect β strands 2–3, 3b–3c, and 6–7. We can see in Figure 12.8 that these loop regions are clustered together in space, at one end of the β sheets. CDR2 and CDR3 are hairpin loops between β strands 3b–3c and 6–7 in the five-strand β sheet. The third hypervariable region CDR1 is a crossover connection from β strand 2 in the four-strand sheet to strand 3 in the five-strand sheet. All strands in the five-strand sheet thus contribute loops to the hypervariable regions, whereas only one strand from the four-strand sheet is involved. The additional loop at this end of the sheet, between β strands 4 and 5 in the four-strand sheet, does not have variable sequence and does not participate in antigen binding. It is thus apparent that loops from the five-strand β sheet contribute more residues to the complete antigen binding site than the four-strand sheet.

The variable domains of immunoglobulins are excellent examples of the important structural principle discussed in Chapter 4 in connection with α/β-barrel structures. Functional residues in protein molecules are frequently provided by loop regions that are attached to one end of a stable structural framework built up mainly by secondary structure elements. In α/β-barrel structures this framework is built from eight parallel β strands surrounded by eight helices, whereas in the variable domains of immunoglobulins it consists of nine antiparallel β strands arranged in two sheets packed against each other. The fact that there exists an almost unlimited number of sequence variations in the loop regions of antibodies demonstrates that the folding of the structural framework is quite insensitive to changes in the functional loop regions. Such loop regions are ideal candidates for attempts to design novel protein molecules with predetermined function by recombinant DNA techniques. It has already been shown by these techniques that stable domains are formed when CDR loops from one species are inserted into a framework from a different species.

However, in order to engineer a predetermined function, it is necessary to know how constant the β-sheet framework is and how accurately one can predict and model the conformations of CDR loops.

Comparison of the known structures of variable domains have shown that about 70 residues in the light chain and about 80 residues in the heavy chain form the β-sheet frameworks that have very similar structures. Therefore, all variable domains would be expected to have the same framework structures for homologous residues. Given the framework, the key question to all attempts to model variable domains of known sequence and unknown structure is then the following: Can the structure of CDR loops vary randomly, or are there certain preferred conformations that can be deduced from the lengths and sequences of these regions?

Cyrus Chothia at MRC, Cambridge, U.K., and Arthur Lesk at EMBL, Heidelberg, have attempted to answer this question by a careful analysis of the CDR-loop conformations in the known structures. For each observed conformation they identified those residues that were mainly responsible for maintaining the conformation, either through interaction with framework residues or through their ability to assume unusual conformations. Examples of the latter are Gly or Pro. For a preserved conformation these residues should be conserved or very conservatively substituted. Examination of the sequences of variable domains of unknown structure showed that many have CDR loops that are similar in size to those of one of the known structures and contain identical residues at the sites responsible for the observed conformation. These results have interesting implications for the molecular mechanisms involved in the generation of antibody diversity, in addition to their value in modeling and design of new antibodies. For five of the six hypervariable regions of most immunoglobulins (CDR3 of the heavy chain is an exception) there seems to be only a small repertoire of main-chain conformations, most of which are known from the set of immunoglobulin structures so far determined. Sequence variations within the hypervariable regions modulate the surface that these canonical structures present to the antigens. Sequence variations within both the framework and the hypervariable regions shift the canonical structures relative to each other by small but significant amounts.

We will give three examples from their analysis, which is further described in Chapter 16. The loop between β strands 3b and 3c in CDR2 of the light chain is three residues long in all known sequences. This loop has very similar conformation in the known structures, and 95% of the 250 known sequences of mouse and human light chains fit the sequence constraints of this canonical structure. There is thus a high probability that this loop has the same conformation in almost all immunoglobulins. Four different conformations of CDR2 from the heavy chain has been found in the 10 x-ray structures that were analyzed. For each of these conformations there are specific constraints at one or two positions in the amino acid sequence of this hypervariable region, which ranges from three to six residues. For 236 of the 300 known sequences of heavy chains the sequence of the CDR2 loop fits one of these constraints and would therefore be expected to have one of these four canonical CDR2-loop structures. The CDR3 loop of the heavy chain, however, varies in length from 6 to 14 residues in the known sequences and forms a long hairpin loop. For such large loops many conformations are possible. The one actually found will depend both on the size of the loop and on the packing against the rest of the protein. This loop, therefore, is difficult to predict and model correctly.

The antigen binding site is formed by close association of the hypervariable regions from both heavy and light chains

Fab fragments contain one arm of the IgG molecule with an intact antigen binding site. In these molecules, as in intact IgG, the light chain is associated with the heavy chain in such a way that C_L associates with C_{H1} and V_L with V_H (Figure 12.9). These associations are very tight and extensive. The segments of the light and heavy polypeptide chains that join V_L to C_L and V_H to C_{H1} have

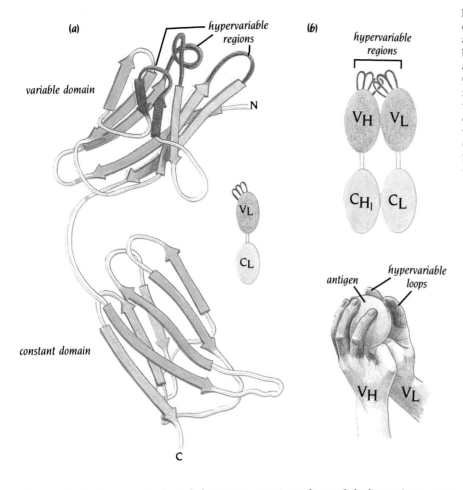

(a) hypervariable regions

variable domain

N

VL

CL

constant domain

C

(b) hypervariable regions

VH   VL

CHI   CL

antigen   hypervariable loops

VH   VL

Figure 12.9 (a) The variable and constant domains in the light chain of immunoglobulins are folded into two separate globular units. In both domains the four-stranded β sheet is blue and the other sheet is green. The hypervariable CDR regions are at one end of this elongated molecule. (b) In the Fab fragment as well as in the intact immunoglobulin molecules the domains associate pairwise so that V_H interacts with V_L and C_{H1} with C_L. By this interaction the CDR regions of both variable domains are brought close to each other and together form the antigen binding site.

few contacts. As a result, the Fab fragment consists of two globular regions, one formed by the two constant domains and the other by the two variable domains. Together, they make an elongated molecule. The hypervariable regions of both the light and heavy chains are close together at one end of this elongated molecule. We will now examine more closely how these domains associate.

The constant domains associate by interactions between the four-stranded β sheets in both C_L and C_{H1}. The sheets are closely packed with mainly hydrophobic residues in the interface. The directions of the β strands in the two domains are almost at right angles to each other (Figure 12.10). This mode of packing β sheets against each other occurs frequently within the core of single domains that are built up from two β sheets, for example, in retinol-binding protein discussed in Chapter 5, but they are rather unusual for domain-domain associations. The effect of this packing is to bury the central regions of the four-strand β sheets away from the solvent. Each four-strand β sheet has the three-strand β sheet of the same domain on one side and the four-strand β sheet from the second domain on its other side (Figure 12.10). This places rather stringent constraints on the sequence of these regions and most of the residues that occur in the interface are, in fact, conserved. In addition, residues that form the hydrophobic core within each domain are in general conserved.

The variable domains associate in a very different manner. It is obvious from Figure 12.11 that if they were associated in the same way as the constant domains, via the four-strand β sheets, the CDR loops, which are linked mainly to the five-strand β sheet, would be too far apart on the outside of each domain to contribute jointly to the antigen binding site. Thus in the variable domains the five-strand β sheets form the domain-domain interaction area. Furthermore, the relative orientation of the β strands in the two domains is closer to parallel than in the constant domains (Figure 12.11), and the curvature of the five-strand β sheets is such that they do not pack tightly against each other. Instead, each sheet forms half a barrel. When the two sheets associate, four of the five

Figure 12.10 Schematic diagrams of the packing of the four-stranded β sheets of the constant domains C_{H1} and C_L in an Fab fragment of IgG. The sheets are viewed perpendicular to the β strands in (a) and end-on in (b), where the four-stranded β sheet is blue.

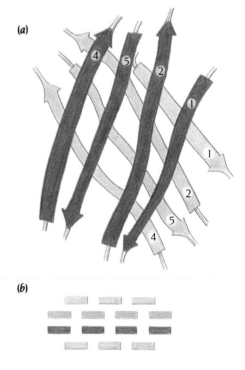

(a)

(b)

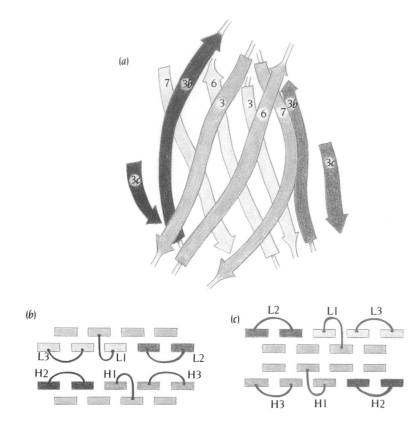

(a)

Figure 12.11 The two variable domains V_H and V_L are packed against each other so that the six hypervariable regions are close to each other. (a) Schematic diagram of the packing of the five-stranded β sheets of V_H and V_L in IgG. Only four of the β strands are involved in packing the variable domains against each other. Strand 3c is not involved. (b) Schematic diagram of the strands viewed end-on in the variable domain. Hypervariable loop regions are red. (c) Diagram illustrating hypothetical packing of the variable domains through their four-stranded sheets. The hypervariable regions would be far apart.

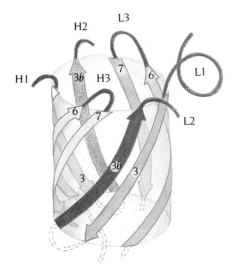

strands from both domains complete a barrel structure of eight antiparallel β strands (Figure 12.12).

The sheets associate in such a way that CDR loops from both sheets are at the same end of the barrel. These six CDR loops form the complete antigen binding site, which extends over a large surface outside the top of the barrel (Figure 12.13).

The antigen binding site binds haptens in crevices and protein antigens through large flat surfaces

The six CDR loops from the two chains at the rim of the eight-strand barrel provide an ideal arrangement for generating antigen binding sites of different shapes depending on the size and sequence of the loops. They can create flat extended binding surfaces for protein antigens or specific deep binding cavities for small hapten molecules.

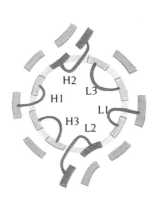

Figure 12.12 Schematic diagram of the barrel arrangement of four β strands from each of the variable domains in Fab. The six hypervariable regions, CDR1–CDR3 from the light chain (L1–L3) and from the heavy chain (H1–H3), are at one end of this barrel. (From J. Novotny et al., *J. Biol. Chem.* 258: 14435, 1983.)

(a)

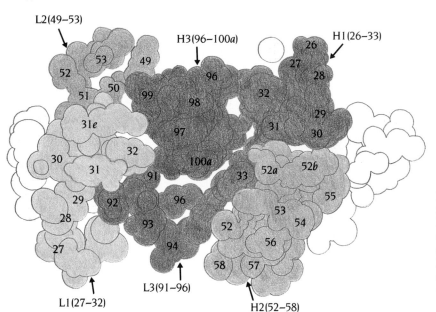

(b)

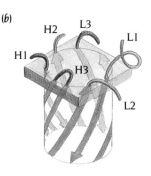

Figure 12.13 Drawing of a space-filling model (a) of the hypervariable regions of an Fab fragment. The superpositions of five sections are shown, cut through a model as shown in (b). It is clearly seen that all six hypervariable regions (L1–L3, H1–H3) contribute to the surface shown here. (From C. Chothia and A. Lesk, *J. Mol. Biol.* 196: 914, 1987.)

Haptens, a special class of antigens, are small molecules that induce specific antibody production when they are attached to a protein that acts as a carrier. Phosphorylcholine is one such. It has been widely used in the investigation of immune responses. The specific binding of this hapten to an Fab fragment has been studied crystallographically by David Davies and co-workers at NIH, U.S.A. The binding cavity is about 15 Å wide at the mouth and 12 Å deep. It is lined by residues from all CDR loops except the short CDR2 from the light chain. A number of side chains from these loops bind phosphorylcholine (Figure 12.14) in a manner that is very similar to that of a substrate or inhibitor binding to an enzyme.

In enzyme-catalyzed reactions substrates in their transition state bind very tightly to the enzyme and are positioned close to polar or charged side chains that participate in catalyzing the enzymatic reaction. It has recently been

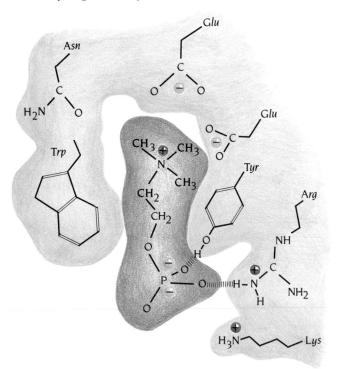

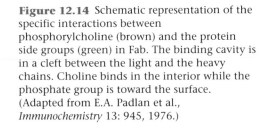

Figure 12.14 Schematic representation of the specific interactions between phosphorylcholine (brown) and the protein side groups (green) in Fab. The binding cavity is in a cleft between the light and the heavy chains. Choline binds in the interior while the phosphate group is toward the surface. (Adapted from E.A. Padlan et al., *Immunochemistry* 13: 945, 1976.)

189

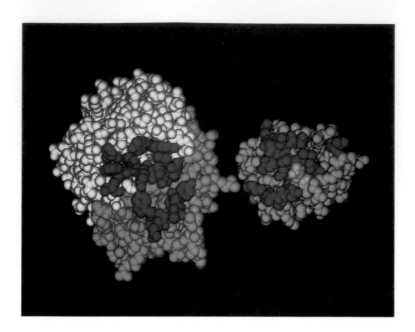

Figure 12.15 Space-filling representation of a complex between lysozyme (green) and the Fab fragment of a monoclonal antilysozyme (blue and yellow). The Fab fragment and the antigen (lysozyme) have been separated in this diagram, and their combining surfaces are viewed end-on. Atoms that are in contact in the complex are colored red both in Fab and lysozyme, except Gln 121 in lysozyme, which is violet. The diagram illustrates the large size of the interaction surfaces. (After A.G. Amit et al., *Science* 233: 749, 1986; courtesy of R. Poljak.)

possible to induce production of antibodies with enzymatic activity by choosing haptens that are analogues of the transition state of the substrate for a chemical reaction. When a transition-state analogue is used as a hapten, it induces an array of different antibodies that bind the hapten specifically. Some of these antibodies happen to have charged or polar side chains suitably placed to promote catalytic reactions of a corresponding substrate. These antibodies can be isolated as monoclonals and improved upon by recombinant DNA techniques. This approach seems to be one promising way to convert antibodies into enzymes with completely novel catalytic functions not previously present in living organisms.

The analysis of hapten binding to antibodies was possible before monoclonal antibodies became available because some myeloma proteins bind haptens. Thus hapten binding was studied as early as 1974. Once monoclonal antibodies with any chosen specificity could be produced, it became possible in principle to study the interactions of antibodies with protein antigens. However, the path from a cell line to a high-resolution three-dimensional antigen-antibody structure is long, and not until 1986 was the first such investigation reported by Roberto Poljak and co-workers at the Pasteur Institute, France. They chose a lysozome-Fab complex because the structure of the enzyme lysozyme was known and its antigenic properties had been extensively studied.

The shape of the interaction area between lysozyme and the CDR loops of the antibody is quite different from the hapten-binding crevice. The interaction extends over a large area with maximum dimensions of about 20×30 Å (Figure 12.15). It is an irregular rather flat surface with small protuberances and depressions that are complementary in the antigen and the antibody. Residues from all six CDR loops contribute to the antibody surface and residues from two stretches of the polypeptide chain of lysozyme, 18–27 and 116–129, form the antigen surface. There are 17 antibody residues that make close contact with 16 lysozyme residues. These residues pack against each other with a density similar to that found in the interior of protein molecules. Exchange of some of these residues would sterically interfere with packing or abolish bonding contacts and therefore cannot be tolerated. For example, Gln 121 in lysozyme protrudes from the lysozyme surface and fits snugly into a cleft between the CDR3 loops of V_H and V_L (Figure 12.16). On the other hand, there are some imperfections in the form of holes in the interaction area. Appropriate changes in the CDR loops could fill these holes and produce an even better fit. This is the sort of adjustment that can presumably occur through somatic mutations in the antibody genes and would account for the observation that a second immunization often gives rise to higher-affinity antibodies.

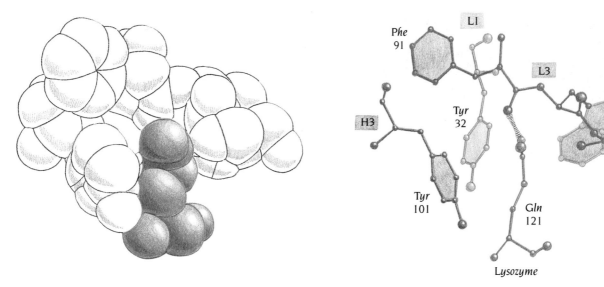

Figure 12.16 Detailed views of the environment of Gln 121 in the lysozyme-antilysozyme complex. Gln 121 in lysozyme is colored green both in the space filling representation to the left and in the ball and stick model to the right. This side chain of the antigen fits into a hole between CDR3 regions of both the heavy (Tyr 101) and the light (Trp 92) chains as well as CDR1 from the light chain (Tyr 32). (After A.G. Amit et al., *Science* 233: 751, 1986.)

Two of the 17 antibody residues that make contact with the lysozyme antigen are from framework regions. These are residues bordering CDR in the sequence, indicating that the functional distinction between framework and CDR residues is not absolute in the border between the two regions. But at the physical center of the interaction area is CDR3 of the heavy chain, which makes a proportionally greater contribution to the formation of this area than any of the other CDR loops. Four residues in the heavy-chain CDR3 participate in interactions with lysozyme, and they are all encoded in the D segment, the special significance of which now becomes clear. CDR3 of the heavy chain is the principal focus of the somatic mechanisms for generating antibody diversity and is central to the specific recognition of protein antigens. It is also the loop that eludes the predictive formula of Chothia and Lesk.

The surface of lysozyme that interacts with the antibody is built up by residues from two different regions of the polypeptide chain and is critically dependent on an intact three-dimensional structure of the protein. Yet immunization with isolated peptides from a protein molecule can induce antibodies that also bind to the intact protein. The peptides that induce these antibodies must presumably adopt the same conformation as in the intact protein; otherwise the known structures of protein-antibody complexes tell us nothing about the mechanism of this recognition.

The structure of lysozyme in the complex is the same as that in crystals of free lysozyme. No conformational changes of the main chain are seen even in the regions that bind to the antibody. The binding area in lysozyme, the antigenic determinant, is in regions exposed to solvent, but these regions are not particularly flexible. The structure of the Fab fragment complexed to lysozyme is very similar to the structure of other Fab fragments not bound to their corresponding antigen. This implies that no gross conformational changes occur upon antigen binding. This is only partly true for a second antigen-antibody complex, the structure of which was determined by Peter Colman and collaborators, CSIRO, Melbourne, Australia, where the antigen was the influenza virus protein neuraminidase, whose structure is described in Chapter 5. In this complex there is a difference within the interaction area between V_H and V_L such that the CDR loops have been displaced about 3 Å compared with other Fab structures. It is not known if this difference is due to antibody variability or to changes induced by antigen binding. In addition, the structure of the antigen, neuraminidase, has been somewhat distorted so that it is no longer catalytically active. On the whole, however, it seems that the classical "lock-and-key" description is an adequate first-order simplification to describe the interaction of antibodies with protein antigens. There is no evidence so far of major conformational changes induced by binding of antigen to antibody, but small local structural changes may occur to increase the complementarity between the interacting surfaces of antigen and antibody.

191

antigen   Fab 2 anti-idiotype   Fab 2 anti-idiotype

Fab 1   Fab 1 idiotype   Fab 3 anti-anti-idiotype

Figure 12.17 Schematic diagram illustrating the idiotype–anti-idiotype reaction.

The structure of an idiotype–anti-idiotype complex has been determined

The characterization of the interaction between antigen and antibody as a flat surface provides a satisfactory structural explanation of idiotype–anti-idiotype reactions in immunology. An idiotope is an antigenic determinant in the variable domains of an antibody. An idiotype in antibody 1 can induce the production of antibody 2 whose CDR loops bind to the CDR loops of antibody 1. This is called an **idiotype–anti-idiotype** reaction (Figure 12.17). Networks of anti-idiotype and anti-anti-idiotype interactions have been proposed as a mechanism for regulating immune responses, but extensive interactions of this sort were hard to envisage on the basis of the cleftlike antigen binding site suggested by the phosphorylcholine complex. However, an interaction area in the form of a flat surface can easily accommodate idiotype–anti-idiotype reactions.

It is also easily imaginable that the anti-idiotype of antibody 2 can structurally mimic the antigenic determinant that induced antibody 1, an idea that has provided the basis for an innovative strategy for isolating receptor molecules by raising antibodies against their ligands and then raising anti-idiotypic antibodies against those antibodies. If the anti-ligand antibody mimics the binding site of the receptor, then an anti-idiotypic antibody may recognize the original binding site in the receptor molecule.

This attractive hypothesis has only partly been validated by a recent structure determination by the group of Roberto Poljak of an idiotype–anti-idiotype complex. Poljak has managed to crystallize the complex between the Fab fragment shown in Figure 12.15 and the Fab fragment of one of its anti-idiotype immunoglobulins. Since the first Fab fragment was derived from a monoclonal antibody against lysozyme, it was expected that its anti-idiotype would mimic lysozyme and show similar interactions with the idiotype. This is not the case.

The structure of the complex shows that the CDR regions of the idiotype interacts with the CDR regions of the anti-idiotype as predicted (Figure 12.18). The details of these interactions are, however, not the same as in the idiotype-lysozyme complex. The anti-idiotype Fab fragment is oriented in the complex in such a way that its L chain occupies the same interaction area with the

Figure 12.18 The three-dimensional structure of a complex between Fab fragments from an idiotype immunoglobulin and its anti-idiotype. The idiotype Fab fragment (L chain, yellow; H chain, orange) is from a monoclonal antibody against lysozyme. The anti-idiotype Fab fragment (L chain, light blue; H chain, dark blue) is from a monoclonal antibody against the idiotype. The interaction between these two Fab fragments takes place largely through the hypervariable regions of both fragments. (Courtesy of R. Poljak.)

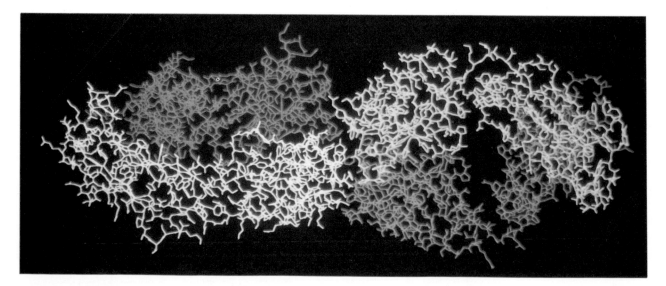

idiotype as lysozyme does in its complex. The interactions between the idiotype and the H chain of the anti-idiotype have no counterpart in the lysozyme complex. The amino acid residues from the anti-idiotype L chain that are involved in this interaction are different from those of lysozyme in its interaction with the idiotype. Details of this interaction are not available at the time of writing this book, since the structure is not refined, but it is quite clear that the binding interactions are different in the lysozyme complex compared to the anti-idiotype complex.

An IgG molecule has several degrees of conformational flexibility

Early physicochemical and electron microscopy studies suggested that the Fab arms were flexible and could move relative to each other and to the Fc stem of intact IgG molecules. The region of the heavy chains that connects C_{H1} with C_{H2} and therefore Fab with Fc is called the **hinge region**. The amino acid sequence of this region includes the cysteine residues that form S-S bridges between heavy chains and has a high proportion of proline residues that frequently are found in flexible regions of proteins. The flexibility of the hinge region is important for the function of immunoglobulins, but it makes immunoglobulins difficult to crystallize. Therefore, it has been expedient to use antibodies from which the hinge has been deleted.

Medium-resolution x-ray studies of an IgG molecule lacking its hinge region, and therefore with Fc rigidly attached to Fab, has provided the picture of an IgG molecule shaped like a T with the Fab arms at right angles to the Fc stem; this structure, however, may not be representative for a functionally intact IgG molecule that includes the hinge region. The C_{H2} and C_{H3} domains of each of the heavy chains associate in $C_{H2}{:}C_{H2}$ and $C_{H3}{:}C_{H3}$ pairs. The C_{H3} pair associate in the same way as the $C_{H1}{:}C_L$ pair discussed previously. In contrast, there is no protein-protein interaction between the C_{H2} domains: instead, a carbohydrate chain is attached to each C_{H2} domain in the interface region and forms a weak bridge between them.

In crystals of IgG molecules with intact hinge regions only the Fab moiety and the beginning of the hinge region are visible: the C-terminal residues of the hinge region and the Fc parts of the molecules are disordered. From a careful study of one such molecule and a proteolytic fragment containing two joined Fab arms, Robert Huber and co-workers in Munich, Germany, have deduced that there are at least four different orientations of the Fc region in the crystal. The crystal thus contains a random mixture of at least four different conformations of the same IgG molecule. Furthermore, the orientation of the two Fab arms is such that the intact IgG molecule must resemble a Y more than a T. This shows that the hinge region is a flexible link that allows the Fab arms to move relative both to each other and to Fc. This flexibility has important functional consequences. It means that the two antigen binding sites at the tips of the Fab arms can move relative to each other to bind antigenic determinants separated by different distances and in different orientations. It also provides flexibility to the Fc region relative to the Fab that facilitates the effector functions of IgG.

There is another flexible region in the IgG molecule that is functionally more obscure. The region of the polypeptide chain that links the variable domain to the adjacent constant domain, the so-called **switch peptide**, can have different conformations. As a result the variable and constant regions of Fab are oriented differently with respect to each other. This orientation can be defined by the angle between the local twofold axes that relates V_H to V_L and C_{H1} to C_L, respectively (Figure 12.19). This "elbow" angle is different in different Fab structures and varies from 132° to 172°. Even the same Fab molecule can have different "elbow" angles in different crystal forms. It has been suggested that antigen binding might trigger a change in the "elbow" angle and thereby transmit a signal through C_{H1} to the Fc effector region. There is, however, no experimental evidence to substantiate this idea. In contrast, no correlation has

Figure 12.19 Schematic diagram illustrating the "elbow" angle between the variable and constant domains in immunoglobulins.

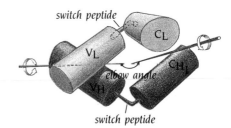

been observed between the size of the "elbow" angle and antigen binding. These results strongly suggest that the observed spread of "elbow" angles reflects an inherent flexibility in Fab with no specific biological function.

The structure of a human MHC molecule has provided insights into the molecular mechanism of T-cell activation

Few protein crystal structures have had such a direct impact as that of the human class I **MHC antigen**, human lymphocyte antigen A2 (HLA-A2), which was determined by Pamela Björkman in the group of Don Wiley at Harvard University. MHC molecules are crucial to the activation of T cells, which play a major role in immune defense. **T-cell activation** is initiated when the antigen receptors on the surface of these cells bind foreign antigens in combination with MHC molecules on the surface of other cells. T-cells are also activated by foreign MHC molecules on the surface of cells from a genetically different individual. An individual's own MHC molecules do not normally activate that individual's T cells because all T cells bearing potentially self-reactive receptors are eliminated during ontogeny.

There have been major questions in understanding these observations at the molecular level, primarily the following: How can a very limited number of MHC molecules recognize and bind an almost unlimited number of antigens and present them to T-cell receptors in such a way that they discriminate between self and nonself? Do the T cells see antigen and MHC molecules, which are firmly attached to the surface of cells, as a single structure or as two separate entities? These questions were immediately answered by the structure of one HLA molecule, in combination with many years of detailed molecular genetic analysis of T-cell–MHC interactions. Furthermore, the structure allows us now to ask very specific questions about the immune system and answer them by site-directed mutagenesis of MHC molecules.

Recognition of antigen is different in MHC compared to immunoglobulins

The class I MHC molecules are plasma membrane proteins expressed in all cells and which are composed of two polypeptide chains: a heavy chain that spans the membrane bilayer and a noncovalently attached light chain. The light chain is invariant, but the class I MHC heavy chain is the most genetically **polymorphic protein** known. Each human individual has three class I genes, and each of those genes has a very large number of alleles so that any given individual is likely to have six different class I molecules out of an estimated total of about 100 in the human population. Exactly how this polymorphism contributes to the recognition of antigen by T cells is a major question to which a general answer emerged from the HLA-A2 structure.

The light chain of the class I heterodimer, which is called β_2 microglobulin (β_2m) has amino acid sequence homology to the immunoglobulin constant domains and essentially the same three-dimensional structure. The extracellular portion of the heavy chain is divided into three domains, $\alpha1$, $\alpha2$, and $\alpha3$, each about 90 amino acids long. The carboxy-terminal domain $\alpha3$ also shows sequence and structural homology to immunoglobulin constant domains, whereas $\alpha1$ and $\alpha2$ have quite different structures. The extracellular portion of HLA-A2 composed of $\alpha1$, $\alpha2$, $\alpha3$, and β_2m was removed from the cells by proteolytic digestion and crystallized for high-resolution x-ray analysis.

The structure is divided into two globular regions (Figure 12.20). One region is built up from the two immunoglobulin-like domains $\alpha3$ and β_2m. Since $\alpha3$ is attached to the membrane in the intact HLA molecule, this region is closest to the cell surface. These domains are paired through their four-strand β sheets but in detail quite differently from the pairing of the constant domains C_{H1}:C_L and C_{H3}:C_{H3} in IgG.

The second region, which presumably is facing away from the surface and which contains the antigen binding site, is composed of domains $\alpha1$ and $\alpha2$.

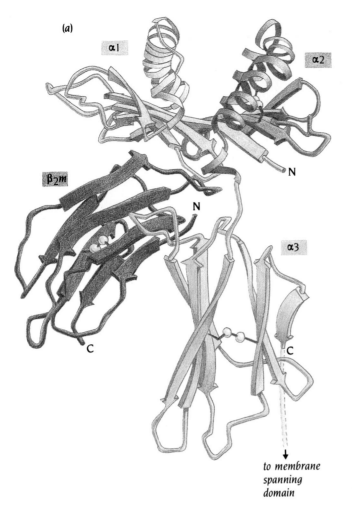

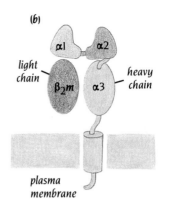

Figure 12.20 (a) Schematic representation of the path of the polypeptide chain in the structure of HLA-A2. Disulfide bonds are indicated as two connected spheres. The molecule is shown with the membrane proximal immunoglobulin-like domains (α3, β_2m) at the bottom and the polymorphic α1 and α2 domains at the top. (b) The domain arrangement in HLA-A2. (Adapted from P.J. Björkman et al., *Nature* 329: 506, 1987.)

Each domain has a very similar structure, which is quite simple (Figure 12.21). Starting from the N terminus the chain forms four up-and-down antiparallel β strands called a "W" followed by a helical region across the β strands on one side of the β sheet. The two domains associate by their "W" regions in such a way that they are hydrogen bonded to each other in an antiparallel fashion. By this association the structure of the complete region has a "floor" of a continuous eight-strand antiparallel β sheet (Figure 12.21). This floor sits on top of the immunoglobulin-like domains α3 and β_2m. The two helical regions are above the floor almost parallel to each other and separated by a large distance, about 18 Å from center to center. A large crevice that faces the solution is thus formed

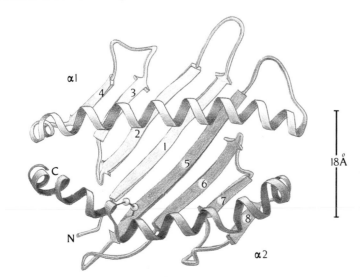

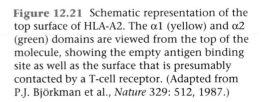

Figure 12.21 Schematic representation of the top surface of HLA-A2. The α1 (yellow) and α2 (green) domains are viewed from the top of the molecule, showing the empty antigen binding site as well as the surface that is presumably contacted by a T-cell receptor. (Adapted from P.J. Björkman et al., *Nature* 329: 512, 1987.)

195

with the floor forming its bottom and the helices its sides. This crevice is the antigen binding site. In the actual structure that was determined this crevice was occupied by an unknown antigen, probably a short peptide.

It is immediately obvious from this structure that the HLA molecule and the antigen bind to the T-cell receptor as one entity. The antigen is presented in the form of a short peptide that binds in the crevice and exposes some of its side chains to the receptor. The dimensions of the crevice, 25 Å long, 10 Å wide, and 11 Å deep, allow the binding of peptides of 8 residues if they are in an extended form and of about 20 residues if they are folded into an α helix.

The constraints of this binding pocket are wholly consistent with immunological studies showing that T cells sensitized to cells bearing a specific class I MHC molecule and infected with influenza virus will recognize cells bearing the same MHC incubated with short peptides from influenza virus proteins. It is believed that intracellular pathogens such as viruses betray their presence when fragments of viral proteins escape from the infected cells and become associated with the MHC molecules on the surface. In other cases, extracellular pathogens or foreign proteins may be internalized by cells, degraded in their lysosomes, and returned to the cell surface for presentation to T cells in association with MHC. In either case, the HLA structure reveals the nature of the antigen-MHC complex, and the effect of MHC polymorphism on their function can be more clearly understood.

The polymorphisms are due to a large number of point mutations at specific positions in the amino acid sequences of the α1 and α2 domains. Insight into the structural basis for this polymorphism was provided in 1989 by Don Wiley's group. They determined the structure of a second MHC molecule, HLA-Aw68, which is related to HLA-A2. The principal differences between the two structures are in the size and precise location of 13 amino acids (Figure 12.22) that have been mutated. Five of these differences (orange in Figure 12.22) are located on the central β strands of the "floor." They all point up between the two helical regions and would be buried if the binding site were occupied. Another five mutations have occurred (red in Figure 12.22) on the sides of the helices facing the binding site and are thus also in a position to bind foreign antigens. Although each of these individual amino acid differences in the binding cleft causes only local structural changes, confined to the actual side-chain difference, the 10 substitutions taken together substantially transform the shape and charge of parts of the cleft but leave other parts unchanged (Figure 12.23). Several pockets that extend from the groove beneath the α helices change drastically and presumably cause major changes in specificity for side chains of the peptide antigen.

One amino acid difference between HLA-A2 and HLA-Aw68 is at residue 62 (blue in Figure 12.22), whose side chain is at the top face of the helix in α1 and

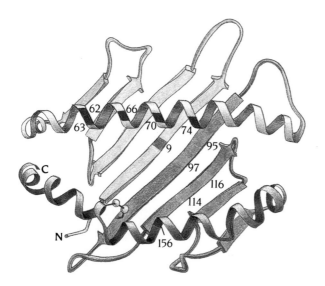

Figure 12.22 Schematic diagram illustrating the location of amino acid differences between two HLA-A molecules that are responsible for their polymorphic difference. Residues that point into the antigenic binding site are orange and red. Residue 62 (blue) may be involved in interactions with the T-cell receptor. (Adapted from P.J. Björkman et al., *Nature* 329: 506, 1987.)

points toward the solvent. This residue is a strong candidate for making direct contacts with the T-cell receptor and may possibly be involved in the discrimination between self and foreign HLA. Thus an immune response involving T cells depends on whether, first, a peptide derived from the pathogen can be successfully bound in the MHC antigen pocket of any of an individual's six class I molecules and, second, whether any of his or her T-cell receptors can recognize the resulting combination of antigen with the remaining exposed polymorphic residues.

We now have the following picture of T-cell activation by HLA. The T-cell receptor, which is composed of four immunoglobulin-like domains, probably has a binding surface similar to the antibody binding surface. This T-cell–receptor surface recognizes a large surface of the HLA molecule. The HLA surface is composed of a pattern of side chains from the top of the two helical regions in HLA as well as side chains from the bound peptide antigen between the helices. The pattern of side chains on this surface that is given by self-HLA and foreign antigens is recognized by the T-cell receptor, which binds the complex with consequent activation of T cells. T cells bearing receptors that bind self-HLA and self-antigen never mature but are destroyed inside the thymus.

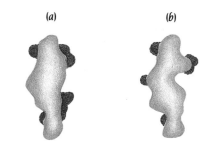

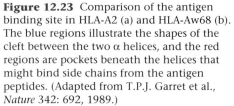

Figure 12.23 Comparison of the antigen binding site in HLA-A2 (a) and HLA-Aw68 (b). The blue regions illustrate the shapes of the cleft between the two α helices, and the red regions are pockets beneath the helices that might bind side chains from the antigen peptides. (Adapted from T.P.J. Garret et al., *Nature* 342: 692, 1989.)

Conclusion

IgG antibody molecules are composed of two light chains and two heavy chains joined together to form a T- or Y-shaped molecule. Each light chain has one variable domain and one constant domain, while each heavy chain has one variable and three constant domains. All domains have a similar three-dimensional structure known as the immunoglobulin fold. The stem of the molecule is formed by association between two constant domains from each of the heavy chains. Each arm is formed by association between a light chain and the amino-terminal half of a heavy chain. The hinge region between the stem and the arms is flexible and provides possibilities for different relative orientations of the arms and stem.

The constant domain has a stable framework structure composed of two antiparallel β sheets comprising seven β strands, four in one sheet and three in the other. The variable domains have a similar framework structure but comprising nine β strands, five in one sheet and four in the other. The three hypervariable regions are in loops at one end of the five-strand β sheet. The variable domains associate at the end of the arms through their five-strand β sheets in such a way that the hypervariable loop regions from both domains are close together at the top of a barrel of β strands from both the light and the heavy chains.

This arrangement of the hypervariable loop regions can form antigen binding sites of different shapes depending on the size and sequence of the loops. Haptens induce antibodies with binding cavities, and protein antigens induce antibodies with extended binding surfaces, which contain grooves or pockets for protruding antigen residues.

The lysozyme-antilysozyme interaction area is a large irregular complementary surface where antibody residues from all six hypervariable regions make close contacts with lysozyme residues from two different regions of the polypeptide chain. No gross conformational changes occur in antibody or antigen when they bind together. The third hypervariable region of the heavy chain is in a central position of the interaction area and makes a large number of contacts with the antigen. This region corresponds to the D gene segment, which, through junctional diversity and addition of nucleotides during the recombination events, has the largest sequence variation of the hypervariable regions.

HLA molecules of class I, which bind foreign antigens and activate T-cell receptors, are composed of one light chain, β2m, and one transmembrane heavy chain. The extracellular portion of the heavy chain is divided into three domains α1, α2, and α3. The light chain and α3, both of which have the immunoglobulin fold, associate to form one region of the HLA molecule close to the cell surface. Domains α2 and α3, which have the same unique fold, associate to form a second region, which contains the antigen binding site. This site has a floor of antiparallel β strands with two α helices on top of the floor separated by 18 Å. Antigens in the form of peptides 8–20 residues long bind in the crevice between these helices. Residues from the β strands on the floor and from the two helices on the sides contribute to antigen binding. HLA and antigen thus present to the T-cell receptors a large binding surface composed of side chains from the peptide antigen flanked by side chains from the two helices of HLA.

Selected readings

General

Alzari, P.N.; Lascombe, M.-B.; Poljak, R.J. Three-dimensional structure of antibodies. *Annu. Rev. Immunol.* 6: 555–580, 1988.

Amzel, L.M.; Poljak, R.J. Three-dimensional structure of immunoglobulins. *Annu. Rev. Biochem.* 48: 961–997, 1976.

Colman, P.M. Structure of antibody-antigen complexes: implications for immune recognition. *Adv. Immunol.* 43: 99–132, 1988.

Davies, D.R.; Metzger, H. Structural basis of antibody function. *Annu. Rev. Immunol.* 1: 87–117, 1983.

Davies, D.R.; Padlan, E.A. Antibody-antigen complexes. *Annu. Rev. Biochem.* 59: 439–473, 1990.

Davies, D.R.; Sheriff, S.; Padlan, E.A. Antibody-antigen complexes. *J. Biol. Chem.* 263: 10541–10544, 1988.

Flavell, R.A., et al. Molecular biology of the H–2 histocompatibility complex. *Science* 233: 437–443, 1986.

Kennedy, R.C.; Melnick, J.L.; Dreesman, G.R. Anti-idiotypes and immunity. *Sci. Am.* 255(1): 40–63, 1986.

Leder, P. The genetics of antibody diversity. *Sci. Am.* 247(5): 72–83, 1982.

Lerner, R.A. Synthetic vaccines. *Sci. Am.* 248(2): 48–62, 1983.

Lerner, R.A.; Tramontano, A. Catalytic antibodies. *Sci. Am.* 258(3): 42–53, 1988.

Mariuzza, R.A.; Phillips, S.E.V.; Poljak, R.J. The structural basis of antigen-antibody recognition. *Annu. Rev. Biophys. Biophys. Chem.* 16: 139–159, 1987.

Marquart, M.; Deisenhofer, J. Three-dimensional structure of antibodies. *Immunol. Today* 3: 160–166, 1982.

Marrack, P.; Kappler, J. The T cell and its receptor. *Sci. Am.* 254(2): 28–37, 1986.

Rees, A.R.; de la Paz, P. Investigating antibody specificity using computer graphics and protein engineering. *Trends Biochem. Sci.* 11: 144–148, 1986.

Riechmann, L., et al. Reshaping human antibodies for therapy. *Nature* 332: 323–327, 1988.

Shokat, K.M.; Schultz, P.G. Catalytic antibodies. *Annu. Rev. Immunol.* 8: 335–364, 1990.

Tonegawa, S. Somatic generation of antibody diversity. *Nature* 302: 575–581, 1983.

Specific structures

Amit, A.G., et al. Three-dimensional structure of an antigen-antibody complex at 2.8 Å resolution. *Science* 233: 747–753, 1986.

Bentley, G.A., et al. Three-dimensional structure of an idiotope-anti-idiotope complex. *Nature* 348: 254–257, 1990.

Bjorkman, P.J., et al. Structure of the human class I histocompatibility antigen, HLA-A2. *Nature* 329: 506–512, 1987.

Bjorkman, P.J., et al. The foreign antigen binding site and T-cell recognition regions of class I histocompatibility antigens. *Nature* 329: 512–518, 1987.

Brown, J.H., et al. A hypothetical model of the foreign antigen binding site of Class II histocompatibility molecules. *Nature* 332: 845–850, 1988.

Bruccoleri, R.E.; Haber, E.; Novotny, J. Structure of antibody hypervariable loops reproduced by a conformational search algoritm. *Nature* 335: 564–568, 1988.

Chothia, C.; Boswell, D.R.; Lesk, A.M. The outline structure of the T-cell αβ receptor. *EMBO J.* 7: 3745–3755, 1988.

Chothia, C., et al. Domain association in immunoglobulin molecules. The packing of variable domains. *J. Mol. Biol.* 186: 651–663, 1985.

Chothia, C., et al. Conformations of immunoglobulin hypervariable regions. *Nature* 343: 877–883, 1989.

Colman, P.M., et al. Three-dimensional structure of a complex of antibody with influenza virus neuraminidase. *Nature* 326: 358–363, 1987.

Cygler, M., et al. Crystallization and structure determination of an autoimmune antipoly (dT) immunoglobulin Fab fragment at 3.0 Å resolution. *J. Biol. Chem.* 262: 643–648, 1987.

Deisenhofer, J. Crystallographic refinement and atomic models of a human Fc fragment and its complex with fragment B of protein A from *Staphylococcus aureus* at 2.9 and 2.8 Å resolution. *Biochemistry* 20: 2361–2369, 1981.

Ely, K.R.; Herron, J.N.; Edmundson, A.B. Three-dimensional structure of a hybrid light chain dimer: protein engineering of a binding cavity. *Mol. Immunol.* 27: 101–114, 1990.

Ely, K.R., et al. Three-dimensional structure of a light chain dimer crystallized in water. Conformational flexibility of a molecule in two crystal forms. *J. Mol. Biol.* 210: 601–615, 1989.

Epp, O., et al. Crystal and molecular structure of a dimer composed of the variable portions of the Bence-Jones protein Rei. *Eur. J. Biochem.* 45: 513–520, 1974.

Furey, W., Jr., et al. Structure of a novel Bence-Jones protein (Rhe) fragment at 1.6 Å resolution. *J. Mol. Biol.* 167: 661–692, 1983.

Garret, T.P.J., et al. Specificity pockets for the side chains of peptide antigens in HLA-Aw68. *Nature* 342: 692–696, 1989.

Herron, J.N., et al. Three-dimensional structure of a fluorescein-fab complex crystallized in 2-methyl–2,4-pentanediol. *Prot.: Struct., Funct., Gen.* 5: 271–280, 1989.

Huber, R., et al. Crystallographic structure studies of an IgG molecule and an Fc fragment. *Nature* 264: 415–420, 1976.

Janda, K.D.; Benkovic, S.J.; Lerner, R.A. Catalytic antibodies with lipase activity and R or S substrate selectivity. *Science* 244: 437–440, 1989.

Lascombe, M.-B., et al. Three-dimensional structure of Fab R 19.9, a monoclonal murine antibody specific for the p-azobenzenearsonate group. *Proc. Natl. Acad. Sci. USA* 86: 607–611, 1989.

Lesk, A.M.; Chothia, C. Elbow motion in the immunoglobulins involves a molecular ball-and-socket joint. *Nature* 335: 188–190, 1988.

Marquart, M., et al. Crystallographic refinement and atomic models of the intact immunoglobulin molecule Kol and its antigen-binding fragment at 3.0 and 1.9 Å resolution. *J. Mol. Biol.* 141: 369–391, 1980.

Novotny, J., et al. Molecular anatomy of the antibody binding site. *J. Biol. Chem.* 258: 14433–14437, 1983.

Padlan, E.A., et al. Structure of an antibody-antigen complex. Crystal structure of the HyHEL–10 Fab-lysozyme complex. *Proc. Natl. Acad. Sci. USA* 86: 5938–5942, 1989.

Poljak, R.J., et al. Three-dimensional structure of the Fab fragment of a human immunoglobulin at 2.8 Å resolution. *Proc. Natl. Acad. Sci. USA* 70: 3305–3310, 1973.

Pollack, S.J.; Jacobs, J.W.; Schultz, P.G. Selective chemical catalysis by an antibody. *Science* 234: 1570–1573, 1986.

Rajan, S.S., et al. Three-dimensional structure of the Mcg IgG1 immunoglobulin. *Mol. Immunol.* 20: 787–799, 1983.

Satow, Y., et al. Phosphorylcholine-binding immunoglobulin Fab McPC603. An x-ray diffraction study at 2.7 Å. *J. Mol. Biol.* 190: 593–604, 1986.

Schiffer, M., et al. Structure of a λ-type Bence-Jones protein at 3.5 Å resolution. *Biochemistry* 12: 4620–4631, 1973.

Segal, D.M., et al. The three-dimensional structure of a phosphorylcholine-binding mouse immunoglobulin Fab and the nature of the antigen binding site. *Proc. Natl. Acad. Sci. USA.* 71: 4298–4302, 1974.

Sheriff, S., et al. Three-dimensional structure of an antibody-antigen complex. *Proc. Natl. Acad. Sci. USA* 84: 8075–8079, 1987.

Shokat, K.M., et al. A new strategy for the generation of catalytic antibodies. *Nature* 338: 269–271, 1989.

Silverton, E.W.; Navia, M.A.; Davies, D.R. Three-dimensional structure of an intact human immunoglobulin. *Proc. Natl. Acad. Sci. USA* 74: 5140–5144, 1977.

Stanfield, R.L., et al. Crystal structures of an antibody to a peptide and its complex with peptide antigen at 2.8 Å. *Science* 248: 712–719, 1990.

Suh, S.W., et al. The galactan-binding immunoglobulin Fab J539: an x-ray diffraction study at 2.6 Å resolution. *Prot.: Struct., Funct., Gen.* 1: 74–80, 1986.

Tramontano, A.; Janda, K.D.; Lerner, R.A. Catalytic antibodies. *Science* 234: 1566–1570, 1986.

Membrane Proteins

13

Cells are bounded, and organelles within them are enveloped by **membranes**, which are extremely thin films of lipids and protein molecules. The lipids form a **bilayered sheet** structure that is hydrophilic on the two surfaces and hydrophobic in between. Protein molecules are embedded in this layer, and in the simplest case they are arranged in such a way that there are three distinct regions: one hydrophobic transmembrane segment and two hydrophilic regions, one on each side of the membrane. **Amphiphilic proteins**, whose polypeptide chain traverses the membrane only once, usually form functional globular domains on at least one side of the membrane (Figure 13.1). Often these can be cleaved off by proteolytic enzymes. The hemagglutinin and neuraminidase of influenza virus (discussed in Chapter 5) and HLA proteins (discussed in Chapter 12) are examples of such cleavage products that can be handled as functional soluble globular domains. The polypeptide chain of other transmembrane proteins passes through the membrane several times (Figure 13.1). The hydrophilic regions on either side of the membrane in these cases comprise the termini of the chain and the loops between the membrane-spanning parts. Proteolytic cleavage produces a number of fragments, and function is not preserved. Other membrane proteins, which we will not discuss further, are associated with the lipid bilayer, not by a membrane-spanning region but by fatty acids that intercalate in the lipid bilayer of the membrane.

The biological membrane functions basically as a permeability barrier that establishes discrete compartments and prevents the random mixing of the contents of one compartment with those of another. However, biological membranes are more than passive containers. The embedded proteins serve as highly active mediators between the cell and its environment or the interior of an organelle and the cytosol. They catalyze specific transport of metabolites and ions across the membrane barriers. They convert the energy of sunlight into chemical and electrical energy, and they couple the flow of electrons to the synthesis of ATP. Furthermore, they act as signal receptors and transduce signals across the membrane. The signals can be, for example, neurotransmitters, growth factors, hormones, light or chemotactic stimuli. The transmembrane proteins of the plasma membrane are also involved in cell-cell recognition.

In this chapter we describe the only structure known to high resolution of a membrane-bound protein—that of the bacterial photosynthetic reaction center complex—and outline how elucidation of this structure has contributed to understanding both photosynthetic mechanisms and the construction of membrane-bound proteins in general. In Chapter 14 we describe current, essentially one-dimensional, knowledge of the domain organization of receptor

Figure 13.1 Two different ways in which transmembrane proteins may be embedded in a lipid bilayer (grey). On the left is a protein whose polypeptide chain traverses the membrane once. The hydrophobic transmembrane helix is shown in green and the hydrophilic regions in red. On the right is a protein whose polypeptide chain crosses the membrane seven times. The transmembrane helices are connected by hydrophilic loop regions.

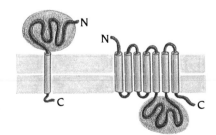

families and one type of signal transduction through the membrane that is mediated by G proteins.

Membrane proteins are difficult to crystallize

Membrane proteins, which have both hydrophobic and hydrophilic regions on their surfaces, are not soluble in aqueous buffer solutions and denature in organic solvents. However, if small amphiphilic molecules—detergents—such as octylglucosides are added to an aqueous solution, these proteins can be solubilized and purified in their native conformation. It is generally assumed that the hydrophobic part of the detergent molecules binds to the protein's hydrophobic surface, while the detergents' polar-head groups face the solution (Figure 13.2a). In this way the protein-detergent complex acquires an essentially hydrophilic surface with the hydrophobic parts buried inside the complex.

Such solubilized protein-detergent complexes are the starting material for purification and crystallization. Empirically, it has been found that for crystallization it is important to add small amphipathic molecules to the detergent-solubilized protein before crystallization. This is probably essential for proper packing interactions between the molecules in all three dimensions in a crystal (Figure 13.2b). Therefore, many different amphipathic molecules are added in separate crystallization experiments until, by trial and error, the correct one is found.

Despite considerable efforts very few membrane proteins have yielded crystals that diffract x-rays to high resolution. In fact, only a handful of such proteins are known at the time of writing this book, among which are porin, which is an outer membrane protein from *E. coli*, the enzyme prostaglandin synthase, and the photosynthetic reaction centers from two bacterial species, namely, *Rhodopseudomonas viridis* and *Rhodobacter sphaeroides*. In contrast, many other membrane proteins have yielded small crystals that diffract poorly, or not at all.

Figure 13.2 (a) Schematic drawing of membrane proteins in a typical membrane and their solubilization by detergents. The hydrophilic surfaces of the membrane proteins are indicated by red color. (b) A membrane protein crystallized with detergents bound to its hydrophobic protein surface. The hydrophilic surfaces of the proteins are indicated by red color. The symbols for detergents are the same as in (a). (Adapted from H. Michel, *Trends Biochem. Sci.* 8: 57, 1983.)

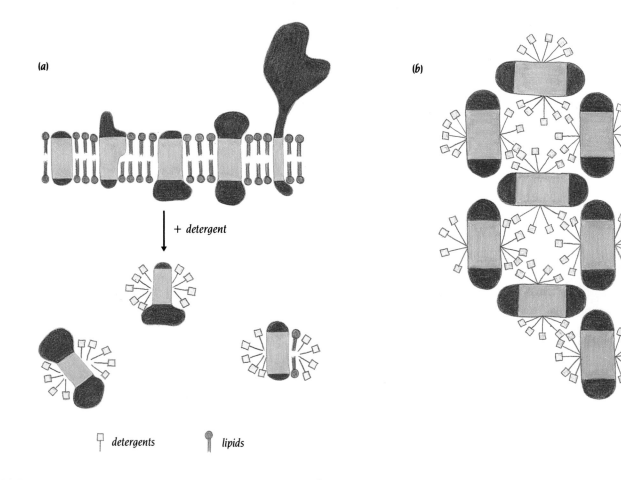

(a)   (b)

+ detergent

⊓ detergents   ☀ lipids

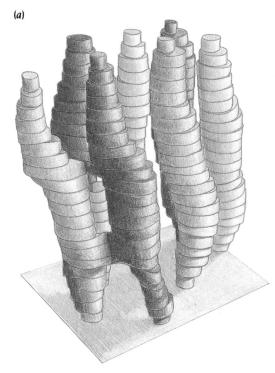

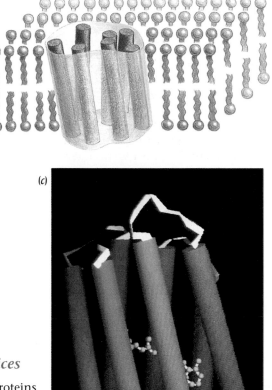

(a)

(b)

(c)

Bacteriorhodopsin contains seven transmembrane α helices

The first really useful information about the structure of membrane proteins came not from x-ray crystallography but from high-resolution electron microscopy of two-dimensional crystals. Such crystals, or ordered sheets of membrane proteins, are easier to obtain than proper crystals, and in some cases they occur *in vivo*, in the plane of the membrane. The purple membrane of *Halobacterium halobium*, for example, contains ordered sheets of a protein, bacteriorhodopsin, that uses the energy of light to pump protons across the membrane. Richard Henderson and Nigel Unwin at MRC, Cambridge, U.K., pioneered high-resolution three-dimensional reconstruction of tilted low-dose electron microscope images using such two-dimensional crystals. The 7 Å model of bacteriorhodopsin (Figure 13.3a) that they obtained in this way in 1975 provided the first glimpse of how membrane proteins are constructed. The fundamental observation that this protein has a number of **transmembrane α helices** (Figure 13.3b) has had great impact on subsequent theories and experiments on membrane proteins. It also provided the first experimental evidence behind the extensively used methods to predict transmembrane helices from amino acid sequences.

This electron microscopy reconstruction has now been extended to high resolution (3 Å) where the connections between the helices and the bound retinal molecule are visible together with the seven helices (Figure 13.3c). The helices are tilted by about 20° with respect to the plane of the membrane. This structure provides the first example of a high-resolution three-dimensional structure determination using electron microscopy. Given the difficulty of obtaining three dimensional crystals of membrane proteins, it is not surprising that the electron microscope technique is now widely used to study large membrane-bound complexes such as the acetylcholine receptor, ion pumps, gap junctions, and enzymes like cytochrome oxidase, which crystallize in two dimensions.

Figure 13.3 Two-dimensional crystals of the protein bacteriorhodopsin were used to pioneer three-dimensional high-resolution structure determination from electron micrographs. An electron density map to 7 Å resolution (a) was obtained and interpreted in terms of seven transmembrane helices (b). In 1989 the resolution was extended to 3 Å, which confirmed the presence of the seven α helices (c). The map also showed how these helices were connected by loop regions and where the retinal molecule was bound to bacteriorhodopsin. (Courtesy of R. Henderson.)

The bacterial photosynthetic reaction center is built up from four different polypeptide chains and many pigments

The crystallographic world was stunned when at a meeting in Erice, Sicily, in 1982, Hartmut Michel of the Max-Planck-Institut in Martinsried, West Ger-

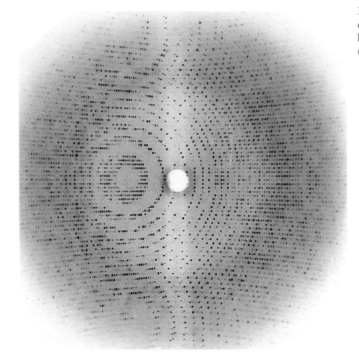

Figure 13.4 X-ray diffraction pattern from crystals of a membrane-bound protein, the bacterial photosynthetic reaction center. (Courtesy of Hartmut Michel.)

many, displayed the x-ray diagram shown in Figure 13.4. Not only was this the first x-ray picture to high resolution of a membrane protein, but the crystal was formed not from a small protein of trivial function but from a large complex of polypeptide chains that represents a class of proteins having a function of central importance for life on earth. The protein complex was a **photosynthetic reaction center** from the photosynthetic bacterium *Rhodopseudomonas viridis*, which converts the energy of captured sunlight into electrical and chemical energy in the first steps of photosynthesis. The structure has subsequently been solved to 2.5 Å resolution by H. Michel in collaboration with Hans Deisenhofer and Robert Huber at the same institute.

The interiors of *Rhodopseudomonad* bacteria are filled with photosynthetic vesicles, which are hollow membrane-enveloped spheres. The photosynthetic reaction centers are embedded in the membrane of these vesicles. One end of the protein complex faces the inside of the vesicles, which is known as the periplasmic side, the other end faces the cytoplasm of the cell. Around each reaction center, on the periplasmic side within the vesicles, there are hundreds of small membrane proteins, the antenna pigment protein molecules with bound chlorophyll. These catch photons over a wide area and funnel them to the reaction center. By this arrangement the reaction center can utilize about a hundred times more photons than those that directly strike the special pair of chlorophyll molecules at the heart of the reaction center.

The reaction center is built up from four polypeptide chains, three of which are called L, M, and H because they have light, medium, and heavy molecular masses as deduced from their electrophoretic mobility on SDS-PAGE. Subsequent amino acid sequence determinations showed, however, that the H chain is in fact the smallest with 258 amino acids, followed by the L chain with 273 amino acids. The M chain is the largest polypeptide with 323 amino acids. This discrepancy between apparent relative masses and real molecular weights underlines the uncertainty in deducing molecular masses of membrane-bound proteins from their mobility in electrophoretic gels.

The L and M subunits show sequence identity of about 25% and are therefore homologous and evolutionarily related proteins. The H subunit, on the other hand, has a completely different sequence. The fourth subunit of the reaction center is a cytochrome that has 336 amino acids with a sequence that is not homologous to any other known cytochrome sequence.

In addition to these polypeptide chains, the reaction center contains a number of pigments. There are four bacteriochlorophyll molecules (Figure

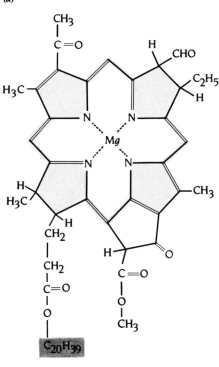

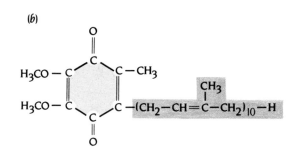

Figure 13.5 Photosynthetic pigments are used by plants and photosynthetic bacteria to capture photons of light and for electron flow from one side of a membrane to the other side. The diagram shows two such pigments that are present in bacterial reaction centers, bacteriochlorophyll (a) and ubiquinone (b).

13.5a), two of which form the strongly interacting dimer called "the special pair." Furthermore, there is one Fe atom, and there are two quinone molecules (Figure 13.5b) and two bacteriopheophytin molecules, which are chlorophyll molecules without the central Mg^{++} atom. Finally, the cytochrome subunit has four bound heme groups. The crystal structure shows how the polypeptide chains bind these pigments into a functional unit allowing electrons to flow from one side of the membrane to the other.

The L, M, and H subunits have transmembrane α helices

The L and the M subunits are firmly anchored in the membrane, each by five hydrophobic transmembrane α helices (yellow and red, respectively, in Figure 13.6). The structures of the L and M subunits are quite similar as expected from their sequence homology; they differ only in some of the loop regions. These loops, which connect the membrane-spanning helices, form rather flat hydrophilic regions on either side of the membrane to provide interaction areas with the H subunit (green in Figure 13.6) on the cytoplasmic side and with the cytochrome (blue in Figure 13.6) on the periplasmic side. The H subunit, in addition, has one transmembrane α helix at the carboxy terminus of its polypeptide chain (Figure 13.6). The carboxy end of this chain is therefore on the same side of the membrane as the cytochrome. In total, eleven transmembrane α helices attach the L, M, and H subunits to the membrane.

No region of the cytochrome penetrates the membrane; nevertheless, the cytochrome subunit is an integral part of this reaction-center complex, held through protein-protein interactions similar to those in soluble globular multisubunit proteins. The protein-protein interactions that bind cytochrome in the reaction center of *R. viridis* are strong enough to survive the isolation procedure. However, when the reaction center of *R. sphaeroides* is isolated, the cytochrome is lost, even though the structures of the L, M, and H subunits are very similar in the two species.

α helices D and E from the L and from the M subunits (Figure 13.6) form the core of the membrane-spanning part of the complex. These four helices are tightly packed against each other in a way quite similar to the four-helix bundle motif in water-soluble proteins. Each of these four helices provides a histidine

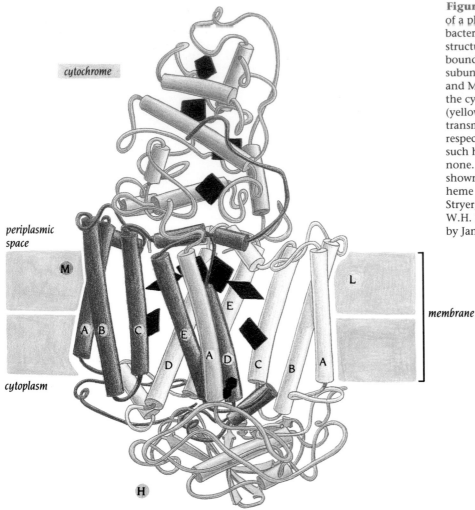

cytochrome

periplasmic
space

M

A B C E D A D C B A

E

L

membrane

cytoplasm

H

Figure 13.6 The three-dimensional structure of a photosynthetic reaction center of a purple bacterium was the first high-resolution structure to be obtained from a membrane-bound protein. The molecule contains four subunits: L, M, H, and a cytochrome. Subunits L and M bind the photosynthetic pigments, and the cytochrome binds four heme groups. The L (yellow) and the M (red) subunits have five transmembrane α helices; AL-EL and AM-EM, respectively. The H subunit (green) has one such helix, AH, and the cytochrome (blue) has none. Approximate membrane boundaries are shown. The photosynthetic pigments and the heme groups appear in black. (Adapted from L. Stryer, *Biochemistry*, 3rd. ed., p. 309 [New York: W.H. Freeman, 1988], after a drawing provided by Jane Richardson.)

side chain as ligand to the Fe atom, which is located between the helices close to the cytoplasm. The role of the Fe atom is probably to stabilize the structure of the four-helix bundle. Since its removal does not change the rate of electron flow through the system, the Fe atom cannot have a functional role in electron transfer. The remaining three helices of each subunit are scattered around the core in a manner that is not found in water-soluble proteins. Presumably, their positions are at least partly determined by the positions of the loop regions outside the membrane and not by close packing of the α helices inside the membrane. It is interesting that none of the α helices are perpendicular to the assumed plane of the membrane; instead, they are all tilted at angles of about 20° to 25°, similar to the tilt of the transmembrane helices in bacteriorhodopsin determined by electron microscopy (see Figure 13.3).

The photosynthetic pigments are bound to the L and M subunits

The structurally similar L and M subunits are related by a pseudo-twofold symmetry axis through the core, between the helices of the four-helix bundle motif. The photosynthetic pigments are bound to these subunits, most of them to the transmembrane helices, and they are also related by the same twofold symmetry axis (Figure 13.7). The pigments are arranged so that they form two possible pathways for electron transfer across the membrane, one on each side of the symmetry axis.

This symmetry is important in bringing the two chlorophyll molecules of the special pair into close contact and conferring on them their unique function in initiating electron transfer. They are bound in a hydrophobic pocket close to

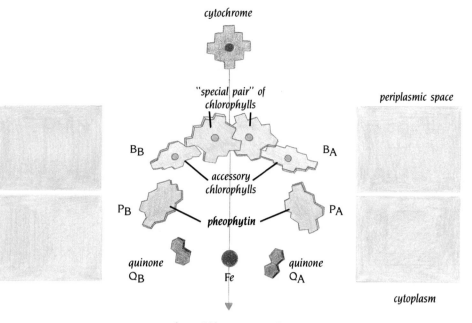

cytochrome

"special pair" of
chlorophylls

periplasmic space

B_B   B_A

accessory
chlorophylls

P_B   pheophytin   P_A

quinone   Fe   quinone
Q_B   Q_A

cytoplasm

pseudo-twofold symmetry axis

Figure 13.7 Schematic arrangement of the photosynthetic pigments in the reaction center of *R. viridis*. The twofold symmetry axis that relates the L and the M subunits is aligned vertically in the plane of the paper. Electron transfer proceeds preferentially along the branch to the right. The periplasmic side of the membrane is near the top, and the cytoplasmic side is near the bottom of the structure. (From B. Furugren, courtesy of the Royal Swedish Academy of Science.)

the symmetry axis between the D and E transmembrane α helices of both subunits. The ring system of these chlorophyll molecules are parallel to each other and interact closely with each other; in particular, two pyrrol rings, one from each molecule, are stacked on top of each other, 3 Å apart (Figure 13.7). The twofold symmetry axis passes between these stacked pyrrol rings.

This pair of chlorophyll molecules, which accepts photons and thereby excites electrons, is close to the membrane surface on the periplasmic side. At the other side of the membrane the symmetry axis passes through the Fe atom. The remaining pigments are symmetrically arranged on each side of the symmetry axis (Figure 13.7). Two bacteriochlorophyll molecules, the accessory chlorophylls, make hydrophobic contacts with the special pair of chlorophylls on one side and with the pheophytin molecules on the other side. Both the accessory chlorophyll molecules and the pheophytin molecules are bound between transmembrane helices from both subunits in pockets lined by hydrophobic residues from the transmembrane helices (Figure 13.8).

The functional reaction center contains two quinone molecules. One of these, Q_B (Figure 13.7), is loosely bound and is lost during purification. The reason for the difference in the strength of binding between Q_A and Q_B is unknown, but as we will see later, it probably reflects a functional asymmetry in the molecule as a whole. Because of the loss of Q_B, only Q_A is seen in the crystal structure. It is positioned between the Fe atom and one of the pheophytin molecules (Figure 13.6). The polar-head group is outside the membrane, bound

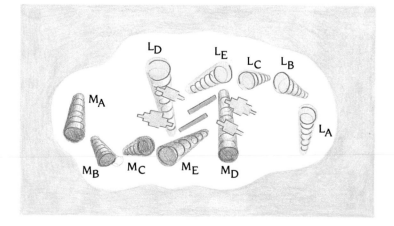

L_D   L_E   L_C   L_B

M_A

L_A

M_B   M_C   M_E   M_D

Figure 13.8 View of the reaction center perpendicular to the membrane illustrating that the pigments are bound between the transmembrane helices. The five transmembrane-spanning α helices of the L (yellow) and the M (red) subunits are shown as well as the chlorophyll (green) and pheophytin (blue) molecules.

207

to a loop region, whereas the hydrophobic tail is bound to the pheophytin molecule and to hydrophobic side chains of transmembrane helices of the L subunit. The corresponding binding site on the other side of the pseudo-twofold axis is empty. Certain weed killers that are inhibitors of photosynthesis in plants bind in the crystal to this empty binding site. The second quinone, therefore, presumably binds in this site in the functional reaction center, as illustrated in Figure 13.7.

Reaction centers convert light energy into electrical energy by electron flow through the membrane

In **photosynthesis** light energy is converted to electrical energy by an electron flow that causes the separation of negatively and positively charged molecules. Many molecules can absorb photons and use the energy of this process to donate an electron to a nearby electron acceptor. The electron donor then becomes positively charged and the electron acceptor negatively charged. In most cases, however, the transfer of electrons back from the acceptor to the donor is as fast as the forward reaction and the absorbed energy is lost, usually as fluorescent radiation. The arrangement of photoreaction centers in both bacteria and green plants results in a very fast forward reaction and a slow back reaction; therefore, the electric charges induced by the absorbed light energy stay separated. This separation of charge represents storage of energy because energy would be released if the charges were able to come together. This is the basic primary process of photosynthesis, the detailed mechanism of which we do not yet understand. However, by interpreting spectroscopic and genetic experiments in terms of the structure of the bacterial reaction center we may come to understand this process and be able to re-create this primary biological function—that of a solar-powered electrolytic battery.

In the bacterial reaction center the photons are absorbed by the special pair of chlorophyll molecules on the periplasmic side of the membrane (Figure 13.6). Spectroscopic measurements have shown that when a photon is absorbed by the special pair of chlorophylls, an electron is moved from the special pair to one of the pheophytin molecules. The close association and the parallel orientation of the chlorophyll ring systems in the special pair facilitates the excitation of an electron so that it is easily released. This process is very fast and occurs within 2 picoseconds. From the pheophytin the electron moves to a molecule of quinone, Q_A, in a slower process that takes about 200 picoseconds. The electron then passes through the protein, to the second quinone molecule, Q_B. This is a comparatively slow process, taking about 100 microseconds.

There are two pheophytin molecules, one on each side of the twofold axis, that in principle could accept the electrons (Figure 13.7). However, only one pathway on the right side of the symmetry axis that is shown in Figure 13.7 is used for electron transfer. Electrons do not pass through the chain of pigments on the left side (Figure 13.7), which appear to have no role in the charge separation. The best guess as to why these pigments are present is that the L and M chains have evolved from an ancestral chain that formed symmetrical homodimers in which both pigment chains were utilized. Presumably, the present-day reaction centers are more efficient for charge separation than the ancestral homodimers.

One apparent discrepancy between the spectroscopic data and the crystal structure is that no spectroscopic signal has been measured for participation of the accessory chlorophyll molecule B_A in the electron-transfer process. However, as seen from Figure 13.7, this chlorophyll molecule is between the special pair and the pheophytin molecule and provides an obvious link for electron transfer in two steps from the special pair through B_A to the pheophytin. This discrepancy has prompted recent, very rapid measurements of the electron-transfer steps, still without any signal from B_A. This means either that the electron bypasses B_A and is transferred directly the very long distance of about 25 Å from the special pair to pheophytin or that the transfer through B is too rapid to detect with current technology, less than 0.01 picosecond. None of these conclusions is compatible with current theories for electron transfer. However, the components for electron transfer are embedded in a protein environment of subunits L and M with special properties that are not taken into account in the theories.

While this electron flow takes place, the cytochrome on the periplasmic side donates an electron to the special pair and thereby neutralizes it. Then the entire process occurs again: another photon strikes the special pair, and another electron travels the same route from the special pair on the periplasmic side of the membrane to the quinone, Q_B, on the cytosolic side, which now carries two extra electrons. This quinone is then released from the reaction center to participate in later stages of photosynthesis. The special pair is again neutralized by an electron from the cytochrome.

The charge separation that stores the energy of the photons is now complete: two positive charges have been left on the cytochrome side of the membrane, and two electrons have traveled through the membrane to a quinone molecule on the cytosolic side. In the photosynthetic reaction centers charge separation is remarkably efficient for capturing light energy. The forward electron transfer from the reaction center to Q_A is more than eight orders of magnitude faster than the back reaction. This large difference allows the reaction center to capture between 98 and 100% of the photons it absorbs. As a solar cell, it is remarkably efficient: the energy stored in separated charges is about half of the energy inherent in the photons. The rest of the energy is lost in the reactions that drive the electrons along the pathway of photosynthetic pigment molecules.

The reaction center is a quantum-mechanical tunneling device

One intriguing feature of the reaction-center molecule is the relatively large distances between the pigments involved in the electron-transfer steps, which are about 10 Å apart. There is, moreover, no evidence that they are connected by molecular "wires" for electrons to flow through. They are rigidly fixed in a nonpolar region of the transmembrane protein matrix. The electron flow must, therefore, occur by electron tunneling between the components. Quantum-mechanical tunneling, a counterintuitive phenomenon, enables electrons to move across relatively large distances without ever being in between. Apparently, primitive biological systems, evolving billions of years ago, incorporated into their metabolism quantum-mechanical devices discovered by physicists only a few decades ago.

Transmembrane α helices can be predicted from the amino acid sequence

We have seen in previous chapters that only short continuous regions of the polypeptide chains contribute to the hydrophobic interior of water-soluble globular proteins. In such proteins α helices in general are arranged so that one side of the helix is hydrophobic and faces the interior while the other side is hydrophilic and at the surface of the protein as discussed in Chapter 3. β strands are usually short with the residues alternating between hydrophobic interior and hydrophilic surface. Loop regions between these secondary structure elements are usually very hydrophilic. Therefore, in soluble globular proteins, regions of more than 10 consecutive hydrophobic amino acids in the sequence are rarely encountered.

In contrast, the transmembrane helices observed in the reaction center are embedded in an hydrophobic surrounding and are built up from continuous regions of predominantly hydrophobic amino acids. To span the lipid bilayer, a minimum of about 20 amino acids are required. In the reaction center these α helices each comprise about 25 to 30 residues, some of which extend outside the hydrophobic part of the membrane. From the amino acid sequences of the polypeptide chains, the regions that comprise the transmembrane helices can be predicted with reasonable confidence.

Naively, one might assume that it should be possible to scan the sequence and pick out regions with about 20 consecutive hydrophobic amino acids. However, no such regions occur in the reaction-center proteins. Whereas in soluble proteins there are hydrophobic side chains at the hydrophilic surface of the molecule, in the transmembrane helices of the reaction center there are hydrophilic side chains among the hydrophobic. However, hydrophobic residues are in a clear majority in transmembrane helices, and such residues occur less frequently in other continuous regions of the polypeptide chain. We, therefore, need some method to measure the amount of hydrophobicity in a segment of the amino acid sequence in order to be able to predict whether or not it is likely to be a transmembrane helix.

Hydrophobicity scales measure the degree of hydrophobicity of different amino acid side chains

Each amino acid side chain within a transmembrane helix has a different hydrophobicity. It is trivial to state that side chains such as Val, Met, and Leu are the most hydrophobic and that charged residues such as Arg and Asp are at the other end of the scale. However, to order all side chains according to hydrophobicity and to assign actual numbers that represent their degree of hydrophobicity is not trivial. Many such **hydrophobicity scales** have been developed over the last decade on the basis of solubility measurements of the amino acids in different solvents, vapor pressures of side-chain analogues, analysis of side-chain distributions within soluble proteins, and theoretical energy calculations. In Table 13.1 two of these hydrophobicity scales are listed. The most frequently used scale, which was introduced by J. Kyte and R.F. Doolittle at La Jolla, California, is based on experimental data. A more refined scale was developed by D.A. Engelman, T.A. Steitz, and A. Goldman at Yale University. They used a semitheoretical approach to calculating the hydrophobicity, taking into account the fact that the side chains are attached to an α-helical framework.

Hydropathy plots identify transmembrane helices

These hydrophobicity scales frequently are used to identify those segments of the amino acid sequence of a protein that have hydrophobic properties consistent with a transmembrane helix. For each position in the sequence a **hydropathic index** is calculated. The hydropathic index is the mean value of the hydrophobicity of the amino acids within a "window," usually 19 residues long, around each position. In transmembrane helices the hydropathic index is high for a number of consecutive positions in the sequence. Charged amino acids are usually absent in the middle region of transmembrane helices because it would cost too much energy to have a charged residue in the hydrophobic lipid environment. It might be possible, however, to have two residues of

Table 13.1
Hydrophobicity scales

Amino acid	Phe	Met	Ile	Leu	Val	Cys	Trp	Ala	Thr	Gly	Ser	Pro	Tyr	His	Gln	Asn	Glu	Lys	Asp	Arg
A	2.8	1.9	4.5	3.8	4.2	2.5	-0.9	1.8	-0.7	-0.4	-0.8	-1.6	-1.3	-3.2	-3.5	-3.5	-3.5	-3.9	-3.5	-4.5
B	3.7	3.4	3.1	2.8	2.6	2.0	1.9	1.6	1.2	1.0	0.6	-0.2	-0.7	-3.0	-4.1	-4.8	-8.2	-8.8	-9.2	-12.3

Row A is from J. Kyte and R.F. Doolittle; row B, from D.A. Engelman, T.A. Steitz, and A. Goldman.

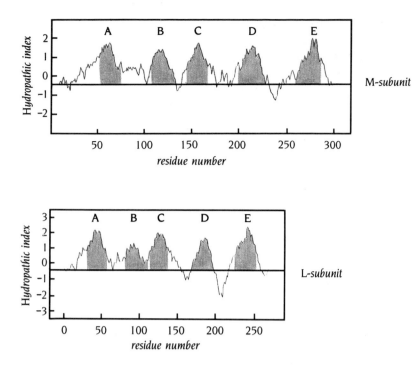

Figure 13.9 Hydropathy plots for the polypeptide chains L and M of the reaction center of *R. sphaeroides*. A window of 19 amino acids was used with the hydrophobicity scales of Kyte and Doolittle. The hydropathic index is plotted against the tenth amino acid of the window. The positions of the transmembrane helices as found by subsequent x-ray analysis by the group of G. Feher, La Jolla, California, are indicated by the green regions.

opposite charge close together inside the lipid membrane because they neutralize each other. Such charge neutralization has been observed in the hydrophobic interior of soluble globular proteins.

When the hydropathic indices are plotted against residue numbers, the resulting curves, hydropathy plots, identify possible transmembrane helices as broad peaks with high positive values. Such hydropathy plots are shown in Figure 13.9 for the L and M chains of the reaction center.

Reaction-center hydropathy plots agree with crystal structural data

The hydropathy plots in Figure 13.9 were calculated and published several years before the x-ray structure of the reaction center was known. It is therefore of considerable interest to compare the predicted positions of the transmembrane-spanning helices with those actually observed in the x-ray structure. These observed positions are indicated in green in Figure 13.9.

It is immediately apparent that these plots correctly predict the number of transmembrane helices, five each in the L and M chains, and also their approximate positions in the polypeptide chain. This gives us confidence in the hydropathy-plot method. Transmembrane helices are the only secondary structure elements that can be predicted from novel amino acid sequences with a reasonable degree of confidence using current knowledge and methods. The exact ends of transmembrane helices, however, cannot be predicted, essentially because they are usually inserted in the polar-head groups of the membrane and therefore contain charged and polar residues. The transmembrane helices in the reaction center, for example, contain a number of charged residues at their ends (Table 13.2), most of which are at the cytoplasmic side. All of the helices, however, have a segment of at least 19 consecutive amino acids that contain no charged side chains. The majority of residues in these segments are hydrophobic, but there are a number of polar residues, such as Ser, Thr, Tyr, and Trp, among them. The presence of histidine residues in the D and E helices of subunits L and M is accounted for by their special function in the reaction center; they are ligands to the magnesium atoms of chlorophyll molecules and to the Fe atom.

Table 13.2

Amino acid sequences of the transmembrane helices of the photosynthetic reaction center in *Rhodobacter sphaeroides*

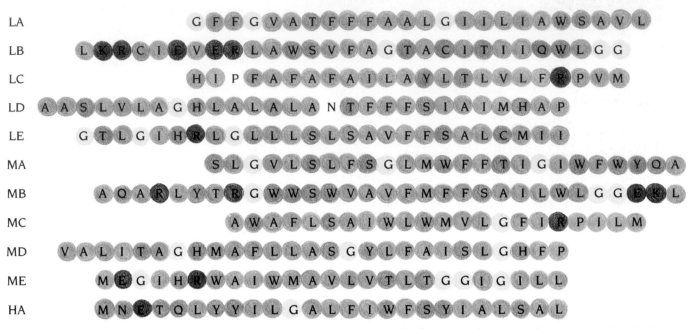

LA	G F F G V A T F F F A A L G I I L I A W S A V L
LB	L K R C I E V E R L A W S V F A G T A C I T I I Q W L G G
LC	H I P F A F A F A I L A Y L T L V L F R P V M
LD	A A S L V L A G H L A L A L A N T F F S I A I M H A P
LE	G T L G I H R L G L L L S L S A V F F S A L C M I I
MA	S L G V L S L F S G L M W F F T I G I W F W Y Q A
MB	A Q A R L Y T R G W W S W V A V F M F F S A I L W L G G E K L
MC	A W A F L S A I W L W M V L G F I R P I L M
MD	V A L I T A G H M A F L L A S G Y L F A I S L G H F P
ME	M E G I H R W A I W M A V L V T L T G G I G I L L
HA	M N T Q L Y Y I L G A L F I W F S Y I A L S A L

The helices are aligned according to approximate positions within the membrane and with respect to the photosynthetic pigments. LA is the first helix of subunit L, ME is the last helix of subunit M, HA is the only transmembrane helix of subunit H. Charged residues are colored red, polar residues are blue, hydrophobic residues are green, and glycine is yellow. (From T.O. Yeates et al., *Proc. Natl. Acad. Sci. USA* 84: 644, 1987.)

Membrane lipids have no specific interaction with protein transmembrane α helices

Comparison of the amino acid sequences of the L and M subunits of the reaction centers from three different bacterial species shows that about 50% of all residues in those two subunits are conserved in all three species. In the transmembrane helices sequence conservation varies. Residues that are buried and have contacts either with pigments or with other transmembrane helices are about 60% conserved. In contrast, residues that are fully exposed to the membrane lipids are only 16% conserved. Clearly, fewer restrictions are placed on residues that are exposed to the membrane lipids than to residues having contact with polypeptide chains or pigments. This implies that there are relatively few specific interactions between these transmembrane helices and the fatty acid side chains of the membrane that require the presence of specific residues. This is consistent with the observation that membrane proteins can move within the plane of the membrane, by lateral diffusion, and are not at fixed positions.

Structural rearrangements convert a water-soluble protein to a membrane-bound form

Colicin A is an antibiotic protein that kills *E. coli* cells by forming channels or pores in their membranes that destroy the energy potential of the cells. A proteolytic fragment of colicin A, comprising around 200 amino acid residues, has pore-forming properties very similar to those of the entire molecule. This fragment is soluble in aqueous media but nevertheless spontaneously inserts itself into lipid bilayers or membranes. Examination of the amino acid sequence of this fragment reveals a region 39 amino acids long of almost entirely hydrophobic residues: 31 residues have hydrophobic side chains, 6 are glycine, and the remaining 2 residues are serine and tyrosine. This hydrophobic region is the major candidate for interactions with the lipid bilayer when these molecules form pores in a membrane. How can this long continuous region of

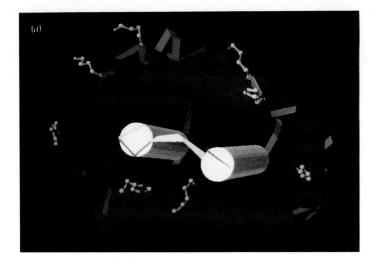

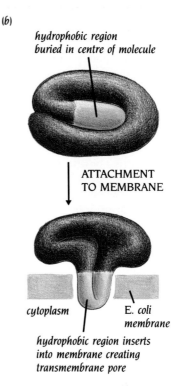

hydrophobic region
buried in centre of molecule

ATTACHMENT
TO MEMBRANE

cytoplasm   E. coli
                   membrane

hydrophobic region inserts
into membrane creating
transmembrane pore

hydrophobic residues be accommodated in the three-dimensional structure of the water-soluble form of the protein?

This question was answered by the groups of Franc Pattus and Demetrius Tsernoglou at EMBL, Heidelberg, who in 1989 determined the x-ray structure of the water-soluble form of the colicin-A fragment to 2.5 Å resolution. The structure provided a neat solution to the paradox of a water-soluble and a membrane-bound form. The fragment folds into 10 α helices, arranged in three layers (Figure 13.10a). Two of the helices (white in Figure 13.10a) are in the middle layer and form a hairpin α-loop-α motif that is completely buried in the interior of the molecule. This motif is built up from the continuous region of hydrophobic residues. Thus, residues that interact with lipid in the membrane-bound form of the protein are buried within the hydrophobic interior of the protein in its water-soluble form.

The mechanism by which this water-soluble structure is converted to a membrane-bound form is unknown as is the structure of the fragment when it is bound to lipid bilayers. It seems highly probable, however, that the hydrophobic helix-turn-helix motif is the first region of the protein that is inserted in the lipid bilayer and that it forms one of the major interaction areas with the lipids (Figure 13.10b). This implies a major rearrangement of the tertiary structure since the motif would be transferred from the interior of the molecule to its surface. Such major structural rearrangements have never been observed, but, on the other hand, we have presently no firm knowledge about how protein structures change when their environment changes from aqueous to lipid.

Figure 13.10 (a) A fragment of the antibiotic membrane-channel forming protein colicin A is stable both in a water-soluble and a membrane-bound form. The water-soluble form has a structure of the α type (α helices are represented as cylinders) with a hydrophobic hairpin helix-turn-helix motif (white) completely buried in the interior of the globular structure. Several charged side chains (connected balls) that are arranged in a ring around the loop of the motif are assumed to associate with the charged surface of the membrane before the helix-turn-helix motif is inserted in the hydrophobic interior of the lipid bilayer. (Courtesy of D. Tsernoglou, drawn by A. Lesk.) (b) One possible model for the conversion of the water-soluble form of colicin A to a membrane-bound form.

Conclusion

The x-ray structure of a photosynthetic bacterial reaction center has given insight into the mechanism for the primary reaction of photosynthesis: conversion of light energy to chemical energy by a stable separation of negatively and positively charged molecules. In addition, the structure has provided geometrical constraints for theoretical calculations on the electron flow through pigments across the membrane during this charge separation.

Important novel information has thus been obtained for the specific biological function of this molecule, but disappointingly few general lessons have been learned that are relevant for other membrane-bound proteins with different biological functions. In that respect the situation is similar to the failure of the structure of myoglobin to provide general principles for the construction of soluble protein molecules as described in Chapter 2.

The most important general lesson is that there are hydrophobic transmembrane helices, the positions of which within the amino acid sequence can be predicted with reasonable accuracy. This applies both to the single transmembrane-spanning helix within the H polypeptide chain and the five trans-membrane helices of the L and M chains that are connected by loop regions. This does not imply, however, that it is equally simple to predict the positions of, for example, channel-forming transmembrane helices that probably contain charged residues facing the ion channel. Such helices would give quite different hydropathy plots. Neither does it imply that all membrane-spanning regions are helical; on the contrary, it has recently been shown by a preliminary x-ray structure determination that porin has transmembrane-spanning β strands.

The structure of the reaction center also establishes that membrane-spanning helices can be tilted with respect to the plane of the membrane and that their relative positions within the membrane might be determined by the way they are anchored to the loop regions. Finally, the structure provides examples of how binding pockets for ligands are formed between such trans-membrane-spanning helices.

The three dimentional structure of one membrane protein, bacteriorhodopsin, has been obtained from electron microscopy of two-dimensional crystals. This method is now being applied to other membrane bound proteins.

A proteolytic fragment of the antibiotic protein colicin A is stable both in a water-soluble and a membrane-bound form. The x-ray structure of the water-soluble form shows that a long continuous region of hydrophobic residues form a hairpin helix-turn-helix motif that is completely buried in the interior of the globular molecule. Thus, residues that interact with lipid in the membrane-bound form of the protein are buried in the hydrophobic interior of the protein in its water-soluble form.

Selected readings

General

Deisenhofer, J.; Michael, H. Nobel lecture. The photosynthetic reaction center from the purple bacterium *Rhodopseudomonas viridis*. *EMBO J.* 8: 2149–2169, 1989.

Engelman, D.M.; Steitz, T.A.; Goldman, A. Identifying nonpolar transbilayer helices in amino acid sequences of membrane proteins. *Annu. Rev. Biophys. Biophys. Chem.* 15: 321–353, 1986.

Fasman, G.D.; Gilbert, W.A. The prediction of transmembrane protein sequences and their conformation: an evaluation. *Trends Biochem. Sci.* 15: 89–95, 1990.

Garavito, R.M.; Rosenbusch, J.P. Isolation and crystallization of bacterial porin. *Methods Enzymol.* 125: 309–328, 1986.

Gust, D.; Moore, T.A. Mimicking photosynthesis. *Science* 244: 35–41, 1989.

Huber, R. Nobel lecture. A structural basis of light energy and electron transfer in biology. *EMBO J.* 8: 2125–2147, 1989.

Michel, H. Crystallization of membrane proteins. *Trends Biochem. Sci.* 8: 56–59, 1983.

Michel, H.; Deisenhofer, J. Relevance of the photosynthetic reaction center from purple bacteria to the structure of photosystem II. *Biochemistry* 27: 1–7, 1988.

Norris, J.R.; Schiffer, M. Photosynthetic reaction centers in bacteria. *Chem. Eng. News* 68(31): 22–37, 1990.

Rees, D.C., et al. The bacterial photosynthetic reaction center as a model for membrane proteins. *Annu. Rev. Biochem.* 58: 607–633, 1989.

Unwin, N.; Henderson, R. The structure of proteins in biological membranes. *Sci. Am.* 250(2): 78–95, 1984.

Warman, J.M. A quantum jump for chemistry. *Nature* 327: 462–464, 1987.

Youvan, D.C.; Marrs, B.L. Molecular mechanisms of photosynthesis. *Sci. Am.* 256(6): 42–49, 1987.

Specific structures

Allen, J.P., et al. Structure of the reaction center from *Rhodobacter sphaeroides* R–26: the cofactors. *Proc. Natl. Acad. Sci. USA* 84: 5730–5734, 1987.

Allen, J.P., et al. Structure of the reaction center from *Rhodobacter sphaeroides* R–26: the protein subunits. *Proc. Natl. Acad. Sci. USA* 84: 6162–6166, 1987.

Brisson, A.; Unwin, P.N.T. Quaternary structure of the acetylcholine receptor. *Nature* 315: 474–477, 1985.

Deisenhofer, J., et al. X-ray structure analysis of a membrane protein complex. Electron density map at 3 Å resolution and a model of the chromophores of the photosynthetic reaction center from *Rhodopseudomonas viridis*. *J. Mol. Biol.* 180: 385–398, 1984.

Deisenhofer, J., et al. Structure of the protein subunits in the photosynthetic reaction center of *Rhodopseudomonas viridis* at 3 Å resolution. *Nature* 318: 618–624, 1985.

Eisenberg, D.; Weiss, R.M.; Terwilliger, T.C. The hydrophobic moment detects periodicity in protein hydrophobicity. *Proc. Natl. Acad. Sci. USA* 82: 140–144, 1984.

Garavito, R.M.; Rosenbusch, J.P. Three-dimensional crystals of an integral membrane protein: an initial x-ray analysis. *J. Cell. Biol.* 86: 327–329, 1980.

Henderson, R.; Unwin, P.N.T. Three-dimensional model of purple membrane obtained by electron microscopy. *Nature* 257: 28–32, 1975.

Henderson, R., et al. Model for the structure of bacteriorhodopsin based on high-resolution electron cryo-microscopy. *J. Mol. Biol.* 213: 899–929, 1990.

Kyte, J.; Doolittle, R.F. A simple method for displaying the hydropathic character of a protein. *J. Mol. Biol.* 157: 105–132, 1982.

Leifer, D.; Henderson, R. Three-dimensional structure of orthorhombic purple membrane at 6.5 Å resolution. *J. Mol. Biol.* 163: 451–466, 1983.

Michel, H. Three-dimensional crystals of a membrane protein complex. The photosynthetic reaction center from *Rhodopseudomonas viridis*. *J. Mol. Biol.* 158: 567–572, 1982.

Michel, H.; Epp, O.; Deisenhofer, J. Pigment-protein interactions in the photosynthetic reaction center from *Rhodopseudomonas viridis*. *EMBO J.* 5: 2445–2451, 1986.

Michel, H., et al. The "heavy" subunit of the photosynthetic reaction center from *Rhodopseudomonas viridis*: isolation of the gene, nucleotide and amino acid sequence. *EMBO J.* 4: 1667–1672, 1985.

Michel, H., et al. The "light" and "medium" subunits of the photosynthetic reaction center from *Rhodopseudomonas viridis*: isolation of the genes, nucleotide and amino acid sequence. *EMBO J.* 5: 1149–1158, 1986.

Parker, M.W., et al. Structure of the membrane-pore-forming fragment of colicin A. *Nature* 337: 93–96, 1989.

Rees, D.C.; DeAntonio, L.; Eisenberg, D. Hydrophobic organization of membrane proteins. *Science* 245: 510–513, 1989.

Valpuesta, J.M.; Henderson, R.; Frey, T.G. Electron cryo-microscopic analysis of crystalline cytochrome oxidase. *J. Mol. Biol.* 214: 237–251, 1990.

Weiss, M.S., et al. The three-dimensional structure of porin from *Rhodobacter capsulatus* at 3 Å resolution. *FEBS Lett.* 267: 268–272, 1990.

Yeates, T.O., et al. Structure of the reaction center from *Rhodobacter sphaeroides* R–26: membrane-protein interactions. *Proc. Natl. Acad. Sci. USA* 84: 6438–6442, 1987.

Receptor Families

14

So far we have discussed various aspects of the three-dimensional structure of proteins determined by x-ray crystallography. This chapter is different. It deals with transmembrane proteins, whose three-dimensional structures are not known, but for which one-dimensional information, namely, the amino acid sequence deduced from the corresponding DNA sequence, is available. As we saw in Chapter 13, the hydrophobic transmembrane helices of membrane proteins can sometimes be predicted with a reasonable degree of confidence from hydropathy plots. We have chosen certain families of transmembrane receptor proteins present in the plasma membrane to exemplify the type of valuable but inevitably limited structural information that can be deduced from sequence studies alone. The hydropathy plots of the membrane proteins that we will discuss in this chapter all show strong signals indicating the position in the sequence of the transmembrane helices.

We must emphasize at the outset, however, that the hydropathy plots of membrane proteins are not universally unambiguous and straightforward to interpret. For some families of important membrane proteins, for example, ion channel proteins, the information from hydropathy plots is often not very reliable. The reason for this is obvious. The membrane-spanning regions of these proteins form channels in the membrane through which water and ions or other solutes can pass. The membrane-spanning regions of these proteins must, therefore, on the one hand contain hydrophobic amino acids so as to be able to insert into the membrane while on the other hand they need hydrophilic residues to line the surface of the channels in contact with the aqueous phase. The hydropathic indices as a result are not high, and additional biochemical information is needed in order to predict with confidence the transmembrane segments of the sequence. Methods such as proteolytic digestion, antibody binding, and chemical modification of transmembrane proteins can provide valuable complementary information. When carefully done, such studies not only reveal which segments of the protein are outside the membrane and hence indirectly which regions span the membrane, but also show which segments are cytosolic and which, extracellular.

Tyrosine kinase growth factor receptors and G-protein linked receptors form two different receptor families

Receptors are plasma membrane proteins that bind specific molecules, such as growth factors, hormones, or neurotransmitters, and then transmit a signal to the cell's interior that causes the cell to respond in a specific manner. These responses are usually cascades of enzymatic reactions that give rise to many different effects within the cell, including changes in gene expression. Interference with these **receptor-signaling systems** therefore can have drastic consequences, for example, the loss of cellular growth control that is caused by those **oncogene products** that simulate the function of receptors or their associated signal transmitters.

Recent DNA sequence studies have shown that many receptors are structurally and evolutionarily related and that they can be grouped into a few families. These proteins provide yet another example of the now familiar story that there are only a limited number of different types of domains and that protein molecules with different functions have evolved by combining these domains in different ways, by gene shuffling during evolution. We will concentrate on just two of these families; the G-protein linked receptors and the **growth factor** receptors that signal through protein **tyrosine kinase** phosphorylation. No three-dimensional structure is yet available for any member of these two receptor families. The protein sequences deduced from DNA sequences have, however, revealed their overall structural organization, as well as their relation to many different oncogenes including one with a known three-dimensional structure, namely, cH-ras p21, the product of the human *ras* protooncogene.

The epidermal growth factor (EGF) receptor folds into distinct domains

The control of cell division is fundamental for the existence of multicellular organisms. Small proteins, growth factors, play a key role in this control process. One such growth factor, the **epidermal growth factor**, EGF, stimulates epidermal and epithelial as well as connective tissue cells to divide by binding to specific EGF receptor molecules inserted in the plasma membrane of these cells (Figure 14.1). There is strong evidence that the receptor molecule dimerizes upon binding EGF and thereby transmits a signal across the plasma membrane from the extracellular growth factor binding domain to an intracellular domain of the receptor. This cytosolic domain, now activated, is an enzyme, protein tyrosine kinase, that catalyzes phosphorylation of tyrosine residues. The primary event is an autophosphorylation of tyrosine residues in the carboxy-terminal region of the receptor chain itself. Subsequently, a number of different intracellular protein molecules are phosphorylated, and these, in an unknown manner, activate transcription of specific genes causing the cell to grow and divide.

The **EGF receptor**, EGFR, is a polypeptide chain 1186 residues long that is anchored in the membrane by one transmembrane helix (Figure 14.1) located in the middle of the chain. Hydropathy plots (Figure 14.2) identify this helix

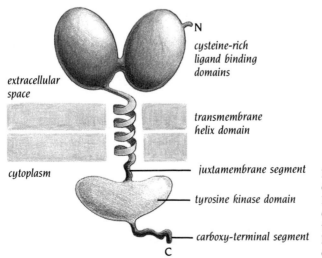

extracellular space

cysteine-rich ligand binding domains

transmembrane helix domain

juxtamembrane segment

tyrosine kinase domain

carboxy-terminal segment

cytoplasm

N

C

Figure 14.1 The 1186 amino acid polypeptide chain of epidermal growth factor receptor, EGFR, folds into distinct domains, two extracellular cysteine-rich domains, which bind the growth factor (blue), one transmembrane helix (green), and a cytosolic tyrosine kinase domain (red).

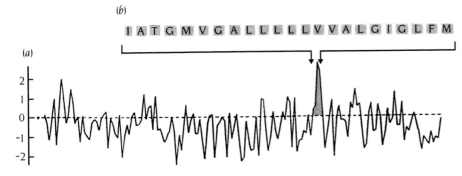

(b)

I A T G M V G A L L L L L V V A L G I G L F M

(a)

Figure 14.2 Hydropathy plots of the EGF receptor amino acid sequence (a), deduced from the nucleotide sequence of the corresponding gene, identify one transmembrane helix (green) with a high degree of confidence. The plot shown here was calculated using a window of 20 residues (see Chapter 13 for description of hydropathy plots). The amino acids of the 23-residue-long transmembrane helix (b) contain no charged side chains, and there is only one polar residue, a threonine. With a rise per residue of 1.5 Å in an α helix, 23 residues are just enough to span the 30-Å-thick hydrophobic portion of the membrane bilayer.

with a high degree of confidence. The extracellular part of the molecule, which binds EGF, is formed by two homologous, cysteine-rich domains (Figure 14.1). The cysteine residues form a network of S-S bridges that stabilize the two domains and render them protease resistant. The cytosolic part of the receptor is divided into three regions (Figure 14.1): a juxtamembrane segment, which may be important for signal transmission; the tyrosine kinase domain, which has an ATP as well as a protein substrate binding site; and the carboxy-terminal segment, which is autophosphorylated by the tyrosine kinase domain.

The v-erb B *oncogene is a coopted EGF receptor gene*

Once a partial amino acid sequence of the EGF receptor had been determined by sequencing peptides from purified receptor, computer-aided comparisons with gene sequences in data banks showed immediately that parts of the EGF receptor were almost identical to the gene product of the *v-erb B* oncogene. This viral oncogene is responsible for the ability of avian erythroblastosis virus to induce erythroleukemia in chickens and to transform both erythroblasts and fibroblasts *in vitro*. However, the *v-erb B* gene product lacks most of the extracellular domain of the EGF receptor (Figure 14.3), and it also lacks 34 amino acids at the cytosolic C terminus, including one of the tyrosine residues that is autophosphorylated. As a consequence of these differences, the *v-erb B* gene product, once it is inserted in the plasma membrane of a cell, has its tyrosine kinase constantly switched on; as a result, cell division becomes constitutive following infection with the virus.

There are strong indications that such **retroviral oncogenes** have arisen by the insertion of copies of normal host cell genes into the viral genome. Mutations to the normal host cell gene within the virus have then led to the viral oncogenes whose protein products transform normal host cells to malignant cells that ultimately are lethal to the organism. Many different oncogenes interfere with the tyrosine kinase activities of cells, including the first retroviral oncogene to be discovered, *pp 60 v-src*, which is derived from a cellular tyrosine kinase gene. The product of this gene is an intracellular tyrosine kinase that has extensive sequence homologies within a stretch of 260 amino acid residues in the tyrosine kinase domains of growth factor receptors, including EGFR, as well as to serine protein kinase enzymes (Figure 14.3b).

Figure 14.3 (a)The gene product of the *v-erb B* oncogene has an amino acid sequence that is highly homologous to parts of the EGF receptor sequence. The cytosolic carboxy-terminal tyrosine kinase domains (red) as well as the transmembrane helices are very similar, but the N-terminal half (blue) of the EGF receptor is truncated in the *v-erb B* gene product. (b) The protein product of the transforming gene of Rous sarcoma virus, known as pp60v-src, has a region of its amino acid sequence that is homologous both to the tyrosine kinase domain (residues 686–951) of the EGF receptor, EGFR, and to residues 41–300 of the catalytic subunit of cyclic-AMP-dependent serine protein kinase from cardiac muscle. The transmembrane helix of EGFR is colored green.

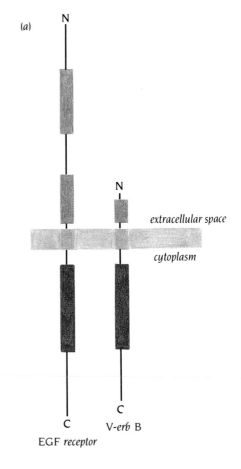

(a)

N

N

extracellular space

cytoplasm

C   C

V-*erb* B

EGF *receptor*

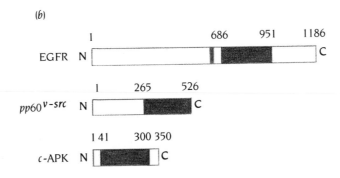

(b)

                1   686   951   1186
EGFR   N [||] C

         1   265   526
pp60 v-src N [   ] C

       1 41   300 350
c-APK N [   ] C

The receptors for insulin and epidermal growth factor are evolutionarily related

Signal transmission through protein kinase phosphorylation, exemplified by EGFR, is a common feature of the receptors of several polypeptide growth factors. Another interesting example is the receptor for one of the **insulinlike growth factors**, IGF-1. This growth factor has an amino acid sequence homologous to that of the polypeptide hormone insulin, hence its name, insulinlike growth factor; and, furthermore, its receptor (IGFR) shows extensive sequence homology to the insulin receptor. The completely different responses of cells to IGF and insulin are mediated by their different but related receptor molecules.

The **insulin receptor** (IR) and the **IGF-1 receptor** are both tetrameric molecules built up from two copies of each of two different polypeptide chains, α and β (Figure 14.4). The chains are joined by S-S bridges. Even though the IGF-1 and insulin receptor molecules are organized differently from the monomeric, single-chain EGF receptor, there are significant homologies at the amino acid sequence level among the three of them.

The β chain of the insulin and IGF-1 receptors contains a cytosolic tyrosine kinase domain, a transmembrane helix, and a short extracellular region with an S-S bridge to the α chain (Figure 14.4). The tyrosine kinase domain of the β chain has significant sequence homology to the corresponding domain in the EGF receptor. The α chains of the two receptors, which bind insulin and IGF-1, respectively, contain one cysteine-rich domain with an arrangement of cysteine residues along the polypeptide chain that is similar to that in each of the two extracellular domains of the EGF receptor. There is almost no sequence homology between these domains of the insulin receptor and the EGF receptor, and

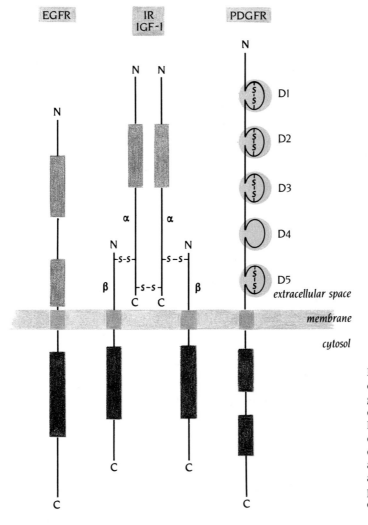

Figure 14.4 Comparison of the domain organization of the receptors for epidermal growth factor (EGFR), insulin (IR), and platelet-derived growth factor (PDGFR). The tyrosine kinase domains are red, the cysteine-rich domains are blue, and the immunoglobulinlike domains of the PDGF receptor are yellow. They all contain one transmembrane helix (green) and a similar tyrosine kinase domain (red) per polypeptide chain, but their ligand-binding cysteine-rich domains are differently organized.

a number of insertions and deletions are necessary to align the cysteine residues (Figure 14.5). Nevertheless, the pattern of cysteine residues is strikingly similar and in all probability reflects a common network of S-S bridges and hence a common fold of these domains. In other words, an evolutionary relationship certainly exists between the β chain of the insulin and IGF-1 receptors and the cytosolic tyrosine kinase domain of the EGF receptor. Furthermore, in all probability the α chain of the insulin and IGF-1 receptors and the two extracellular domains of the EGF receptor are also evolutionarily related.

The PDGF receptor is also a protein tyrosine kinase receptor

The EGF and insulin receptors represent two subclasses of the family of tyrosine kinase receptors. The receptor for **platelet-derived growth factor**, PDGF, is a member of a third subclass of this family. A single polypeptide chain is organized in an N-terminal extracellular segment, one transmembrane helix, and a cytosolic tyrosine kinase domain (Figure 14.4). The N-terminal, ligand-binding segment is quite different from the cysteine-rich domains of the other two subclasses. It contains about 500 amino acids and is organized into five domains, each of which shows low (about 20%) but significant sequence homology to immunoglobulin domains. Four of these PDGF receptor domains have the internal S-S bridge that is characteristic for immunoglobulin domains as described in Chapter 12. One receptor domain, D4, however, lacks this S-S bridge, which is quite unusual for members of the immunoglobulin superfamily. The tyrosine kinase domain of the PDGF receptor contains an insertion of about 100 residues in the middle of the chain, but the remaining regions are homologous to the tyrosine kinase domains of the other two subclasses of this receptor family. Extensive reshuffling of genes must have occurred during the evolution of this family of receptors, but, that notwithstanding, their kinship is still discernible.

Similar mechanisms are used for signal transduction across the membrane

Do these receptors use the same signal mechanism through the membrane despite different architectures and different cellular responses? To answer that question, the group of Axel Ullrich at Genentech, San Francisco, constructed a **chimeric receptor** using recombinant DNA techniques. The α subunit of the

Figure 14.5 Amino acid sequences of the cysteine-rich domains in insulin receptor, IR, and the two domains of the epidermal growth factor receptor, EGFR1 (residues 134–312 of EGFR) and EGFR2 (residues 446–604 of EGFR). Residue numbers are given at the beginning of each line. The sequences are aligned to maximize overlap of cysteine residues, which illustrates similarities in the arrangement of cysteine bridges in the folds of these domains.

```
IR     126  C Y L A T   I D W S R I L D S V   E D N Y I V L N K D   D N E E - C G D I - - - C P
EGFR1  134  C N V E S   I Q W R D I V S S D   F L S N M S M D F Q   N H L G S C Q K - - - - C D
EGFR2  446  C Y A N T   I N W K K L F G T S   G Q K T K I I S N R   G E N S - C K A T G Q V C H

IR     161  G T A K G K T N C A   T V I N G Q F V E R   C W T H S H C Q K V   C P T I C K S H G C
EGFR1  169  P S C P N G S - C W   G A G E E N - - - -   C Q K L T K I I - -   C A Q Q C S G R - C
EGFR2  484  A L C S P E G - C W   G P E P R D C V S -   C R N V S R G R E -   C V D K C N L L E G

IR     201  T A E G L - - - - - - - C C H S E   C L G N - - - - - C S Q P D D   P T K C V A C R N F
EGFR1  201  R G K S P S D - - - - - C C H N Q   C A A G - - - - - C T G P R E   S D - C L V C R K F
EGFR2  521  E P R E F V E N S E C I N C H P E   C L P Q A M N I T C T G R G P   D N - C I Q C A H Y

IR     231  Y L D G R C V E T C   P P P Y Y H F Q D W   R C V N F S F C Q D   L H H K C K N S R R
EGFR1  232  R D E A T C K D T C   P P L M L Y N P T T   Y Q M D V N P E G K   Y S F G A T C V K K
EGFR2  562  I D G P H C V K T C   P A G V M G E N N T   L V - - - - - - -   - - - - - - - - -

IR     271  Q G C H Q Y V I H N   N K C I P E C P S G   Y T M N S S N L L - - C   T P C L G P C P K V C -
EGFR1  272  C P R N Y V V T D H   G S C V R A C G A D   S Y E M E E D G V R K C   K K C E G P C R K V C -
EGFR2  584  - - - - - - - - -   - - - - - - - - -   - - W K Y A D A G H V C   H L C H P N C T Y G C -
```

insulin receptor was fused to the transmembrane and cytosolic portions of the EGF receptor. This hybrid receptor was expressed and integrated into the cell membrane. It bound insulin, and this binding produced autophosphorylation of the EGF tyrosine kinase domain and the phosphorylation of cytosolic proteins that is characteristic of the activated EGF receptor. This chimeric receptor thus responds to insulin in the same way that the intact EGF receptor responds to EGF.

In summary, despite differences, all members of the tyrosine kinase growth factor receptor family contain an evolutionarily related cytosolic tyrosine kinase domain and one transmembrane helix that is connected to an extracellular hormone-binding domain. Different subclasses of this family have quite different extracellular domains. Nevertheless, they have similar mechanisms for signal transduction across the membrane, as well as for signal transmission through protein phosphorylation leading to activation of gene expression.

G proteins are molecular amplifiers

Several important physiological responses like vision, smell, and stress response produce large metabolic effects from a small number of input signals. The receptors for these signals have two things in common. First, they are transmembrane proteins with seven helices spanning the lipid bilayer of the plasma membrane. Second, the signals received by these receptors are amplified and the amplifiers are members of a common family of proteins called **G proteins**. Most G proteins are heterotrimers consisting of one copy of α (45kd), β (35kd), and γ (7kd) subunits. They are called G proteins because they bind guanine-containing nucleotides on their α subunits. The α subunits have GTPase activity and slowly hydrolyze GTP to GDP and inorganic phosphate. Phosphate diffuses away from the active site, but GDP remains tightly bound. The GTP-bound form of the protein is the active form and sends a signal from the receptor to some cellular effector protein. The GDP-bound form is inactive. GTP hydrolysis is the switch between the active and inactive forms. Dissociation of GDP from the inactive form occurs by simple chemical exchange with GTP. There are a number of other proteins that can modulate either the rate of exchange of GTP for GDP or the rate of GTP hydrolysis. The cytoplasmic domain of the receptor for epinephrine is one such modulator.

Table 14.1
Examples of physiological processes mediated by G proteins

Stimulus	Receptor	Effector	Physiological response
Epinephrine	β-adrenergic receptor	adenylate cyclase	glycogen breakdown
Light	rhodopsin	c-GMP phosphodiesterase	visual excitation
IgE-antigen complexes	mast cell IgE receptor	phospholipase C	histamine secretion in allergic reactions
Acetylcholine	muscarinic receptor	potassium channel	slowing of pacemaker activity that controls the rate of the heartbeat

When **epinephrine** (adrenaline) binds to the membrane-bound **adrenergic receptor**, the latter is activated by an allosteric change (Figure 14.6). As a result, the cytoplasmic moiety of the receptor binds an inactive G protein and stimulates the exchange of bound GDP for GTP. The α subunit of the G protein, which now has bound GTP, dissociates from the heterotrimer and is released from the receptor–G protein complex. The Gα-GTP complex then activates adenylate cyclase, a membrane-bound enzyme that catalyzes the formation of cyclic AMP. Cyclic AMP, a second messenger for many hormones, in turn affects a wide range of cellular processes.

As long as epinephrine remains bound to the adrenergic receptor, the cytoplasmic domain can continue to trigger the production of many active molecules of Gα-GTP. Each active Gα-GTP, by complexing with adenylate cy-

222

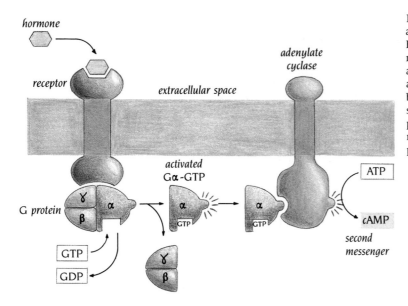

Figure 14.6 G-protein–mediated activation of adenylate cyclase by hormone binding. Hormone binding on the extracellular side of a receptor such as the β-adrenergic receptor activates a G protein in the cytoplasm. The activated form of the G protein (α subunit with bound GTP) dissociates from the β and γ subunits and activates adenylate cyclase to produce cyclic AMP, which is a second messenger that affects a diverse range of cellular processes.

clase, in turn triggers the production of many cyclic AMP molecules. Thus one molecule of the first messenger, the hormone, in this way produces a cascade of molecules of the second messenger, cyclic AMP, through the action of G proteins and adenylate cyclase. The GTPase activity of the Gα subunit determines the length of time that the signal remains "on." Once hydrolysis to GDP has occurred, the α subunit is once again able to bind to the β and γ subunits and the cycle can recommence.

G proteins are homologous in sequence to elongation factor Tu and cH-ras p21

At present the three-dimensional structure of the receptor-mediated G proteins is unknown. However, a comparison of the amino acid sequences of the GTP binding region of the α subunit of G proteins with other, monomeric GTP-binding proteins of known structure, such as the **elongation factor Tu** and different **ras proteins**, reveals three small regions of conserved sequences. These conserved regions are intimately involved in nucleotide binding by Tu and ras proteins. It is, therefore, highly probable that the structure of the GTP-binding region in the receptor-mediated G proteins, is very similar to that of the elongation factor Tu and the ras proteins.

Point mutation generates transforming ras oncogenes

The oncogenes most commonly isolated from human tumors or transformed human cell lines belong to the ***ras gene family***. The proteins encoded by these genes are called **ras p21**. All have a single polypeptide chain of 188 or 189 residues with very similar amino acid sequences. The difference between normal *ras* protooncogenes (*c-ras*) in healthy cells and the transforming *ras* oncogenes in cancer cells is usually a single base change at one of a very few critical positions, resulting in single amino acid substitutions in p21. Most of these substitutions occur in the regions conserved between the ras proteins and the G proteins, i.e., the regions that form or border the GTP binding site. Despite numerous studies it has not yet been possible to pinpoint the function of the normal ras p21 proteins in human cells, although in yeast a ras like protein activates adenyl cyclase. But in all probability human ras proteins are involved in signal transduction from hormone-receptor complexes to effector molecules; they might act simply as a G protein or as a relay from tyrosine kinase receptors. In either case, it appears that the mutations result in a constitutively activated protein. Most of the substitutions reduce the rate of GTP hydrolysis and thus leave the protein in the "on" state for a longer period of time.

A deletion mutant gene is used to produce truncated p21

The three-dimensional structure of a truncated form of the human *c-ras* protooncogene product complexed to GDP was published in 1988 by the group of Sung-Hou Kim in Berkeley, University of California, who subsequently corrected their structure in the light of a structure published a year later by the laboratory of Ken Holmes, Max Planck Institute, Heidelberg. Holmes and his colleagues determined the structure of a truncated cH-ras p21 complexed with a GTP analogue. In the following we describe the results from the Heidelberg group.

In mammalian cells the concentration of normal p21 is too low for purification of the protein in the amounts needed for crystallization. Instead, both the Berkeley and the Heidelberg groups used a truncated form of the protein that was purified from *E. coli* cells harboring an expression system plasmid that contained a deletion mutant of H-ras cDNA, which coded for amino acid residues 1 to 166 of the p21 protein. These 166 residues contain the catalytic domain of the protein. Why was not a gene for the complete polypeptide chain of p21 used? The missing residues at the carboxy terminus of the complete p21 molecule form a region that includes a covalently attached phospholipid that anchors p21 to the cell membrane. This region was known to be very flexible and was therefore deleted from the cloned *H-ras* gene and hence its protein product to improve the chances of crystallizing the protein. The successful crystallization of the protein and determination of its structure proved the wisdom of this decision.

The crystal structures of p21 and EF-Tu have the same fold of their polypeptide chains

p21 has an α/β-type structure (Figure 14.7) like that of the nucleotide-binding proteins discussed in Chapter 10. The central pleated sheet comprises six β strands, five of which are parallel. There are five α helices positioned on both sides of the β sheet. The GTP is bound in a pocket at the carboxy ends of the β strands in a way similar to the binding of nucleotides to other nucleotide-binding proteins. Loop regions that connect the β strands with the α helices form the binding pocket.

The topology of the fold of the polypeptide chain in p21 (Figure 14.7b) is exactly the same as that observed earlier by Jens Nyborg in Brian Clark's laboratory at Aarhus University, Denmark, and Frances Jurnak at Riverside,

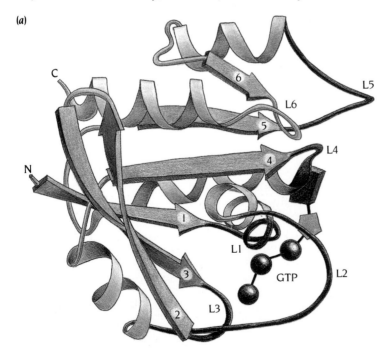

(a)

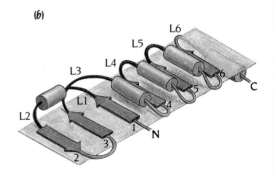

(b)

Figure 14.7 Schematic diagram (a) and topology diagram (b) of the polypeptide chain of cH-ras p21. The central β sheet of this α/β structure comprises six β strands, five of which are parallel. α helices are green, β strands are blue, and the adenine ribose and phosphate parts of the GTP analogue are blue, green, and red, respectively. The loop regions that are involved in the activity of this protein are red and labeled L1–L5. (Adapted from E.F. Pai et al., *Nature* 341: 209, 1989.)

University of California, in their crystal structures of elongation factor Tu. Furthermore, the mode of nucleotide binding, which involves the three conserved peptide regions, is very similar in these two GTP-binding proteins. Consequently, it is highly probable that the α subunits of receptor-mediated G proteins, which also contain these conserved peptide regions, have very similar three-dimensional structures and mode of GTP binding.

Regions of conserved amino acid sequence bind GTP

Five of the six loop regions (L1–L6 in Figure 14.7a) that are present at the carboxy end of the β sheet in the p21 structure form the GTP binding site. Three of these loops, L1 (residues 10–17), L3 (57–60), and L5 (116–119), contain the regions of conserved amino acid sequence that have been identified among small GTP-binding proteins and the α subunits of receptor-mediated G proteins. These regions have the consensus sequences Gly-X-X-X-X-Gly-Lys-Ser/Thr, Asp-X-X-Gly, and Asn-Lys-X-Asp, where X is any residue.

The first loop region, L1, is wrapped around the triphosphate part of GTP (Figure 14.8) and seems to be essential for proper positioning of the phosphate groups. The side chain of the conserved lysine residue, K16 in p21, forms a strong hydrogen bond (or ion-pair interaction) with the terminal (γ) phosphate group of GTP (Figure 14.8). In addition, the same lysine side chain stabilizes the position of the beginning of this loop region by forming hydrogen bonds to the main chain carbonyl groups of amino acids 11 and 12.

Almost all nucleotide triphosphate hydrolyzing enzymes, including p21, require magnesium ions for their catalytic activities. Mg++ is presumably required in these enzymes both for proper positioning of the γ phosphate and for weakening the P–O bond that is split during catalysis. In the structure of the p21–GTP analogue complex a magnesium ion is coordinated to two phosphate oxygen atoms, one each from the β and γ phosphates (Figure 14.8). In addition the Mg++ in p21 has two protein oxygen ligands, one from the side chain of Ser 17 in loop L1 and one from Thr 35 in loop L2. Furthermore, Asp 57 in loop L3 participates indirectly in metal binding in the triphosphate or "active" form of ras, since it binds a water molecule that is ligated to the magnesium.

The third conserved loop region, L5, has been called the guanine specificity region, since *in vitro* mutations of Asp 119 in this region drastically reduce the affinity of p21 for guanine nucleotides. In Chapter 6 we mentioned that the features that distinguish guanine from other common purine bases, including adenine, are a hydrogen bond donor substituent, NH₂, in position S2 and an oxygen hydrogen bond acceptor in position W2′. The three-dimensional structure of p21 clearly demonstrates the molecular basis of the specificity of G proteins for GTP. Both the NH₂ and the oxygen substituents are specifically recognized by the protein. NH₂ forms a hydrogen bond to the side chain of Asp 119, which is conserved in ras proteins. The oxygen substituent forms another

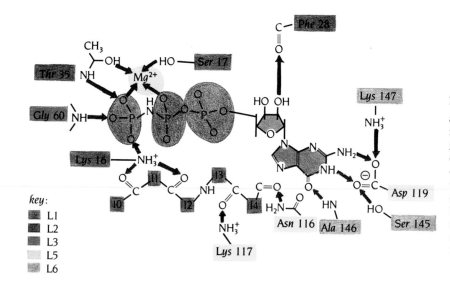

Figure 14.8 Schematic view of the binding of a GTP analogue to cH-ras p21. A magnesium ion, Mg²⁺, positions the terminal phosphate groups of the GTP analogue by binding to two of their oxygen atoms. Oxygen atoms from side chains of Ser 17 and Thr 35 anchor the Mg ion to the protein. Residue Asp 119, which is conserved in ras proteins, forms a hydrogen bond to the NH₂ substituent of the guanine moiety, which accounts for the specificity of ras for GTP. (Adapted from E.F. Pai et al., *Nature* 341: 209, 1989.)

225

hydrogen bond to the main-chain NH group of residue 146. None of these hydrogen bonds could be formed with ATP, since adenine has a hydrogen atom instead of the NH_2 substituent in guanine and an NH_2 hydrogen donor in place of the oxygen hydrogen acceptor in guanine. These differences are more than sufficient to account for the specificity of ras for GTP rather than ATP.

Oncogenic activation is caused by mutations in the GTP-binding loops

In a high percentage of human tumors of several types the p21 protein has been found to be mutated at amino acid positions 12/13 or 61. Furthermore, additional mutations at these and other sites causing p21 to become oncogenic have been generated experimentally by either random or site-directed mutagenesis. Biochemical studies had shown that these mutated forms of p21 have either altered binding affinity for GDP/GTP or altered rates of GTPase activity *in vitro*. The three-dimensional structure of p21 now allows us to explain the biochemical and biological properties of these oncogenic mutants in terms of the atomic structure of the protein.

Mutations of Gly 12 or Gly 13 would be expected to affect the conformation of the phosphate-binding loop region, L1. Only glycines can be tolerated in these positions because of the close spatial interaction between this loop and the phosphate residues. The large side chain of Gln 61 in loop L3 points toward the γ phosphate of GTP but does not seem to form direct hydrogen bonds. Mutations where Gln 61 is replaced by an amino acid with a large hydrophobic side chain, such as Leu or Val, have the highest transforming efficiency of all the H-ras mutants changed at this position. One possibility is that such substitutions could displace either the γ-phosphate group or amino acid side chains in its vicinity, thereby changing GTP binding and GTPase activity. Finally, mutations of K117 in loop L5 have been found in chemically induced mouse tumors. This residue in normal p21 interacts with the main chain oxygen atom of residue 13 (Figure 14.8), thereby stabilizing the phosphate-binding loop region, L1. In short, it appears that most mutations that render p21 oncogenic affect either GTP hydrolysis or the binding of GTP or GDP. Conversely, it is possible that all mutations that change the binding of the guanine nucleotides are potentially transforming.

The molecular basis of autophosphorylation of viral p21

The structure of p21 also provides some clues to the mechanism of **autophosphorylation** that is observed with viral p21 proteins. Viral p21 proteins differ at two amino acid residues from the normal cH-ras p21 specified by the cellular protooncogene. Gly 12 is replaced by either Arg or Ser and Ala 59 is replaced by Thr, which is phosphorylated. This last residue is located in loop L3, which is close to the Mg atom as well as to the γ phosphate of GTP (Figure 14.8). Changing Ala 59 to a threonine would place its side-chain hydroxyl in a position close to the γ phosphate. Instead of a water molecule attacking the γ phosphate during GTP hydrolysis by the cellular enzyme (Ala 59), the hydroxyl group of Thr 59 could attack the phosphate in the viral mutant protein leading to autophosphorylation and a decrease in GTPase activity. In this case autophosphorylation occurs because of the close proximity of the γ phosphate of bound GTP and a suitable hydroxyl group in the enzyme. It seems quite possible that autophosphorylation of tyrosine kinase domains, triggered by ligand binding to the receptor, as described in the beginning of this chapter, has an equally simple mechanism. Ligand binding to the extracellular domain may trigger a conformational change of the cytosolic tyrosine kinase domain, such that the hydroxyl group of a specific tyrosine residue is brought close to the γ-phosphorous atom of a bound nucleotide triphosphate.

Protein-protein interactions can modulate the GTP-binding and hydrolysis properties of ras p21

The normal rate of hydrolysis of GTP catalyzed by cellular ras p21 is too slow for proper control of cell growth and division. In fact, the normal level of GTPase activity would lead to transformation of cells. This paradox was resolved by the discovery by Frank McCormick, of Cetus Corporation, Emeryville, California, of a high molecular weight protein that binds to p21 and accelerates its rate of GTP hydrolysis over tenfold. This protein is called **GAP** (*GTPase Activating Protein*). It is the ras-GAP complex that functions as the "on" signal in normal cells. Synthetic peptides with sequences corresponding to loop L2 (Figure 14.7) in the ras structure compete with ras for GAP binding, so it seems likely that this loop, which also borders the triphosphate binding site and contains one of the magnesium ligands, Thr 35, is involved in GAP binding to ras. The mechanism by which the GAP-ras interaction increases the catalytic activity of ras is unknown, but it is important to remember that a factor of 10 in catalytic rate can arise from any one of a large number of small effects. A single extra hydrogen bond between the protein and the transition state, or even a change in the strength of an existing hydrogen bond is all that might be needed, as will be discussed in Chapter 15.

One might wonder if a protein exists that modulates the other part of the ras "on-off" switch, the exchange of GTP for bound GDP. Thus far, no such protein has been conclusively identified for human ras, but yeast cells, which possess ras like proteins that stimulate adenylate cyclase, appear to have proteins that can accelerate the rate of GTP binding and/or GDP release. Thus, at least in some cells, the duration of the cell growth signal sent by the *ras* protooncogene product is modulated by protein-protein interactions. It may be that proteins such as GAP and the yeast GTP-GDP exchange catalyst are ras counterparts of the β and γ subunits as well as the cytoplasmic receptor domain in the receptor-mediated G-protein signaling systems.

Receptors that utilize G proteins contain seven transmembrane helices

The genes for the receptors that use G proteins listed in Table 14.1 have all been cloned and sequenced. These receptors are all single polypeptide chains of around 450 amino acids, except rhodopsin, which is about 100 residues shorter. They all exhibit significant sequence homology to one another and can therefore be expected to have the same topology. In contrast to the tyrosine kinase receptors, which have continuous domains on each side of the membrane joined by a single transmembrane helix, hydropathy plots strongly indicate that the **G protein receptors** have seven transmembrane helices (Figure 14.9). In the case of bacteriorhodopsin, the bacterial analogue of

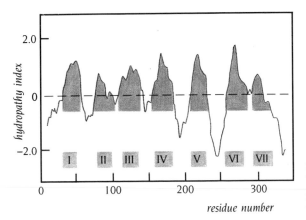

Figure 14.9 Hydropathy plots of the rhodopsin amino acid sequence identifies seven transmembrane helices (green). Hydrophobicity increases with increasing values of the hydropathy index.

rhodopsin, structural data to 3 Å resolution in projection (see Chapter 13) directly establishes the presence of seven transmembrane helices; this fact reinforces the interpretation of the hydropathy plots of the other molecules. Biochemical studies have shown that the NH$_2$ terminus is extracellular and that the carboxy terminus is cytoplasmic. The polypeptide chain therefore goes back and forth through the lipid bilayer seven times leaving the two termini on opposite sides of the membrane (Figure 14.10).

The extracellular loop regions of the adrenergic and muscarinic receptors (Figure 14.10) are short, and it seems unlikely that the few residues can form a complete ligand-binding domain. This correlates well with compelling biochemical evidence that epinephrine binds within the lipid of the adrenergic receptor matrix of the membrane to residues of at least two transmembrane helices. Such a situation is analogous to that of the photon receptor retinal in rhodopsin, which is covalently attached to a lysine residue in the middle of one of the predicted transmembrane helices.

The G-protein linked receptors, with their seven membrane-spanning helices, have been modified and adapted during the course of evolution to fulfill the role of signal receivers and transducers in a wide variety of cellular communication systems. The signals they recognize range from photons of light to hormones and neurotransmitters. These receptors all utilize members of another family of proteins, the G proteins, to amplify the signals that they receive. The challenge now is to solve the three-dimensional structures of some of these receptors and G proteins to atomic resolution so that we can begin to understand the details of these signaling systems. As we have seen, a considerable amount can be learned from analyses of one-dimensional amino acid sequences. But, as this chapter also shows, in the absence of three-dimensional structural information at a high resolution our understanding of protein function remains tantalizingly superficial.

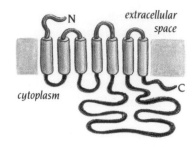

Figure 14.10 Schematic diagram of the arrangement of the seven transmembrane helices of the muscarinic M2 receptor.

Conclusion

Receptor proteins in the plasma membrane bind specific molecules, such as growth factors, hormones, or neurotransmitters, and then transmit a signal to the cell interior that causes the cell to respond in a specific manner. Many receptors are structurally and evolutionarily related and can be grouped into a few families.

The receptors for insulin and several growth factors belong to one such family, the members of which contain an evolutionarily related cytosolic tyrosine kinase domain in the cytoplasm and a single transmembrane helix that is connected to an extracellular hormone binding domain. Certain oncogenes, such as *v-erb B*, are truncated versions of this receptor family that constitutively send signals to the cells to multiply.

Rhodopsin and β adrenergic receptors belong to a different family with seven transmembrane helices. The signals received by this receptor family are amplified through the GTPase activity of G proteins to give a cascade of second messengers, including cyclic AMP.

The three-dimensional structure of the cellular protooncogene protein cH-ras p21 has provided insight into the mechanism of action of GTP-binding proteins. In addition, it has shown how point mutations that alter GTP binding or hydrolysis can convert protooncogenes to oncogenes and result in cell transformation.

Selected readings

General

Barbacid, M. *ras* genes. *Annu. Rev. Biochem.* 56: 779–827, 1987.

Bos, J.L. *ras* oncogenes i human cancer: a review. *Cancer Res.* 49: 4682–4689, 1989.

Carpenter, G. Receptors for epidermal growth factor and other polypeptide mitogens. *Annu. Rev. Biochem.* 56: 881–914, 1987.

Changeux, J.-P.; Giraudat, J.; Dennis, M. The nicotinic acetylcholine receptor: molecular architecture of a ligand-regulated ion channel. *Trends Pharm. Sci.* 8: 459–465, 1987.

Gilman, A.G. G proteins: transducers of receptor-generated signals. *Annu. Rev. Biochem.* 56: 615–650, 1987.

Hanks, S.K.; Quinn, A.M.; Hunter, T. The protein kinase family: conserved features and deduced phylogeny of the catalytic domains. *Science* 241: 42–52, 1988.

Heldin, C.-H.; Westermark, B. Growth factors: mechanism of action and relation to oncogenes. *Cell* 37: 9–20, 1984.

Heldin, C.-H.; Westermark, B. Platelet-derived growth factor: mechanism of action and possible *in vivo* function. *Cell Regulation* 1: 555–566, 1990.

Lefkowitz, R.J.; Caron, M.G. Adrenergic receptors. *J. Biol. Chem.* 263: 4993–4996, 1988.

Mc Cormick, F. *ras* GTPase-activating protein: signal transmitter and signal terminator. *Cell* 56: 5–8, 1989.

Neer, E.J.; Clapham, D.E. Roles of G protein subunits in transmembrane signaling. *Nature* 333: 129–134, 1988.

Ovchinnikov, Yu.A. Rhodopsin and bacteriorhodopsin: structure-function relationships. *FEBS Lett.* 148: 179–191, 1282.

Ramachandran, J.; Ullrich, A. Hormonal regulation of protein tyrosine kinase activity. *Trend Pharm. Sci.* 8: 28–31, 1987.

Stryer, L. Cyclic GMP cascade of vision. *Annu. Rev. Neurosci.* 9: 87–119, 1986.

Stryer, L. The molecules of visual excitation. *Sci. Am.* 257(1): 42–50, 1987.

Stryer, L.; Bourne, H.R. G proteins: a family of signal transducers. *Annu. Rev. Cell. Biol.* 2: 391–419, 1986.

Williams, L.T. Signal transduction by the platelet-derived growth factor receptor. *Science* 243: 1564–1570, 1989.

Yarden, Y.; Ullrich, A. Growth factor receptor tyrosine kinases. *Annu. Rev. Biochem.* 57: 443–478, 1988.

Specific structures

Brünger, A.T., et al. Crystal structure of an active form of ras protein, a complex of a GTP analog and the H-ras p21 catalytic domain. *Proc. Natl. Acad. Sci. USA* 87: 4849–4853, 1990.

Dixon, R.A.F., et al. Cloning of the gene and cDNA for mammalian β-adrenergic receptor and homology with rhodopsin. *Nature* 321: 75–79, 1986.

Dixon, R.A.F., et al. Ligand binding to the β-adrenergic receptor involves its rhodopsin-like core. *Nature* 326: 73–77, 1987.

Doolittle, R.F., et al. Simian sarcoma virus onc gene, v-*cis*, is derived from the gene (or genes) encoding a platelet-derived growth factor. *Science* 221: 275–277, 1983.

Downward, J., et al. Close similarity of epidermal growth factor receptor and v-*erb*-B oncogene protein sequences. *Nature* 307: 521–527, 1984.

Grenningloh, G., et al. The strychnine-binding subunit of the glycine receptor shows homology with nicotinic acetylcholine receptors. *Nature* 328: 215–220, 1987 198.

Holbrook, S.R.; Kim, S.-H. Molecular model of the G protein, a subunit based on the crystal structure of the H-ras protein. *Proc. Natl. Acad. Sci. USA* 86: 1751–1755, 1989.

Jurnak, F. Structure of the GDP domain of EF-Tu and location of the amino acids homologous to *ras* oncogene proteins. *Science* 230: 32–36, 1985.

Krengel, U., et al. Three-dimensional structures of H-*ras* p21 mutants: molecular basis for their inability to function as signal switch molecules. *Cell* 62: 539–548, 1990.

Kubo, T., et al. Cloning, sequencing and expression of complementary DNA encoding the muscarinic acetylcholine receptor. *Nature* 323: 411–416, 1986.

LaCour, T.F.M., et al. Structural details of the binding of guanosine diphosphate to elongation factor Tu from *E. coli* as studied by x-ray crystallography. *EMBO J.* 4: 2385–2388, 1985.

Milburn, M.V., et al. Molecular switch for signal transduction: structural differences between active and inactive forms of protooncogenic ras proteins. *Science* 247: 939–945, 1990.

Mishina, M., et al. Location of functional regions of acetylcholine receptor α subunit by site-directed mutagenesis. *Nature* 313: 364–369, 1985.

Nathans, J.; Hogness, D.S. Isolation, sequence analysis, and intron-exon arrangement of the gene encoding bovine rhodopsin. *Cell* 34: 807–814, 1983.

Pai, E.F., et al. Structure of the guanine-nucleotide-binding domain of the Ha-*ras* oncogene product p21 in the triphosphate conformation. *Nature* 341: 209–214, 1989.

Pai, E.F., et al. Refined crystal structure of the triphosphate conformation of H-*ras* at 1.35 Å resolution: implications for the mechanism of GTP hydrolysis. *EMBO J.* 9: 2351–2360, 1990.

229

Riedel, H., et al. A chimaeric receptor allows insulin to stimulate tyrosine kinase activity of epidermal growth factor receptor. *Nature* 324: 68–70, 1986.

Schlichting, I., et al. Time-resolved x-ray crystallographic study of the conformational change in Ha-ras p21 protein on GTP hydrolysis. *Nature* 345: 309–315, 1990.

Schofield, P.R., et al. Sequence and functional expression of the GABA$_A$ receptor shows a ligand-gated receptor super-family. *Nature* 328: 221–227, 1987.

Tong, L., et al. Structure of ras protein. *Science* 245: 244, 1989.

Ullrich, A., et al. Human epidermal growth factor receptor cDNA sequence and aberrant expression of the amplified gene in A431 epidermoid carcinoma cells. *Nature* 309: 418–425, 1984.

Ullrich, A., et al. Human insulin receptor and its relationship to the tyrosine kinase family of oncogenes. *Nature* 313: 756–761, 1985.

Ullrich, A., et al. Insulin-like growth factor I receptor primary structure: comparison with insulin receptor suggests structural determinants that define functional specificity. *EMBO J.* 5: 2503–2512, 1986.

Waterfield, M.D., et al. Platelet-derived growth factor is structurally related to the putative transforming protein p28[sis] of simian sarcoma virus. *Nature* 304: 35–39, 1983.

Yarden, Y., et al. Structure of the receptor for platelet-derived growth factor helps define a family of closely related growth factor receptors. *Nature* 323: 226–232, 1986.

An Example of Enzyme Catalysis: Serine Proteinases

15

Linus Pauling first formulated in 1946 the basic principle underlying enzyme catalysis, namely, that an enzyme increases the rate of a chemical reaction by strongly binding the **transition state** of the specific substrate. However, for many years it was not generally appreciated that the high affinity of an enzyme for the transition state of a substrate plays a major role in determining **substrate specificity** as well as the **rate of catalysis**. In the past few years, kinetic studies of site-directed mutants, combined with x-ray structures, have made it possible to identify unambiguously the role of particular amino acids in both the substrate specificity and the catalytic reaction of an enzyme as well as providing information on the energetic basis of catalysis itself. The full consequences of Pauling's principle emerged only when it was found that mutants designed to change an enzyme's catalytic rate also changed its substrate specificity and vice versa.

In this chapter we will illustrate some fundamental aspects of enzyme catalysis using as an example the **serine proteinases**, a group of enzymes that hydrolyze peptide bonds in proteins. We also examine how the transition state is stabilized in this particular case.

Proteinases form four functional families

Proteinases are widely distributed in nature, where they perform a variety of different functions. Viral genes code for proteinases that cleave the precursor molecules of their coat proteins, bacteria produce many different extracellular proteinases to degrade proteins in their surroundings, and higher organisms use proteinases for such different functions as food digestion, cleavage of signal peptides, and control of blood pressure as well as blood clotting. Many proteinases occur as domains in large multifunctional proteins, but others are independent smaller polypeptide chains. *In vivo* the activity of many proteinases is controlled by endogenous protein inhibitors that complex with the enzymes and block them. The three-dimensional structures of a relatively large number of the smaller proteinases and of their complexes with protein inhibitors have been determined, and this wealth of data allows some general conclusions to be drawn.

All the well-characterized proteinases belong to one or other of four families: serine, cysteine, aspartic, or metallo proteinases. This classification is based on a functional criterion, namely, the nature of the most prominent functional group in the active site. Members of the same functional family are usually evolutionarily related, but there are exceptions to this rule. We have chosen two serine proteinases, mammalian **chymotrypsin** and bacterial **subtilisin**, as representative examples to illustrate one of the catalytic mechanisms leading to proteolysis. Before describing the structures, mechanism, and engineering of these two enzymes, however, we shall define some basic enzymological concepts.

The catalytic properties of enzymes are reflected in K_m and k_{cat} values

Leonor Michaelis and Maud Menten laid the foundation for enzyme kinetics as early as 1913 by proposing the following scheme:

$$E + S \underset{K_m}{\rightleftarrows} ES \underset{k_{cat}}{\rightleftarrows} E + P$$

Enzyme and substrate first combine to give a reversible **enzyme-substrate** (ES) complex. Chemical processes then occur in a second step with a rate constant called **k_{cat}**, or the **turnover number**, which is the maximum number of substrate molecules converted to product per active site of the enzyme per unit time. The k_{cat} is, therefore, a rate constant that refers to the properties and reactions of the ES complex. For simple reactions k_{cat} is the rate constant for the chemical conversion of the ES complex to free enzyme and products.

These definitions are valid only when the concentration of the enzyme is very small compared with that of the substrate. Moreover, they apply only to the initial rate of formation of products: in other words, the rate of formation of the first few percent of the product, before substrate has been depleted and products that can interfere with the catalytic reaction have accumulated.

The Michaelis-Menten scheme nicely explains why a maximum rate, **V_{max}**, is always observed when the substrate concentration is much higher than the enzyme concentration (Figure 15.1). V_{max} is obtained when the enzyme is saturated with substrate. There are then no free enzyme molecules available to turn over additional substrate. Hence, the rate is constant, V_{max}, and is independent of further increase in the substrate concentration.

The substrate concentration when the half maximal rate, ($V_{max}/2$), is achieved is called the **K_m**. For simple reactions it can easily be shown that the K_m is equal to the dissociation constant of the ES complex. The K_m, therefore, describes the affinity of the enzyme for the substrate. For more complex reactions, K_m may be regarded as the overall dissociation constant of all enzyme-bound species.

The quantity **k_{cat}/K_m** is a rate constant that refers to the properties and reactions of the free enzyme and free substrate. The ultimate limit to the value of k_{cat}/K_m is therefore set by the rate constant for the initial formation of the ES complex. This rate cannot be faster than the diffusion-controlled encounter of an enzyme and its substrate, which is between 10^8 to 10^9 per mole per second. The quantity k_{cat}/K_m is sometimes called the **specificity constant** because it describes the specificity of an enzyme for competing substrates. As we will see, it is a useful quantity for kinetic comparison of mutant proteins.

Enzymes decrease the activation energy of chemical reactions

The Michaelis complex, ES, undergoes rearrangement to one or several transition states before product is formed. Energy is required for these rearrangements. The input energy required to bring free enzyme and substrate to the highest transition state of the ES complex is called the **activation energy** of the reaction (Figure 15.2). In the absence of enzyme, spontaneous conversion of substrate to product also proceeds through transition states that require activation energy. The more than 1-million-fold enhancement of rate achieved by enzyme catalysis results from the ability of the enzyme to decrease the activation energy of the reaction (Figure 15.2) since the rate of a chemical reaction is strictly dependent on its activation energy.

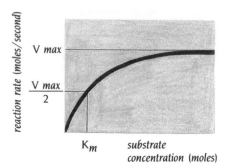

Figure 15.1 A plot of the reaction rate as a function of the substrate concentration for an enzyme catalyzed reaction. V_{max} is the maximal velocity. The Michaelis constant, K_m, is the substrate concentration at half V_{max}. The rate v is related to the substrate concentration, [S], by the Michaelis-Menten equation:

$$v = \frac{V_{max} \times [S]}{K_m + [S]}$$

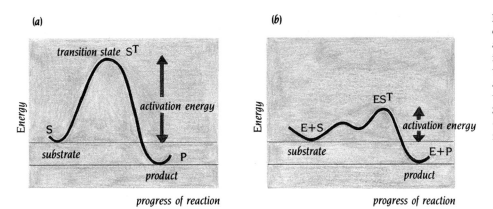

Figure 15.2 Enzymes accelerate chemical reactions by decreasing the activation energy. The activation energy is higher for a noncatalyzed reaction (a) than for the same reaction catalyzed by an enzyme (b). Both reactions proceed through one or several transition states, S^T. Only one transition state is shown in (a), whereas the two bumps in (b) represent two different transition states.

This decrease in activation energy is achieved by enzymes in several different ways: for example, by providing catalytically competent groups for a specific reaction mechanism, by binding several substrates in an orientation appropriate to the reaction catalyzed, and, most importantly, by using the differential binding energy of the substrate in its transition state compared with its normal state. The activation energy for the conversion of ES to E + P is lower if the enzyme binds more tightly to the transition state of S than to its normal structure (Figure 15.3). The higher affinity of the enzyme for the transition state makes the transition energetically favorable and thus decreases the activation energy. If, on the other hand, the enzyme were to bind the unaltered substrate more strongly than the transition state, the decrease in binding energy on the formation of the transition state would increase the activation energy and catalysis would not be achieved (Figure 15.3). It is therefore catalytically advantageous for the enzyme to be complementary to the transition state of the substrate rather than to the normal structure of the substrate.

Since this differential binding energy relates to the complete substrate molecule, including groups that determine the substrate specificity, it is obvious that specificity and catalytic rate are interdependent. The importance of this differential binding energy for catalysis is nicely illustrated by the recent production of antibodies with catalytic activity. Such antibodies were raised against small hapten molecules that simulate a transition state structure for a specific chemical reaction, such as ester hydrolysis. These antibodies not only bound the transition state more tightly than the normal structure of the ester, but also they exhibited significant catalytic activity even though they had not been selected to have any catalytically competent residues in the binding site.

Figure 15.3 One of the most important factors in enzyme catalysis is the ability of an enzyme to bind substrate more tightly in its transition state than in its normal state. The difference in binding energy between these states lowers the activation energy of the reaction. This is illustrated by energy profiles for an enzyme in its wild-type form (a), for a mutant that stabilizes the substrate in its transition state and therefore decreases the activation energy from ES to the transition state ES^T giving higher rates (b), and for a mutant that stabilizes the substrate in its normal state giving lower rates (c). (Adapted from A. Fersht, Enzyme Structure and Mechanism, 2nd ed., pp. 314–315. New York: W.H. Freeman, 1984.)

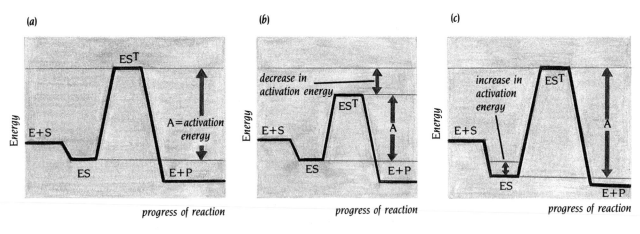

233

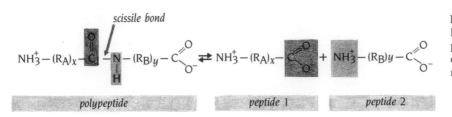

scissile bond

polypeptide ⇌ peptide 1 + peptide 2

Figure 15.4 Serine proteinases catalyze the hydrolysis of peptide bonds within a polypeptide chain. The bond that is cleaved is called the scissile bond. $(R_A)_x$ and $(R_B)_y$ represent polypeptide chains of varying lengths.

Serine proteinases cleave peptide bonds by forming tetrahedral transition states

The serine proteinases have been very extensively studied, both by kinetic methods in solution and by x-ray structural studies to high resolution. From all these studies the following reaction mechanism has emerged.

A serine proteinase cleaves peptide bonds within a polypeptide to produce two new smaller peptides (Figure 15.4). The reaction proceeds in two steps. The first step produces a covalent bond between C_1 of the substrate and the hydroxyl group of a reactive Ser residue of the enzyme (Figure 15.5a). Production of this acyl-enzyme intermediate proceeds through a negatively charged transition state intermediate (Figure 15.5a) where the bonds of C_1 have tetrahedral geometry in contrast to the planar triangular geometry in the peptide group. During this step the peptide bond is cleaved, one smaller product peptide is attached to the enzyme in the acyl-enzyme intermediate, and the other peptide product rapidly diffuses away. In the second step of the reaction (Figure 15.5b), deacylation, the acyl-enzyme intermediate is hydrolyzed by a water molecule to release the second product peptide with a complete carboxy terminus and to restore the Ser-hydroxyl of the enzyme. This step also proceeds through a negatively charged tetrahedral transition state intermediate (Figure 15.5b). What are the structural requirements for the enzyme to perform these reactions?

Figure 15.5 (a) Formation of an acyl-enzyme intermediate involving a reactive Ser residue of the enzyme is the first step in hydrolysis of peptide bonds by serine proteinases. A transition state is first formed where the peptide bond is cleaved and where the C_1 carbon has a tetrahedral geometry with bonds to four groups, including the reactive Ser residue of the enzyme and a negatively charged oxygen atom. (b) Deacylation of the acyl-enzyme intermediate is the second step in hydrolysis. This is essentially the reverse of the acylation step, with water in the role of the NH_2 group of the polypeptide substrate. The base shown in the figure is a His residue of the protein that can accept a proton during the formation of the tetrahedral transition state.

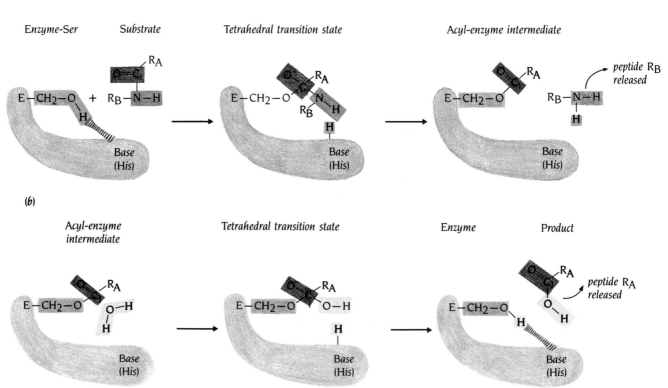

(a)

Enzyme-Ser   Substrate   Tetrahedral transition state   Acyl-enzyme intermediate

(b)

Acyl-enzyme intermediate   Tetrahedral transition state   Enzyme   Product

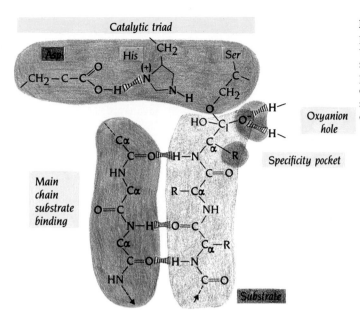

Catalytic triad

Oxyanion hole

Specificity pocket

Main chain substrate binding

Substrate

Figure 15.6 A schematic view of the presumed binding mode of the tetrahedral transition state intermediate for the deacylation step. The four essential features of the serine proteinases are highlighted in yellow: the catalytic triad, the oxyanion hole, the specificity pocket, and the unspecific main-chain substrate binding.

Four essential structural features are required for the catalytic action of serine proteinases

X-ray studies have shown that the serine proteinases have four essential structural features that facilitate this mechanism of catalysis (Figure 15.6).

1. The enzyme provides a general base, a His residue, that can accept the proton from the hydroxyl group of the reactive Ser thus facilitating formation of the covalent tetrahedral transition state. This His residue is part of a **catalytic triad** (Figure 15.6) consisting of three side chains from Asp, His, and Ser, which are close to each other in the active site, although they are far apart in the amino acid sequence of the polypeptide chain.

2. Tight binding and stabilization of the tetrahedral transition state intermediate is accomplished by providing groups that can form hydrogen bonds to the negatively charged oxygen atom attached to C_1. These groups are in a pocket of the enzyme called the **oxyanion hole** (Figure 15.6). The positive charge that develops on the His residue after it has accepted a proton also stabilizes the negatively charged transition state. These features also presumably destabilize binding of substrate in the normal state.

3. Most serine proteinases have no absolute substrate specificity. They can cleave peptide bonds with a variety of side chains adjacent to the peptide bond to be cleaved (the scissile bond). Thus, polypeptide substrates exhibit a nonspecific binding to the enzyme through their main-chain atoms, which form hydrogen bonds in a short antiparallel β sheet with main-chain atoms of a loop region in the enzyme (Figure 15.6). One of these hydrogen bonds is long (3.6 Å) in enzyme-substrate complexes but short in complexes that simulate the transition state. This nonspecific binding therefore also contributes to stabilization of the transition state.

4. Even though these enzymes have no absolute specificity, many of them show a preference for a particular side chain before the scissile bond as seen from the amino end of the polypeptide chain. The preference of chymotrypsin to cleave after large aromatic side chains and of trypsin to cleave after Lys or Arg side chains is exploited when these enzymes are used to produce peptides suitable for amino acid sequence determination and fingerprinting. In each case, the preferred side chain is orientated so as to fit into a pocket of the enzyme called the **specificity pocket**.

235

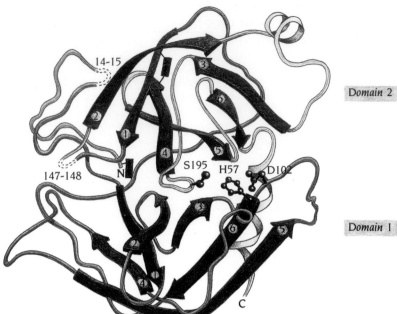

Figure 15.7 Schematic diagram of the structure of chymotrypsin, which is folded into two domains of the antiparallel β type. The six β strands of each domain are red, the side chains of the catalytic triad are blue, and the S-S bridges that join the three polypeptide chains are marked in purple. Chain A (green, residues 1–13) is linked to chain B (blue, residues 16–146) by an S-S bridge between Cys 1 and Cys 122. Chain B is in turn linked to chain C (yellow, residues 149–245) by an S-S bridge between Cys 136 and Cys 201. Dotted lines indicate residues 14–15 and 147–148 in the inactive precursor, chymotrypsinogen. These residues are excised during the conversion of chymotrypsinogen to the active enzyme chymotrypsin.

Convergent evolution has produced two different serine proteinases with similar catalytic mechanisms

These four features all occur in an almost identical fashion in all members of the chymotrypsin superfamily of homologous enzymes, which includes among other enzymes chymotrypsin, trypsin, elastase, and thrombin. Naively, one might imagine that such a combination of four characteristic features had arisen only once during evolution to give an ancestral molecule from which all serine proteinases diverged. However, subtilisin, a bacterial serine proteinase with an amino acid sequence, and as we will see, a three-dimensional structure quite different from the mammalian serine proteinases, exhibits these same four characteristic features. Subtilisin is not evolutionarily related to the chymotrypsin family of enzymes; nevertheless, the atoms in subtilisin that participate in the catalytic triad, in the oxyanion hole, and in substrate binding are in almost identical positions relative to one another in the three-dimensional structure as they are in chymotrypsin and its relatives. Starting from unrelated ancestral proteins, convergent evolution has resulted in the same structural solution to achieve a particular catalytic mechanism. The serine proteinases, in other words, provide a spectacular example of convergent evolution at the molecular level, which we can best appreciate by explaining in detail the structures of chymotrypsin and subtilisin.

The chymotrypsin structure has two antiparallel β-barrel domains

In 1967 the group of David Blow at the MRC Laboratory of Molecular Biology in Cambridge, England, determined the three-dimensional structure of chymotrypsin. This was one of the very first enzyme structures known at high resolution. Since then a large number of serine proteinase structures, complexed both with small peptide inhibitors and large endogenous polypeptide inhibitors, have been determined to high resolution mainly by the groups of Michael James, Edmonton, and Robert Huber, Munich.

The polypeptide chain of chymotrypsinogen, the inactive precursor of chymotrypsin, comprises 245 amino acids. During activation of chymotrypsinogen residues 14–15 and 147–148 are excised. The remaining three polypeptide chains are held together by S-S bridges to form the active chymotrypsin molecule.

The polypeptide chain is folded into two domains (Figure 15.7), each of which contains about 120 amino acids. The two domains are both of the antiparallel β-barrel type, each containing six β strands that have the same topology (Figure 15.8). Even though the actual structure looks complicated, the topology is very simple, a Greek key motif (strands 1–4) followed by an antiparallel hairpin motif (strands 5 and 6).

The active site is formed by two loop regions from each domain

The active site is situated in a crevice between the two domains. Domain 1 contributes two of the residues in the catalytic triad, His 57 and Asp 102, whereas the reactive Ser 195 is part of the second domain (Figure 15.7).

Inhibitors as well as substrates bind in this crevice between the domains. From the numerous studies of different inhibitors bound to serine proteinases we have chosen as an illustration the binding of a small peptide inhibitor, Ac-Pro-Ala-Pro-Tyr-COOH to a bacterial chymotrypsin (Figure 15.9). The enzyme-peptide complex was formed by adding a large excess of the substrate Ac-Pro-Ala-Pro-Tyr-CO-NH$_2$ to crystals of the enzyme. The enzyme molecules within the crystals catalyze cleavage of the terminal amide group to produce the products Ac-Pro-Ala-Pro-Tyr-COOH and NH$_3^+$. The ammonium ions diffuse away, but the peptide product remains bound as an inhibitor to the active site of the enzyme.

This inhibitor does not form a covalent bond to Ser 195 but one of its carboxy oxygen atoms is in the oxyanion hole forming hydrogen bonds to the main chain NH groups of residues 193 and 195. The tyrosyl side chain is positioned in the specificity pocket, which derives its specificity mainly from three residues, 216, 226, and 189, as we will see later. The main chain of the inhibitor forms a short stretch of antiparallel β sheet with residues 215–216 of the enzyme forming hydrogen bonds to the NH group of residue 215 and the CO group of residue 216.

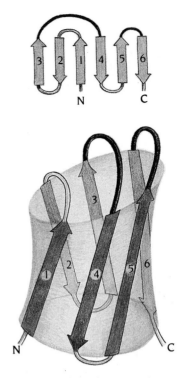

Figure 15.8 Topology diagrams of the domain structure of chymotrypsin. The chain is folded into a Greek key motif followed by a hairpin motif that are arranged as a six-stranded antiparallel β barrel.

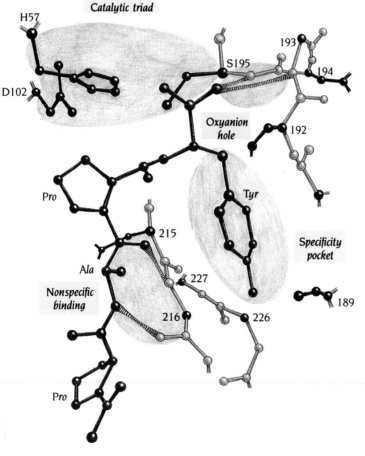

Figure 15.9 A diagram of the active site of chymotrypsin with a bound inhibitor, Ac-Pro-Ala-Pro-Tyr-COOH. The diagram illustrates how this inhibitor binds in relation to the catalytic triad, the substrate specificity pocket, the oxyanion hole and the nonspecific substrate binding region. The inhibitor is red. Hydrogen bonds between inhibitor and enzyme are striped. (Adapted from M.N.G. James et al., *J. Mol. Biol.* 144: 43–88, 1980.)

A closer examination of these essential residues, including the catalytic triad, reveals that they are all part of the same two loop regions in the two domains (Figure 15.10). The domains are oriented in such a way in the molecule that the ends of the two barrels that contain the Greek key crossover connection (described in Chapter 5) between β strands 3 and 4 face each other along the active site. The essential residues in the active site are in these two crossover connections and in the adjacent hairpin loops between β strands 5 and 6. Most of these essential residues are conserved between different members of the chymotrypsin superfamily. They are, of course, surrounded by other parts of the polypeptide chains, which provide minor modulations of the active site, specific for each particular serine proteinase.

His 57 and Ser 195 are within loop 3–4 of domains 1 and 2, respectively. The third residue in the catalytic triad, Asp 102, is within loop 5–6 of domain 1. The rest of the active site is formed by two loop regions (3–4 and 5–6) of domain 2. As in so many other protein structures described previously, the barrels apparently provide a stable scaffold to position a few loop regions that constitute the essential features of the active site.

Did the chymotrypsin molecule evolve by gene duplication?

Although the two domains of chymotrypsin have similar three-dimensional structures there is no amino acid sequence homology between them. Nevertheless, based on the argument that three-dimensional structure is more conserved than amino acid sequence, it has been suggested that the members of the chymotrypsin superfamily evolved by gene duplication of a single ancestral proteinase domain. The putative ancestral domain, obviously, could not have had the catalytic triad in present-day serine proteinases since the contemporary triad is derived from both domains. However, this is less of an obstacle to the gene-duplication hypothesis than it seems at first sight. The ancestral domain could have been a barrel structure similar to the second domain of chymotrypsin, which contains most of the essential features of the active site, including the reactive serine residue. We also now know from experiments with genetically engineered mutants in which the triad has been abolished that the catalytic triad is not absolutely essential for catalytic activity. As we will see later, these mutants retain some proteinase activity. It is, therefore, quite possible that there was a single ancestral gene specifying a single domain with some catalytic activity. This activity could then have been enhanced by a gene-duplication event followed by mutation and evolution leading to the catalytic triad of today. The fact that the active-site residues that comprise the catalytic triad of chymotrypsin and its relatives are clustered in the same two loop regions of domains 1 and 2 supports such an evolutionary history.

Different side chains in the substrate specificity pocket confer preferential cleavage

The serine proteinases all have the same substrate, namely, polypeptide chains of proteins. However, different members of the family cleave polypeptide chains preferentially at sites adjacent to different amino acid residues in the substrate. The structural basis for this preference lies in the side chains that line the substrate specificity pocket in the different enzymes.

This is nicely illustrated by members of the chymotrypsin superfamily; the enzymes chymotrypsin, trypsin, and elastase, have very similar three-dimensional structures but different specificity. They preferentially cleave adjacent to bulky aromatic side chains, positively charged side chains, and small uncharged side chains, respectively. Three residues, numbers 189, 216, and 226, are responsible for these preferences (Figure 15.11). Residues 216 and 226 line the sides of the pocket. In trypsin and chymotrypsin these are both glycine residues that allow side chains of the substrate to penetrate into the interior of the specificity pocket. In elastase they are Val and Thr, respectively, that fill up most

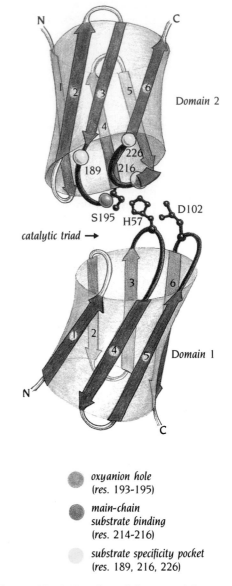

oxyanion hole
(res. 193-195)

main-chain
substrate binding
(res. 214-216)

substrate specificity pocket
(res. 189, 216, 226)

Figure 15.10 Topological diagram of the two domains of chymotrypsin, illustrating that the essential active-site residues are part of the same two loop regions (3–4 and 5–6) of the two domains (colored red). These residues form the catalytic triad (red), the oxyanion hole (green), and the substrate binding regions (yellow and blue) including essential residues in the specificity pocket.

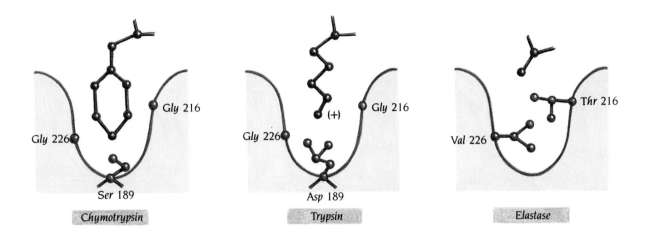

Figure 15.11 Schematic diagrams of the specificity pockets of chymotrypsin, trypsin and elastase, illustrating the preference for a side chain adjacent to the scissile bond in polypeptide substrates. Chymotrypsin prefers aromatic side chains and trypsin prefers positively charged side chains that can interact with Asp 189 at the bottom of the specificity pocket. The pocket is blocked in elastase, which therefore prefers small uncharged side chains.

of the pocket with hydrophobic groups (Figure 15.11). Consequently, elastase does not cleave adjacent to large or charged side chains but rather adjacent to small uncharged side chains.

Residue 189 is at the bottom of the specificity pocket. In trypsin the Asp residue at this position interacts with the positively charged side chains Lys or Arg of a substrate. This accounts for the preference of trypsin to cleave adjacent to these residues. In chymotrypsin there is a Ser residue instead at position 189, which does not interfere with the binding of the substrate. Bulky aromatic groups are therefore preferred by chymotrypsin since such side chains fill up the mainly hydrophobic specificity pocket. It has now become clear, however, from site-directed mutagenesis experiments that this simple picture does not tell the whole story.

Engineered mutations in the substrate specificity pocket change the rate of catalysis

How would substrate preference be changed if the glycine residues in trypsin at positions 216 and 226 were changed to alanine rather than to the more bulky valine and threonine groups that are present in elastase? This question was addressed by the groups of Charles Craik, William Rutter, and Robert Fletterick in San Francisco, who have made and studied three such trypsin mutants: one in which Ala is substituted for Gly at 216, one in which the same substitution is made at Gly 226, and a third containing both substitutions.

Model building shows that both Arg- and Lys-containing substrates can be accommodated by the substrate specificity pocket after these Gly to Ala changes but that some details of the binding mode at the bottom of the pocket would be altered. An enzyme with only the Ala 226 substitution would introduce a methyl group in the region where the end of the side chain binds (Figure 15.12) and would therefore be expected to accommodate Lys better than Arg, since the latter has a longer and more bulky side chain. Based on these steric arguments alone, one would therefore predict that the K_m for an Arg-containing substrate would be larger (less favorable binding) and that the K_m for Lys would be essentially unaltered. The specificity constant, k_{cat}/K_m, would decrease more for an Arg-containing substrate than for one with Lys.

Model building also predicts that the Ala 216 mutant would displace at the bottom of the specificity pocket a water molecule that in the wild type binds to the NH_3+ group of the substrate Lys side chain (Figure 15.12). This extra CH_3 group of the mutant is not expected to disturb the binding of the Arg side chain. One would therefore expect that the K_m for Lys substrates would increase and therefore k_{cat}/K_m would decrease more for Lys- than for Arg-containing substrates. For the double mutant where both Gly 216 and Gly 226 are changed to Ala, one would predict an increase in the K_m values for both Lys- and Arg-containing substrates.

239

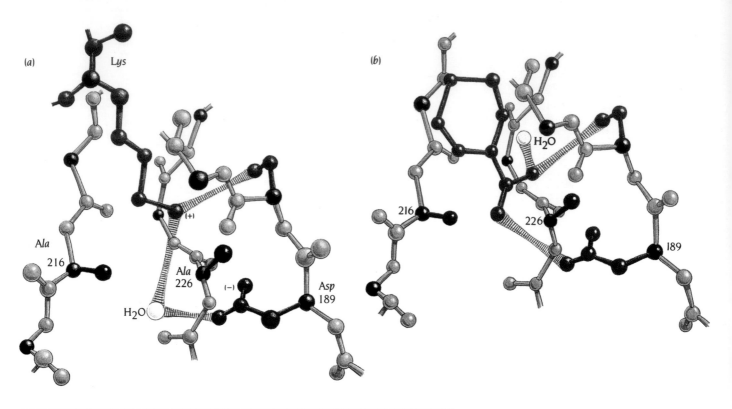

Lys

(+)

Ala
216

Ala
226

(−)

H₂O

Asp
189

(b)

H₂O

216

226

189

Table 15.1
Kinetic data for wild-type and mutant trypsins

Enzyme	Arg			Lys			
	k_{cat}	K_m	k_{cat}/K_m	k_{cat}	K_m	k_{cat}/K_m	$\dfrac{(k_{cat}/K_m)\ \text{Arg}}{(k_{cat}/K_m)\ \text{Lys}}$
Wild type	1	1	1	0.9	10	0.1	10
Gly216, Gly226—Ala	0.001	15	0.0001	0.0005	25	0.00002	25
Gly 226—Ala	0.01	35	0.0003	0.1	250	0.0005	0.5
Gly 216—Ala	0.7	30	0.02	0.2	280	0.001	20

The substrates used were D-Val-Leu-Arg-amino fluorocoumarin (Arg) and D-Val-Leu-Lys-amino fluorocoumarin (Lys). For clarity the K_m and k_{cat} values have been normalized to those of the wild-type enzyme for the Arg substrate.

Figure 15.12 Schematic diagram of the specificity pocket of mutant trypsin with Ala (purple) at positions 216 and 226. (a) The position of a Lys side chain (red) in this pocket as observed in the structure of a complex between trypsin (green) and a peptide inhibitor. The NH₃⁺ group of the Lys side chain interacts with the COO⁻ group of Asp 189 through a water molecule. (b) No structure is available for an Arg side chain in the substrate specificity pocket of trypsin. It is assumed that the complex of trypsin (green) with benzamide (red) is a good model for arginine binding in this pocket. One NH₂ group of benzamide interacts directly with the COO⁻ group of Asp 189 and the second NH₂ group interacts with a water molecule and the OH group of Ser 190. (Adapted from C.S. Craik et al., *Science* 228: 293, 1985.)

The kinetic results are shown in Table 15.1. The dramatic kinetic effects of these mutants are best illustrated for the Arg substrates. The three mutants have roughly similar K_m values 15–35 times higher than wild type, but the k_{cat} values decrease by factors of 10 to about 1000. The mutants were designed to change the specificity, but by far the largest changes occur in the catalytic rates. Apparently, these mutations affect the structure of the enzyme in additional ways, possibly by conformational changes outside the specificity pocket, so that the stabilization of the transition state is reduced and consequently the activation energies for the reactions are different.

The changes in the specificity constants, on the other hand, were as expected from the predictions. The ratio of the k_{cat}/K_m values for the Arg and Lys substrates (last column in Table 15.1) gives a measure of the relative specificities. This ratio decreases for the Ala 226 mutant and increases for the Ala 216 mutant as predicted. However, the changes in these values depend not only on changes in the K_m values, which reflect binding of substrate, but even more on changes in the k_{cat} values, which reflect catalytic rate. It can, therefore, be argued that the agreement with prediction is fortuitous.

The simple lesson to be learnt from these experiments is that critical amino acid residues can have pleiotropic roles in determining a protein's structure and therefore its function.

The Asp 189-Lys mutant in trypsin has unexpected changes in substrate specificity

Asp 189 at the bottom of the substrate specificity pocket interacts with Lys and Arg side chains of the substrate, and this is the basis for the preferred cleavage sites of trypsin (Figure 15.11 and 15.12). It is almost trivial to infer, from these observations, that a replacement of Asp 189 with Lys would produce a mutant that would prefer to cleave substrates adjacent to negatively charged residues, especially Asp. On a computer display, similar Asp-Lys interactions between enzyme and substrate can be modeled within the substrate specificity pocket but reversed compared with the wild-type enzyme.

The results of experiments in which the mutation was made were, however, a complete surprise. The Asp 189-Lys mutant was totally inactive with both Asp and Glu substrates. It was, as expected, also inactive toward Lys and Arg substrates. The mutant was, however, catalytically active with Phe and Tyr substrates, with the same low turnover number as wild-type trypsin. On the other hand, it showed a more than 5000-fold increase in k_{cat}/K_m with Leu substrates over wild type. The three-dimensional structure of this interesting mutant has not yet been determined, but one possible explanation for its kinetic behavior, which was not predicted before the mutant was made, is that the tip of the Lys 189 side chain is directed away from the substrate specificity pocket to avoid having the positively charged group in the hydrophobic pocket.

As these experiments with engineered mutants of trypsin prove, we still have far too little knowledge of the functional effects of even single point mutations to be able to make really accurate and comprehensive predictions of the properties of a point-mutant enzyme, even in the case of such well-characterized enzymes as the serine proteinases. Predictions of the properties of mutations based on computer modeling are not infallible. Once constructed, the mutant enzymes often exhibit properties that are entirely surprising, but they may be correspondingly informative.

The structure of the serine proteinase subtilisin is of the α/β type

Subtilisins are a group of serine proteinases that are produced by different species of bacilli. These enzymes are of considerable commercial interest because they are added to the detergents in washing powder to facilitate removal of proteinaceous stains. Numerous attempts have therefore recently been made to change by protein engineering such properties of the subtilisin molecule as its thermal stability, pH optimum, and specificity. In fact, in 1988 subtilisin mutants were the subject of the first U.S. patent granted for an engineered protein.

The subtilisin molecule is a single polypeptide chain of 275 amino acids with no similarities in the amino acid sequence to chymotrypsin. The three-dimensional structure of subtilisin BPN′ from *Bacillus amyloliquefaciens* was determined in 1969 by the group of Joseph Kraut in San Diego, California, and that of subtilisin Novo in 1971 by the group of Jan Drenth in Groningen, the Netherlands. The main feature of the subtilisin structure is a region of five parallel β strands (blue in Figure 15.13) surrounded by four helices, two on each side of the parallel β sheet. This α/β structure is thus quite different from the double antiparallel β-barrel structure of chymotrypsin (cf. Figure 15.7).

The active sites of subtilisin and chymotrypsin are similar

The active site of subtilisin is outside the carboxy ends of the central β strands analogous to the position of the binding sites in other α/β proteins as discussed in Chapter 4. Details of this active site are surprisingly similar to those of chymotrypsin, in spite of the completely different folds of the two enzymes (Figures 15.14 and 15.9). A catalytic triad is present that comprises residues Asp

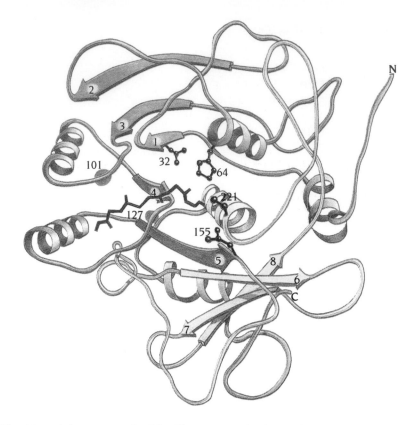

Figure 15.13 Schematic diagram of the three-dimensional structure of subtilisin viewed down the central parallel β sheet. The N-terminal region that contains the α/β structure is blue. It is followed by a yellow region, which ends with the fourth α helix of the α/β structure. The C-terminal part is green. The catalytic triad Asp 32, His 64, and Ser 221 as well as Asn 155, which forms part of the oxyanion hole are shown in purple. The main chain of part of a polypeptide inhibitor is shown in red. Main-chain residues around 101 and 127 (orange circles) form the nonspecific binding regions of peptide substrates.

32, His 64, and the reactive Ser 221. The negatively charged oxygen atom of the tetrahedral transition state binds in an oxyanion hole, where the amide group of the side chain of Asn 155 provides a hydrogen bond in addition to the main-chain nitrogen atom of Ser 221. Peptide substrates and inhibitors bind nonspecifically by forming a small antiparallel pleated sheet, which in subtilisin comprises three β strands (Figure 15.14). There is also a hydrophobic specificity pocket adjacent to the scissile bond.

All the four essential features of the active site of chymotrypsin are thus also present in subtilisin. Furthermore, these features are spatially arranged in the same way in the two enzymes, even though different framework structures bring

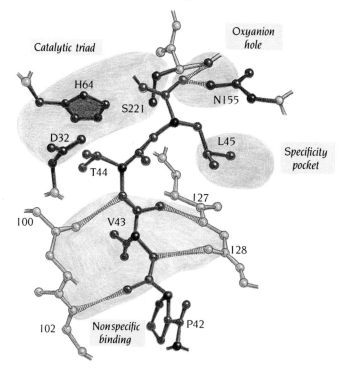

Figure 15.14 Schematic diagram of the active site of subtilisin. A region (residues 42–45) of a bound polypeptide inhibitor, eglin, is shown in red. The four essential features of the active site—the catalytic triad, the oxyanion hole, the specificity pocket, and the region for nonspecific binding of substrate—are highlighted in yellow. Important hydrogen bonds between enzyme and inhibitor are striped. This figure should be compared to Figure 15.9, which shows the same features for chymotrypsin. (Adapted from W. Bode et al., *EMBO J.* 5: 813, 1986.)

different loop regions into correct positions in the active site. This is a classical example of convergent evolution at the molecular level.

A structural anomaly in subtilisin has functional consequences

There is one anomalous and puzzling feature of the subtilisin structure. We mentioned in Chapter 4 that all β-α-β motifs were of the same hand, they were right-handed. Subtilisin contains the one exception to this general rule, which is illustrated in the topology diagram Figure 15.15. There are three β-α-β motifs in subtilisin, β_2-α_B-β_3, β_3-α_C-β_4, and β_4-α_D-β_5. If these motifs were of the same hand, the three α helices α_B, α_C, and α_D would be on the same side of the β sheet. However, α_B is beneath the sheet in the topology diagram in contrast to the other two helices because β_2-α_B-β_3 is left-handed. Why has this exception to the general rule of right-handed β-α-β motifs evolved?

The answer is quite clear. His 64, which is part of the catalytic triad, is in the first turn of helix α_B (Figure 15.13). This helix would be on the other side of the β sheet, far removed from the active site if the motif β_2-α_B-β_3 were right-handed. Therefore, to produce a proper catalytic triad of Asp 32, His 64, and Ser 221, helix α_B must be on the same side of the β sheet as Ser 221; consequently, the motif has evolved to be left-handed.

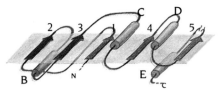

Figure 15.15 Topology diagram of the α/β region of subtilisin illustrating that β_2-α_B-β_3 has a different hand than the other β-α-β units.

Transition-state stabilization in subtilisin is dissected by protein engineering

Two essential features are required to stabilize the covalent tetrahedral transition state in serine proteinases—the oxyanion hole, which provides hydrogen bonds to the negatively charged oxygen atom in the transition state, and the histidine residue of the catalytic triad, which provides a positive charge. The charge on this histidine is, in turn, stabilized by the aspartic acid side chain of the catalytic triad (Figure 15.6). The histidine residue also plays a second role in the catalytic mechanism by accepting a proton from the reactive serine residue and then donating that proton to the nitrogen atom of the leaving group. The effects on the catalytic rate of the different side chains involved in the catalytic triad and the oxyanion hole have been examined by P. Carter, J.A. Wells, and D. Estell at Genentech, USA, by analyses of mutants of subtilisin where one or several of these side chains have been changed.

Catalysis occurs without a catalytic triad

By changing Ser 221 in subtilisin to Ala the reaction rate (both k_{cat} and k_{cat}/K_m) is reduced by a factor of about 10^6 compared with the wild-type enzyme. The K_m value and, by inference, the initial binding of substrate are essentially unchanged. This mutation prevents formation of the covalent bond to the substrate and therefore abolishes the reaction mechanism outlined in Figure 15.5. When the Ser 221 to Ala mutant is further mutated by changes of His 64 to Ala or Asp 32 to Ala or both, there is no effect on the catalytic reaction rate as expected, since the reaction mechanism that involves the catalytic triad is no longer in operation. However, the enzyme still has an appreciable catalytic effect, peptide hydrolysis is still about 10^3–10^4 times the nonenzymatic rate. Whatever the reaction mechanism used by these mutants, it is apparent that the remaining parts of the active site must bind more tightly to the substrate in its transition state than in its initial state and thereby increase the reaction rate above that in the absence of enzyme.

Substrate molecules provide catalytic groups in substrate-assisted catalysis

The single mutation His 64-Ala decreases the reaction rate of subtilisin for most substrates by the same factor (~10^6) as the mutation of Ser 221. This histidine (His 64), therefore, seems to be as important as Ser 221 for the formation of a

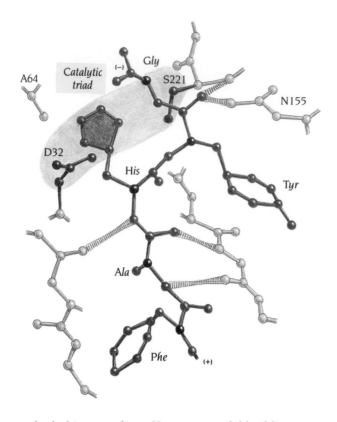

Figure 15.16 Substrate-assisted catalysis. Schematic diagram from model building of a substrate NH$_2$-Phe-Ala-His-Tyr-Gly-COOH (red) bound to the subtilisin mutant His 64-Ala. The diagram illustrates that the His residue of the substrate can occupy roughly the same position in this mutant as His 64 in wild-type subtilisin (Figure 15.14) and thereby partly restore the catalytic triad.

covalent tetrahedral intermediate. However, model building suggested that it might be possible at least partly to compensate for the loss of this histidine in the catalytic triad of the mutant protein with a histidine side chain from a peptide substrate (Figure 15.16). Experiments confirmed this prediction and showed that the mutant His 64-Ala catalyzes hydrolysis of a peptide substrate about 400 times faster when the peptide has histidine at the appropriate position in the sequence. The rate is, however, still several orders of magnitude below the rate of the wild-type enzyme, presumably because of the slightly different position and orientation of the histidine side chain. Nevertheless, the principle of **substrate-assisted catalysis** has been demonstrated; an essential group that is lacked by a mutant enzyme can be replaced by a similar group from the substrate. One consequence of substrate-assisted catalysis is that the mutant enzyme is highly specific for substrates containing the essential group. The His 64-Ala mutant of subtilisin, for example, has a specificity factor (ratios of k_{cat}/K_m) of about 200 for substrates containing histidine.

The single mutation Asp 32-Ala reduces the catalytic reaction rate by a factor of about 10^4 compared with wild type. This rate reduction reflects the role of Asp 32 in stabilizing the positive charge that His 64 acquires in the transition state. A similar reduction of k_{cat} and k_{cat}/K_m (2.5×10^3) is obtained for the single mutant Asn 155-Thr. Asn 155 provides one of the two hydrogen bonds to the substrate transition state in the oxyanion hole of subtilisin. Model building shows that the OH group of Thr in the mutant is too far away to provide such a hydrogen bond. The loss of this feature of the stabilization of the transition state thus reduces the rate by more than 1000-fold.

The subtilisin mutants described here illustrate the power of protein engineering as a tool to allow us to identify the specific roles of side chains in the catalytic mechanisms of enzymes. In Chapter 16 we will discuss the utility of protein engineering in other contexts, such as design of novel proteins and the elucidation of the energetics of ligand binding to proteins.

Conclusion

Enzymes increase the rate of chemical reactions by decreasing the activation energy of these reactions. This is achieved primarily by preferential binding to the transition state of the substrate. Catalytic groups of the enzyme are required to achieve a specific reaction path for the conversion of substrate to product.

Serine proteinases such as chymotrypsin and subtilisin catalyze cleavage of peptide bonds. Four features essential for catalysis are present in the three-dimensional structures of all serine proteinases: a catalytic triad, an oxyanion binding site, a substrate specificity pocket, and a nonspecific binding site for polypeptide substrates. These four features in a very similar arrangement are present in both chymotrypsin and subtilisin even though they are achieved in the two enzymes in completely different ways by quite different three-dimensional structures. Chymotrypsin is built up from two β-barrel domains, whereas the subtilisin structure is of the α/β type. These two enzymes provide an example of convergent evolution where completely different loop regions, attached to different framework structures, form similar active sites.

The catalytic triad consists of the side chains of Asp, His, and Ser close to each other. The Ser residue is reactive and forms a covalent bond with the substrate, thereby providing a specific pathway for the reaction. His has a dual role: first, it accepts a proton from Ser to facilitate formation of the covalent bond; and, second, it stabilizes the negatively charged transition state. The proton is subsequently transferred to the N atom of the leaving group. Mutations of either of these two residues decrease the catalytic rate by a factor of 10^6 because they abolish the specific reaction pathway. Asp, by stabilizing the positive charge of His, contributes a rate enhancement of 10^4.

The oxyanion binding site stabilizes the transition state by forming two hydrogen bonds to a negatively charged oxygen atom of the substrate. Mutations that prevent formation of one of these bonds in subtilisin decrease the rate by a factor of about 10^3.

Mutations in the specificity pocket of trypsin, designed to change the substrate preference of the enzyme, also have drastic effects on the catalytic rate. These mutants demonstrate that the substrate specificity of an enzyme and its catalytic rate enhancement are tightly linked to each other because both are affected by the difference in binding strength between the transition state of the substrate and its normal state.

Selected readings

General

Blow, D.M. Structure and mechanism of chymotrypsin. *Acc. Chem. Res.* 9: 145–152, 1976.

Fersht, A. Enzyme structure and mechanism, 2nd ed. New York: W.H. Freeman, 1984.

Huber, R.; Bode, W. Structural basis of the activation and action of trypsin. *Acc. Chem. Res.* 11: 114–122, 1978.

James, M.N.G. An x-ray crystallographic approach to enzyme structure and function. *Can. J. Biochem.* 58: 251–270, 1980.

Jencks, W.P. Binding energy, specificity, and enzymatic catalysis: the Circe effect. *Adv. Enzymol.* 43: 219–410, 1975.

Knowles, J.R. Tinkering with enzymes: what are we learning? *Science* 236: 1252–1258, 1987.

Kraut, J. Serine proteases: structure and mechanism of catalysis. *Annu. Rev. Biochem.* 46: 331–358, 1977.

Kraut, J. How do enzymes work? *Science* 242: 533–540, 1988.

Neurath, H. Evolution of proteolytic enzymes. *Science* 224: 350–357, 1984.

Pauling, L. Nature of forces between large molecules of biological interest. *Nature* 161: 707–709, 1948.

Steitz, T.A.; Shulman, R.G. Crystallographic and NMR studies of the serine proteases. *Annu. Rev. Biochem. Biophys.* 11: 419–444, 1982.

Stroud, R.M. A family of protein-cutting proteins. *Sci. Am.* 231(1): 74–88, 1974.

Walsh, C. Enzymatic reaction mechanisms. New York: W.H. Freeman, 1979.

Warshel, A., et al. How do serine proteases really work? *Biochemistry* 28: 3629–3637, 1989.

Wells, J.A., et al. On the evolution of specificity and catalysis in subtilisin. *Cold Spring Harbor Symp. Quant. Biol.* 52: 647–652, 1987.

Specific structures

Bode, W., et al. Refined 1.2 Å crystal structure of the complex formed between subtilisin Carlsberg and the inhibitor eglin c. Molecular structure of eglin and its detailed interaction with subtilisin. *EMBO J.* 5: 813–818, 1986.

Bode, W., et al. The refined 1.9 Å crystal structure of human α-thrombin: interaction with D-Phe-Pro-Arg chloromethylketone and significance of the Tyr-Pro-Pro-Trp insertion segment. *EMBO J.* 8: 3467–3475, 1989.

Bryan, P., et al. Site-directed mutagenesis and the role of the oxyanion hole in subtilisin. *Proc. Natl. Acad. Sci. USA* 83: 3743–3745, 1986.

Carter, P.; Wells, J.A. Engineering enzyme specificity by "substrate-assisted catalysis." *Science* 237: 394–399, 1987.

Carter, P.; Wells, J.A. Dissecting the catalytic triad of a serine protease. *Nature* 332: 564–568, 1988.

Craik, C.S., et al. Redesigning trypsin: alteration of substrate specificity. *Science* 228: 291–297, 1985.

Craik, C.S., et al. The catalytic role of the active site aspartic acid in serine proteases. *Science* 237: 909–913, 1987.

Cunningham, B.C.; Wells, J.A. Improvement in the alkaline stability of subtilisin using an efficient random mutagenesis and screening procedure. *Prot. Eng.* 1: 319–325, 1987.

Drenth, J., et al. Subtilisin novo. The three-dimensional structure and its comparison with subtilisin BPN. *Eur. J. Biochem.* 26: 177–181, 1972.

Estell, D.A.; Graycar, T.P.; Wells, J.A. Engineering an enzyme by site-directed mutagenesis to be resistant to chemical oxidation. *J. Biol. Chem.* 260: 6518–6521, 1985.

Fehlhammer, H.; Bode, W.; Huber, R. Crystal structure of bovine trypsinogen at 1.8 Å resolution. II. Crystallographic refinement, refined crystal structure and comparison with bovine trypsin. *J. Mol. Biol.* 111: 415–438, 1977.

Fujinaga, M., et al. Crystal and molecular structures of the complex of α-chymotrypsin with its inhibitor turkey ovomucoid third domain at 1.8 Å resolution. *J. Mol. Biol.* 195: 397–418, 1987.

Graf, L., et al. Selective alteration of substrate specificity by replacement of aspartic acid 189 with lysine in the binding pocket of trypsin. *Biochemistry* 26: 2616–2623, 1987.

Grütter, M.G., et al. Crystal structure of the thrombin-hirudin complex: a novel mode of serine protease inhibition. *EMBO J.* 9: 2361–2365, 1990.

James, M.N.G., et al. Structures of product and inhibitor complexes of *Streptomyces griseus* protease A at 1.8 Å resolution. A model for serine protease catalysis. *J. Mol. Biol.* 144: 43–88, 1980.

Krieger, M.; Kay, L.M.; Stroud, R.M. Structure and specific binding of trypsin: comparison of inhibited derivatives and a model for substrate binding. *J. Mol. Biol.* 83: 209–230, 1974.

Matthews, B.W.; Sigler, P.B.; Henderson, R.; Blow, D.M. Three-dimensional structure of tosyl-a-chymotrypsin. *Nature* 214: 652–656, 1967.

McLachlan, A.D. Gene duplications in the structural evolution of chymotrypsin. *J. Mol. Biol.* 128: 49–79, 1979.

Poulos, T.L., et al. Polypeptide halomethyl ketones bind to serine proteases as analogs of the tetrahedral intermediate. *J. Biol. Chem.* 251: 1097–1103, 1976.

Read, R.J.; James, M.N.G. Refined crystal structure of *Streptomyces griseus* trypsin at 1.7 Å resolution. *J. Mol. Biol.* 200: 523–551, 1988.

Rühlman, A., et al. Structure of the complex formed by bovine trypsin and bovine pancreatic trypsin inhibitor. *J. Mol. Biol.* 77: 417–436, 1973.

Shotton, D.M.; Watson, H.C. Three-dimensional structure of tosyl-elastase. *Nature* 225: 811–816, 1970.

Sigler, P.B., et al. Structure of crystalline α-chymotrypsin II. A preliminary report including a hypothesis for the activation mechanism. *J. Mol. Biol.* 35: 143–164, 1968.

Smith, S.O., et al. Crystal versus solution structures of enzymes: NMR spectroscopy of a crystalline serine protease. *Science* 244: 961–964, 1989.

Sprang, S., et al. The three-dimensional structure of Asn[102] mutant of trypsin: role of Asp[102] in serine protease catalysis. *Science* 237: 905–909, 1987.

Thomas, P.G.; Russel, A.J.; Fersht, A. Tailoring the pH dependence of enzyme catalysis using protein engineering. *Nature* 318: 375–376, 1985.

Tsukada, H.; Blow, D.M. Structure of α-chymotrypsin refined at 1.68 Å resolution. *J. Mol. Biol.* 184: 703–711, 1985.

Wang, D.; Bode, W.; Huber, R. Bovine chymotrypsinogen A. X-ray crystal structure analysis and refinement of a new crystal form at 1.8 Å resolution. *J. Mol. Biol.* 185: 595–624, 1985.

Wells, J.A., et al. Designing substrate specificity by protein engineering of electrostatic interactions. *Proc. Natl. Acad. Sci. USA* 84: 1219–1223, 1987.

Wells, J.A., et al. Recruitment of substrate-specificity properties from one enzyme into a related one by protein engineering. *Proc. Natl. Acad. Sci. USA* 84: 5167–5171, 1987.

Wright, C.S.; Alden, R.A.; Kraut, J. Structure of subtilisin BPN' at 2.5 Å resolution. *Nature* 221: 235–242, 1969.

Prediction, Engineering, and Design of Protein Structures

16

Over a period of more than 3 billion years a large variety of protein molecules has evolved to run the complex machinery of present-day cells and organisms. Most of us believe that these molecules have evolved by random mutation of genes and natural selection for those gene products that have conferred some functional advantage contributing to the survival of individual organisms.

Long before Darwin and Wallace proposed the theory of evolution and Mendel discovered the laws of genetics, plant and animal breeders had begun to interfere with the process of evolution in the species that gave rise to domesticated animals and cultivated plants. Considering their total lack of knowledge of both evolutionary theory and genetics, their achievements, brought about by forcing the pace of and subverting natural selection, were impressive albeit very gradual. With the advent of molecular genetics and in particular techniques for gene cloning and gene insertion, we are now entering an era of genetic exploitation of other organisms undreamed of only 50 years ago. We can now begin to design genes to produce in other organisms novel gene products for the benefit of human beings; we are no longer restricted to selecting useful genes that arise by mutation. We are, however, only at the beginning of this new era, and so far we have only scratched the surface of the knowledge that is required for true engineering and design of protein molecules. We distinguish **protein engineering**, by which we mean mutating the gene of an existing protein in an attempt to alter its function in a *predictable* way, from **protein design**, which has the more ambitious goal of designing *de novo* a protein to fulfill a desired function.

Protein engineers frequently have been surprised by the range of effects caused by single mutations that they hoped would change only one specific and simple property in enzymes; some examples are described in Chapter 15. The often surprising results of such experiments reveal how little we know about the rules of protein stability and the energetics of ligand binding and catalytic efficiency; they also serve to emphasize how difficult it is to design *de novo* stable proteins with specific functions. However, by using the methods of engineering and design, we are now at least increasing rapidly our basic knowledge of the function of protein molecules. For example, we now know that the difference in energetic terms between the transition states of a naturally evolved useful enzyme and an engineered useless mutant corresponds to less than the energy of a single hydrogen bond, even for such important life-sustaining enzymes as the CO_2-fixing enzyme in green plants, rubisco (ribulose–1,5-bisphosphate carboxylase/oxygenase).

Knowledge of a protein's tertiary structure is a prerequisite for the proper engineering of its function. Unfortunately, in spite of recent significant technological advances, the experimental determination of tertiary structure is still slow compared with the rate of accumulation of amino acid sequence data. This makes the folding problem, the successful prediction of a protein's tertiary structure from its amino acid sequence, central to rapid progress in protein engineering and design. We will, therefore, in this chapter first examine the current status of methods for the prediction of tertiary structure before giving some examples of protein engineering and protein design.

Prediction of protein structure from sequence is an unsolved problem

How to predict the three-dimensional structure of a protein from its amino acid sequence is the major unsolved problem in structural molecular biology. We would like to have a computer program that could simulate the action of the simple physical laws that operate in a test tube or a living cell when a polypeptide chain with a specific amino acid sequence folds into a precise three-dimensional structure. Why is this prediction of protein folding so difficult? The answer is usually formulated in terms of the complexity of the task of searching through all the possible conformations of a polypeptide chain to find those with low energy. It requires enormous amounts of computing time.

Others have argued that there must be a pattern of amino acid side chains at a limited number of positions along the polypeptide chain that determines a specific domain structure, and they are searching for such patterns among the known structures. Only limited success has been achieved so far by this method, and no general solution seems to be in sight.

Without a general method for predicting the tertiary structure of an amino acid sequence, what can we do when we want to learn something about the structure and function of a protein whose gene has been newly cloned and sequenced? The obvious first step is to try to obtain information by comparing the new sequence with the sequences of other already characterized proteins. Several sequence comparison methods are available and later some of them will be described, as well as a few selected examples of predicted three-dimensional structures, but first we should discuss the difficulties inherent in the methods of **structure prediction** based upon **pattern recognition**.

Many different amino acid sequences give similar three-dimensional structures

How many completely different amino acid sequences might give a similar three-dimensional structure for an average-sized domain of 150 amino acid residues? Simple combinatorial calculations show that there are a total of 20^{150} or roughly 10^{200} possible amino acid sequences for such a domain, given the 20 different amino acids in natural proteins. This number is much larger than the number of atoms in the known universe. A more laborious calculation shows that out of these 10^{200} possible combinations we can extract about 10^{38} members that have less than 20% amino acid sequence identity with each other and that therefore can be considered to have completely different sequences. In other words, there are 10^{38} nonrelated ways of constructing a domain of 150 amino acids using the 20 standard amino acids as building blocks. We do not know how many of these can form a stable three-dimensional structure, but let us assume that one out of a hundred can fold into a stable structure; we are left with 10^{36} folded proteins. In the previous chapters we have seen that simple structural motifs arrange themselves into a limited number of topologically different domain structures. It has been estimated on reasonable grounds that there are fewer than 1000 topologically different domain structures. Since there are 10^{36} possible nonhomologous sequences that might fold into 10^3 different structures, it follows that there are of the order of 10^{33} completely different side-chain arrangements that can give similar polypeptide folds. This is a very large number that happens to be roughly equal to the number of spheres of 0.1 mm in radius that could be packed into the volume of the earth. Even if we assume that only one in a million of the nonhomologous sequences can fold into a stable structure, there are still 10^{29} totally different side-chain arrangements for a domain of 150 amino acids.

For each of the about 100 different domain structures that have so far been observed, we might at best know about a dozen of these different possible sequences. It is not trivial to recognize the general sequence patterns that are common to specific domain structures from such a limited knowledge base.

Amino acid sequence homology implies similarity in structure and functions

Proteins with homologous amino acid sequences have similar three-dimensional structures. Usually, they also have similar functions although there are some exceptions known, where genes for ancient enzymes have been recruited at a later stage in evolution to produce proteins with quite different functions. An example is provided by one of the structural components in the eye lens that is homologous to the ancient glycolytic enzyme lactate dehydrogenase. Once a novel gene has been cloned and sequenced, a search for amino acid sequence homology between the corresponding protein and other known protein sequences should be made. Usually, this is done by comparison with databases of known protein sequences using one of the standard sequence alignment computer programs.

If amino acid sequence homology is found with a protein of known crystal structure, a three-dimensional model of the novel protein can be constructed in a computer display, on the basis of the sequence alignment and the known three-dimensional structure. This model can then serve as an excellent basis for identifying amino acid residues involved in the active site or in antigenic epitopes, and the model can be used for protein engineering, drug design, or immunological studies.

Since the sequence databases are large, currently comprising around 17,000 known protein sequences, the standard sequence alignment programs have been designed to provide a compromise between the speed and the accuracy of the search. As a result, they work well only when there is a reasonably high degree of sequence identity, usually of the order of 30% or more. Much more sensitive programs have been written which search for both identity and conserved structural properties and also for relatedness in different physical properties, but these inevitably require far more computing time. Carefully used, such programs can identify structural and functional homology where the standard programs fail to do so.

Homologous proteins have conserved structural cores and variable loop regions

Homologous proteins always contain a core region where the general folds of the polypeptide chains are very similar. This core region contains mainly the secondary structure elements that build up the interior of the protein: in other words, the **scaffolds** of homologous proteins have similar three-dimensional structures. Even distantly related proteins with low-sequence homology have similar scaffold structures although minor adjustments occur in the positions of the secondary structure elements to accommodate differences in the arrangements of the hydrophobic side chains in the interior of the protein. The greater the sequence homology, the more closely related are the scaffold structures (Figure 16.1). This has important implications for model building of homologous proteins; the more distantly related two proteins are, the greater the adjustments required to the scaffold in the model structure.

Loop regions that connect the building blocks of scaffolds can vary considerably both in length and in structure. The problem of predicting the three-dimensional structure of a protein whose sequence is homologous to a

Figure 16.1 The relation between the divergence of amino acid sequence and three-dimensional structure of the core region of homologous proteins. Known structures of 32 pairs of homologous proteins such as globins, serine proteinases, and immunoglobulin domains have been compared. The root mean square deviation of the main-chain atoms of the core regions is plotted as a function of amino acid homology (red dots). The curve represents the best fit of the dots to an exponential function. Pairs of high-sequence homology are almost identical in three-dimensional structure, whereas deviations in position for pairs of low homology are of the order of 2 Å. (From C. Chothia and A. Lesk, *EMBO J.* 5: 823, 1986.)

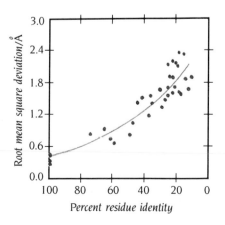

protein of known three-dimensional structure is therefore mainly a question of predicting the structure of loop regions, after the scaffold has been adjusted. As mentioned in Chapter 2, loop regions do not have random structures, and their main-chain conformations cluster in sets of similar structures. The conformation of each set depends more on the number of amino acids in the loop and the type of secondary structure elements that it connects, whether they are α-α, β-β, α-β, or β-α connections, than on the actual amino acid sequences. Therefore it is possible to use a **database of loop regions** from proteins of known structure to obtain a preliminary model of the loops of an unknown structure. To model a protein structure, suitable main-chain loop conformations from this database are attached to the scaffold modeled to have a structure similar to that of the known homologous protein. Finally, the conformations of the side chains of the loops in the model are predicted by a combination of model building and energy refinement.

The most clear-cut example of the use of such procedures has been in modeling antigenic binding sites in immunoglobulins. These binding sites are built up from three hypervariable loop regions, CDR1–CDR3, from the variable domains of both the light and the heavy chains of immunoglobulins as described in Chapter 12. There is usually high-sequence homology within the scaffolds of the variable domains in different immunoglobulin molecules. Consequently, the scaffold of variable domains of known three-dimensional structure can be used in modeling a new monoclonal antibody with a known amino acid sequence. However, the CDR regions are usually very different in sequence from those of any other known antibody, and their three-dimensional structures must be predicted. By comparing known antibody structures and sequences, it has been shown that there is only a small repertoire of main-chain conformations for at least five of the six CDR regions and that the particular conformation adopted is determined by a few key conserved residues for each loop conformation. For example, three different conformations were found for the CDR3 regions of the light chains in the nine known x-ray structures. More than 90% of the 204 known sequences of light-chain CDR3 regions obey the sequence constraints of one or other of these three conformations. By using this repertoire of loop conformations, considerable success has been achieved in correctly predicting the structure of antigen-binding surfaces. An example of such a prediction compared with the actual structure, subsequently determined, is given in Figure 16.2.

Figure 16.2 An example of prediction (*top row*) of the conformations of three CDR regions of a monoclonal antibody compared with the unrefined x-ray structure (*bottom row*). L1 and L2 are CDR regions of the light chain, and H1 is from the heavy chain. The amino acid sequences of the loop regions in the modeled antibody (*top*) are compared with the sequences of the loop regions selected from a database of known antibody structures (*bottom*). The three-dimensional structure of two of the loop regions, L1 and L2, were in good agreement with the preliminary x-ray structure, whereas H1 was not. However, during later refinement of the x-ray structure it was found that there were errors in the conformations of H1, and in the refined x-ray structure this loop was found to agree with the predicted conformations. In fact, all six loop conformations were correctly predicted in this case. (From C. Chothia et al., *Science* 233: 755–758, 1986.)

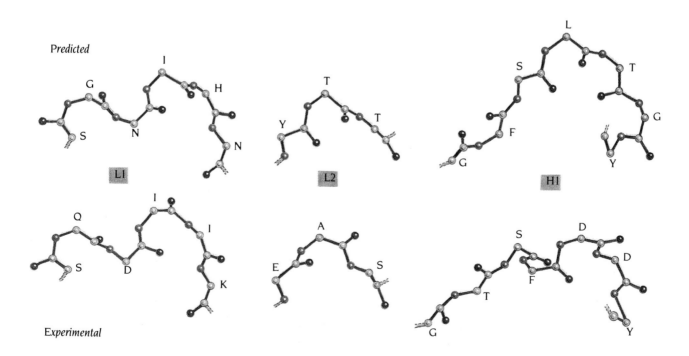

Knowledge of secondary structure is necessary for prediction of tertiary structure

What can be done by predictive methods if the homology search fails to reveal any sequence homology with a protein of known tertiary structure? Is it possible to model a tertiary structure from the amino acid sequence alone? There are no methods available today to do this and obtain a model detailed enough to be of any use, for example, in drug design and protein engineering. This is, however, a very active area of research and quite promising results are being obtained; in some cases it is possible to predict correctly the type of protein, α, β, or α/β, and even to derive approximations to the correct fold.

Today's **predictive methods** all rely on prediction of secondary structure: in other words, which amino acid residues are α helical and which are in β strands. We have previously emphasized that secondary structure cannot in general be predicted with a high degree of confidence with the possible exception of transmembrane helices. This imposes a basic limitation on the prediction of tertiary structure. Once the correct secondary structure is known, we know enough about the rules for packing elements of secondary structure against each other (see Chapter 2 for helix packing) to derive a very limited number of possible stable globular folds. Consequently, secondary structure prediction lies at the heart of the prediction of tertiary structure from the amino acid sequence.

Unfortunately for predictive methods, secondary and tertiary structures are closely linked in the sense that global tertiary structure imposes local secondary structure at least in some regions of the polypeptide chain. The ability of a specific short sequence of amino acids to form an α helix, a β strand, or a loop region is dependent not only on the sequence of that region but also on its environment in the three-dimensional structure. For example, by analyzing all the known tertiary structures, it has been shown that peptide regions of up to five residues long with identical amino acid sequences are α-helical in one structure and a β strand or a loop in other structures. While this interdependence of secondary and tertiary structure complicates secondary structure predictions, it can, sometimes, be used to improve such predictions, by an iterative scheme in which a preliminary assignment of secondary structure is used to predict the type of domain structure, for example, four helix bundle, α/β barrels, etc. The structure type of the domain imposes additional constraints on possible secondary structure, which can be used to refine the secondary structure prediction, as we will see.

Prediction methods for secondary structure have low accuracy

Over 20 different methods have been proposed for **predictions of secondary structure**; they can be categorized in two broad classes. The empirical statistical methods use parameters obtained from analyses of known sequence and tertiary structure. All such methods are based on the assumption that the local sequence in a short region of the polypeptide chain determines local structure; as we have seen, this is not a universally valid assumption. The second group of methods is based on stereochemical criteria, such as compactness of form with a tightly packed hydrophobic core and a polar surface. The three most frequently used methods are the empirical approaches of P.Y. Chou and G.D. Fasman, Palo Alto and Brandeis University, USA, respectively, and of B. Robson and co-workers (the GOR-method), Manchester University, UK, and third, the stereochemical method of V.I. Lim, Moscow, USSR.

Although these three methods use quite different approaches to the problem, the accuracy of their secondary structure prediction is about equally low. All three methods can be used to give a prediction of either of three states to each residue: α helix, β strand, or loop. Random assignment of these three states to residues in a polypeptide chain will give an average score of 33% correctly predicted states. The methods have been assessed in an analysis of a large number of known x-ray structures comprising more than 10,000 residues.

For the three-state definition of secondary structure, the overall accuracy of prediction did not exceed 56% for the best of these methods, Lim's, and was only 50% for the Chou and Fasman method. Other objective assessments have given similar results. This is not quite as bad as it looks: first, because a large fraction of the errors occurs at the ends of α helices and β strands whereas their central regions are correctly predicted and, second, some errors occur because of occasional difficulties in distinguishing between α helices and β strands. These latter errors can be corrected if the structural class, α, β, or α/β, can be deduced from a combination of physical studies, for example, circular dichroism spectra, and the general features of the secondary structure prediction. For example, if the prediction scheme assigns one or two short α helices among many β strands in a protein of the β class, there is a high probability that the regions of secondary structures are essentially correctly predicted but that they should all be β strands.

In spite of their limited accuracy, these predictive methods are very useful in many contexts, for example, in the design of novel polypeptides for the identification of possible antigenic epitopes, in the analysis of common motifs in sequences that direct proteins into specific organelles (for instance, mitochondria), and to provide starting models for tertiary structure predictions. We will, therefore, describe in more detail the **Chou and Fasman method** of secondary structure prediction because it is conceptually the simplest method and the most widely used. Chou and Fasman based their method on a statistical analysis of an x-ray structure database. The number of occurrences in the database of a given amino acid in α helix, β strand, and loop (turn) conformations was used as early as 1974 to calculate the conformational parameters P_α, P_β, and P_t, which essentially are a measure of the frequency of a given amino acid's occurrence in an α helix, β strand, and turn, respectively. In the case of turns, which in this method means short regular loop regions, a significant difference was observed in the frequency of residues in the first, second, third, and fourth position of the turn. In the prediction of turns this positional preference is therefore taken into account.

The effect of neighbors in the amino acid sequence is treated by computing average values for the probabilities of a residue and its nearest neighbors, three neighbors for the helical and two for the β-strand probability. The computer output with an unmodified Chou and Fasman program is a list of such **average probability values**, four for each residue, one each for the probability of it being α helical and β strand, and two for the turn probability, one of which takes the positional preference into account. This computer list should then be converted to a prediction of secondary structure by the human brain. Many users of the Chou and Fasman method have modified their program to automate this last step, which is not to be recommended unless very sophisticated algorithms are used. The list is first analyzed for stretches of consecutive amino acids that have a high probability for α helices or β strands. These form starting regions for secondary structure elements that are extended at both ends until residues with a high probability of turns occur or residues with a low probability of both α and β structures are reached, in which case they are by default classified as coil regions (loops of irregular structure). Often there are regions with high probabilities for both α helix and β strands, it is then useful to know that α helices in general contain more amino acids than β strands and that there are almost always several turn residues between an α helix and a β strand. Proline requires special consideration; with very few exceptions it occurs in α helices only in the first three positions at the amino end.

The tertiary fold of an enzyme has been successfully predicted

The tertiary structure of the α subunit of tryptophan synthase was predicted independently in two different laboratories before the x-ray structure was determined, and this example illustrates both the power and the pitfalls of current prediction methods. Tryptophan synthase is an enzyme that catalyzes the final reactions in the synthesis of the amino acid tryptophan. The bacterial

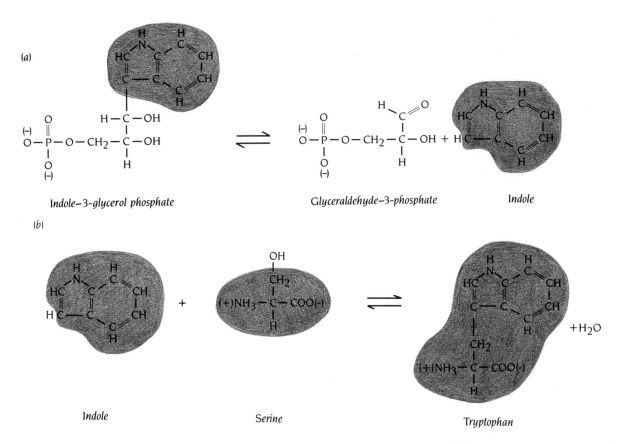

(a)

Indole–3-glycerol phosphate   Glyceraldehyde–3-phosphate   Indole

(b)

Indole   Serine   Tryptophan

enzyme is composed of two polypeptide chains, α (Mw = 29,000 D) and β (Mw = 43,000 D), that combine to form a tetrameric molecule α₂ β₂. The α subunit catalyzes formation of indole from indole–3-glycerol phosphate, and the β subunit combines indole and serine to yield tryptophan (Figure 16.3).

The indole–3-glycerol phosphate used in this reaction is the product of the reactions catalyzed by PRA-isomerase:IGP-synthase, which is the double-headed α/β-barrel enzyme shown in Figure 4.5.

The x-ray structure of the α₂β₂ tryptophan synthase molecule from *Salmonella typhinurium* was determined in 1988 to 2.5 Å resolution by Craig Hyde in the laboratory of David Davies, National Institutes of Health, Washington, D.C. The structure showed that the α subunit was also an eight-stranded α/β

Figure 16.3 The enzyme tryptophan synthase catalyzes the final reactions in the biosynthesis of tryptophan. (a) One subunit, α, catalyzes the formation of indole from indole–3-glycerol phosphate. (b) The second subunit, β, forms tryptophan by coupling the indole ring to C_β of serine.

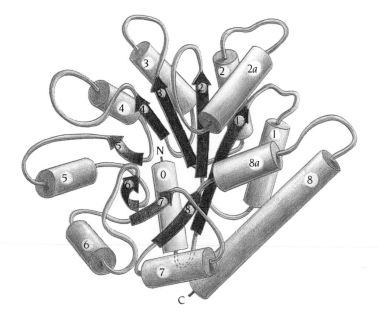

Figure 16.4 Schematic diagram of the three-dimensional structure of the α subunit of tryptophan synthase. The structure is of the α/β-barrel type with eight parallel β strands (purple arrows labeled 1–8) and eight α helices (orange cylinders labeled 1–8). There are three additional α helices, one (green) at the amino end of the polypeptide chain (labeled 0) and two (blue) preceding helices α2 and α8 (labeled 2a and 8a, respectively). Loop regions at the amino end of the β strands are green, and those at the carboxy end are blue. Some of these latter loops form the active site. (Adapted from C.C. Hyde et al., *J. Biol. Chem.* 263: 17861, 1988.)

253

barrel (Figure 16.4), similar to the two domains of PRA-isomerase:IGP-synthase. In addition to the eight α helices of the α/β barrel, there are three extra α helices in the α subunit called α_0, α_{2A}, and α_{8A}.

Prior to the determination of this structure, the groups of Kasper Kirschner at the Biozentrum in Basel and Fred Cohen in San Francisco Medical School each predicted a tertiary fold for the α subunit. Kirschner's group based the secondary structure prediction on nine carefully aligned amino acid sequences of the α subunit from different species. In the alignment procedure the number of gaps was minimized, while the strictly conserved residues in all sequences were properly aligned. Nevertheless, gaps had to be introduced at seven regions, four of which were present in the sequences from more than one species. The presence of insertions or deletions in these regions strongly indicates that they are loop regions and not secondary structure elements.

The prediction of secondary structure was then performed for each sequence individually, using both the GOR and Chou-Fasman methods. At an early stage in these predictions it was apparent that the structure was of the α/β type with alternating α helices and β strands along the sequence. This assignment of the structure class was reinforced by the discovery of several conserved active site residues at or immediately after the carboxy ends of the predicted β strands, in agreement with the observed position of the active sites in α/β domains as discussed in Chapter 4. Constraints for the α/β structure type were, therefore, introduced in the prediction schemes, and each residue was assigned an average predicted state from all sequences. The predicted secondary structures using the two methods are given in Table 16.1 along with the observed secondary structure from the 2.5 Å resolution x-ray model. Additional information in the form of hydropathy plots (Chapter 13) was included when these two predictions were merged into one joint prediction of the secondary structure (Table 16.1).

Table 16.1
Observed and predicted secondary structure in the α subunit of tryptophan synthase

1	2	3	4	5	6	7
	X-ray	GOR	CHF	COM	OPE	BAR
α_0	1–12	1–14	1–14	1–14	—	1–12
β_1	15–24	19–25	18–25	19–24	18–24	18–24
α_1	33–42	30–48	31–52	30–40	30–41	30–41
β_2	47–52	49–51	—	46–50	—	56–60
α_{2A}	63–73	(64–68)β	(64–68)β	(64–68)β	—	—
α_2	79–89	79–89	83–91	79–89	—	—
β_3	95–102	95–107	97–107	95–104	82–88	82–88
α_3	111–120	112–124	111–121	112–124	93–107	93–107
β_4	124–128	125–129	125–129	125–130	125–132	128–132
α_4	138–145	134–145	130–143	134–144	—	138–143
β_5	148–154	149–154	147–154	149–154	148–154	148–154
α_5	160–168	160–167	161–167	160–167	161–168	161–168
β_6	173–177	173–179	173–177	173–179	172–178	172–178
α_6	192–202	188–201	185–198	188–201	192–200	192–200
β_7	208–212	208–210	208–212	208–212	208–215	208–215
α_7	217–225	218–230	214–226	217–227	221–232	221–232
β_8	229–232	231–234	236–241	237–241	—	235–242
α_{8A}	236–242	—	242–244	—	—	—
α_8	248–266	248–268	250–268	249–268	—	250–261

Column 1 gives the nomenclature of the secondary structure elements, where $\alpha_1 \ldots \alpha_8$ and $\beta_1 \ldots \beta_8$ are the eight α helices and β strands of the α/β barrel and α_0, α_{2A}, and α_{8A} are additional α helices in the subunit. Column 2, X-ray, shows the residues assigned to the α helices and β strands in the x-ray structure, and the remaining columns, those that were assigned in the different prediction methods. Columns 3–5 are from Kirschner's group, where GOR is the result of Robson's method, CHF of the Chou and Fasman method, and COM the combination of these two methods. Columns 6 and 7 are from Cohen's group, where OPE is the result of predictions assuming an open α/β structure and BAR assuming a closed α/β barrel.

The results of Kirschner's prediction proved to be in astonishingly good agreement with the x-ray structure. All the eight α helices and seven of the eight β strands of the α/β barrel had been essentially correctly predicted, with only occasional slight errors of one or two residues at the ends of the α helices and β strands. There was only one major error: one of the additional α helices, α_{8A} (residues 236–242), was incorrectly predicted to form β strand 8, which in the crystal structure is formed by residues 229–232.

To obtain a three-dimensional model from these secondary structure predictions, a decision must be made between an open twisted α/β structure or a closed α/β barrel. The predicted eightfold repeat of the β-loop-α-loop unit favored the α/β barrel as did genetic information. The properties of double mutants had shown that a structural interaction must occur between residue 22 in β_1 and residue 211 in β_7. The side chains of these residues could be close to each other and involved in packing interactions in the interior of a closed α/β barrel (Chapter 4), but they would be far apart in an open α/β sheet because they belong to β strands far apart in the sequence. On the other hand, a tryptic fragment (residues 1–188) had been isolated and shown to adopt a stable folded conformation in solution. This result is consistent with an open α/β structure for the complete α subunit with this fragment forming an independent domain having a structure similar to the one that it adopts in solution. If, however, the complete α subunit had a closed α/β-barrel structure, then residues 1 through 188 would be part of the barrel and not an independent domain. In other words, in this case the structure of the tryptic fragment, residues 1 through 188, in solution must be quite different from the structure of these same residues when they are part of the intact α subunit.

Kirschner decided that on balance the observations in favor of a closed α/β structure were stronger than the folding arguments against the barrel, and he was right. A three dimensional α/β-barrel model based on his prediction has an essentially correct tertiary fold but with errors in the detailed position of side chains.

The prediction methods used by Cohen's group are quite different and are based on **artificial intelligence algorithms**. The method attempts to recognize specific patterns of residues that form β strands, α helices, and turns in different types of domain structures. Cohen's group, therefore, first determined the domain type from an analysis of the circular dichroism spectrum of the α subunit, which showed an α/β structure with 12% parallel β strands, 41% α helix, and 0% antiparallel β strands. Unlike Kirschner's group they considered the arguments in favor of an open α/β structure strong enough to assume that that was the way the protein folded. Using specific algorithms for a combination of secondary and tertiary structure prediction in open α/β structures, they concluded that the polypeptide chain contained six β strands alternating with five α helices (Table 16.1) with a topology shown in Figure 16.5.

Figure 16.5 Predicted three-dimensional structure of tryptophan synthase from the amino acid sequence. The molecule was predicted to fold into an open twisted α/β structure with the topology shown in (a) instead of the correct closed α/β-barrel structure (Figure 16.4). A schematic diagram of the predicted structure is shown in (b). (Adapted from M.R. Hurle et al., *Proteins: Structure, Function and Genetics* 2: 217, 1987.)

(a)

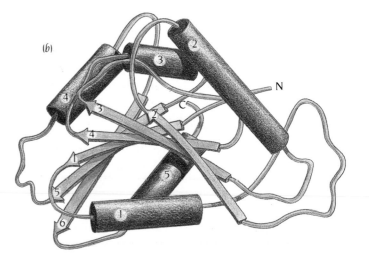

(b)

As well as predicting the fold incorrectly, this method predicted only five α helices and four β strands correctly. When Cohen's group learned that the structure was a closed α/β barrel, they repeated their predictions now based on the algorithms for a closed α/β barrel (Table 16.1). Like Kirschner's group they failed to predict β_8 correctly; but there were also large errors in the prediction of β_2, β_3, and β_8, and they failed to predict α_2 of the barrel.

These exercises in the prediction of tertiary structures demonstrate that successful prediction is still a hit-and-miss business, which only occasionally produces a close to correct model. Even Cohen's program, which is currently the most sophisticated computer algorithm for prediction of tertiary folds, failed to produce an acceptable model of an actual fold known from crystallography.

Folded and unfolded proteins are almost equally stable

Even though we are far from understanding the rules for protein folding, we can engineer, nevertheless, stable proteins with desired functions. The central problem in understanding **protein stability** is that there is only a small difference in total energy between a protein's folded structure and the immense number of different, rapidly interconverting, unfolded structures. The unfolded state has a more favorable **entropy** (amount of disorder) because of its large number of possible structures, and the folded state must compensate for this through various types of chemical interactions, particularly by bringing hydrophobic side chains from the solvent to the interior of the protein, but also through specific hydrogen bonds and electrostatic interactions as well as covalent cross-links through S-S bridges. The observed stability of a native protein structure is the result of a small difference, about 5–15 kcal/mol, between the total energies of the folded and unfolded states, each of which is of the order of 10 million kcal/mol. Clearly, it is an almost impossible task to derive a native folded structure by calculations from first principles, which would require not only extremely well-defined energy terms for all interactions, but also applying them to all possible structures of both the folded and the unfolded polypeptide chain.

If it is impossible to calculate accurately the energetic difference between the folded and the unfolded states, it is, fortunately, relatively easy to measure this difference experimentally. Folded proteins can be made to unfold (denature) by different methods, such as heating the protein or exposing it to denaturing agents like urea or guanidine hydrochloride. The amount of heat released when a protein denatures can be determined by calorimetric methods, which gives a measure of the difference in energy between the folded and the unfolded states of the protein. To compare the stability of wild-type and mutant proteins, a simpler approach is frequently used, such as comparing the melting temperatures, Tm, at which the proteins denature. Spectrophotometric methods, in particular circular dichroism, are used in these measurements to follow changes in protein conformation as the temperature increases. It has been shown that a difference of one degree in Tm values between wild-type and mutant proteins corresponds to a difference in energetic stability of roughly 0.5 kcal/mol. For comparison, each hydrogen bond in liquid water is estimated to have a bond energy, i.e., the energy required to break the bond, of 4.5 kcal/mol compared with 110 kcal/mol for the covalent H-O bond in the water molecule.

When the amino acid sequences of thermostable proteins from thermophilic organisms became available, it was hoped that important factors that determine protein stability would be found by comparing the sequences of homologous proteins from thermophiles and mesophiles living at room temperature. However, it was soon realized that these factors were hidden by a large number of unimportant amino acid substitutions caused by genetic drift. The main conclusion from such comparisons was that the increased stability of proteins in thermophiles results from a large number of different effects, each contributing a small increase.

256

Proteins can be made more stable by engineering

Protein engineering, via site-directed mutagenesis, can be used to answer very specific questions about protein stability, and the results of these studies are now being used to increase the stability of industrially important enzymes. To illustrate some of the factors of importance for protein stability that have been revealed by **protein engineering studies**, we have chosen the extensive work on the enzyme lysozyme from bacteriophage T4 that has been done by the group of Brian Mathews, University of Oregon, Eugene.

Lysozyme from bacteriophage T4 is a 164 amino acid polypeptide chain that folds into two domains (Figure 16.6) There are no S-S bridges; the two cysteine residues in the amino acid sequence, Cys 54 and Cys 97, are far apart in the folded structure. The stability of both the wild-type and mutant proteins is expressed as the melting temperature, Tm, which is obtained by studying reversible heat denaturation. For the wild-type T4 lysozyme the Tm is 41.9°.

We will discuss three different approaches to engineer a more thermostable protein than wild-type T4 lysozyme, namely, (1) reducing the difference in entropy between folded and unfolded protein, which in practice means reducing the number of conformations in the unfolded state, (2) stabilizing the α helices, and (3) increasing the number of hydrophobic interactions in the interior core.

Disulfide bridges increase protein stability

The most obvious way to decrease the number of unfolded conformations is to introduce novel **S-S bridges**, based on knowledge of the tertiary structure of the folded protein. The longer the loop between the cysteine residues, the more restricted is the unfolded polypeptide chain giving more stabilization of the folded structure. To design such bridges is, however, not a simple task, since the geometry of an unstrained -CH$_2$-S-S-CH$_2$- bridge in proteins is confined to rather narrow conformational limits, and deviations from this geometry will

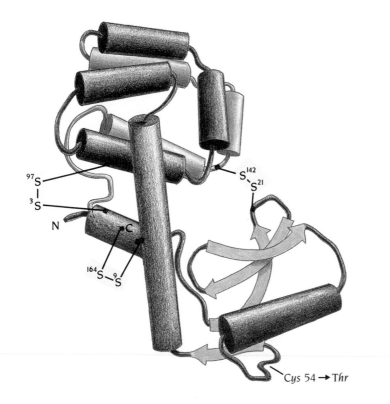

Figure 16.6 The polypeptide chain of lysozyme from bacteriophage T4 folds into two domains. The N-terminal domain is of the α + β type, built up from two α helices (red) and a four-stranded antiparallel β sheet (green). The C-terminal domain comprises seven short α helices (brown and blue) in a rather irregular arrangement. (The last half of this domain is colored blue for clarity.) One long α helix connects the two domains (purple). Thermostable mutants of this protein were constructed by introducing S-S bridges at three different places (yellow). The position of Cys 54, which was mutated to Thr, is also shown. (Adapted from M. Matsumura et al., *Nature* 342: 290, 1989.)

introduce strains into the folded structure and hence reduce rather than increase its stability. It is, therefore, not sufficient to choose at random two residues close together in space to make such a bridge, rather the protein engineer must carefully select pairs of residues with main-chain conformations that fulfill the conditions needed for an unstrained S-S bridge.

Mathews made a very careful comparison between the geometry of the 295 disulfide bridges in known x-ray structures and all possible pairs of amino acid residues close enough to each other in the refined T4 lysozyme structure to accommodate a disulfide bridge. This was followed by energy minimization of the most likely candidates and an analysis of stabilizing interactions that were present in the wild-type structure but would be lost by mutation to a Cys residue. Such losses should be minimized. Three candidate S-S bridges remained after this filtering, one of which, Cys 3-Cys 97, contained one of the cysteine residues (Cys 97) that is present in the wild type. The five amino acid residues—Ile 3, Ile 9, Thr 21, Thr 142, and Leu 164 (Figure 16.6)—were mutated to Cys residues in separate experiments so that all single (3–97, 9–164, and 21–142) as well as combinations of double and triple S-S bonds could be formed. In addition, the second Cys residue of the wild-type enzyme, Cys 54, was mutated to Thr to avoid the formation of incorrect S-S bonds during folding.

The results of this careful design of novel S-S bridges were very encouraging (Figure 16.7). All the mutants were more stable in their oxidized forms than wild-type protein. The longer the loop between the cysteine residues of the mutants with single S-S bonds, the larger was the effect on stability. Furthermore, the effects were additive so that the increase in Tm of 23° for the mutant with three S-S bonds was approximately equal to that of the sum of the increases in Tm values for the three mutants with single S-S bonds $(4.8 + 6.4 + 11.0 \approx 22°)$. The effect on the stability of the protein from reducing the number of possible unfolded structures through introduction of S-S bridges, the entropic effect, is even larger than these values show because the reduced forms of the mutants had a lower Tm than wild type, which indicates that favorable contacts in the folded structure had been lost by the mutations. These experiments show that engineered S-S bridges can be combined together to enhance stability dramatically. Needless to say, knowledge of the three-dimensional structure of the protein is a prerequisite to engineer increased stability in this way.

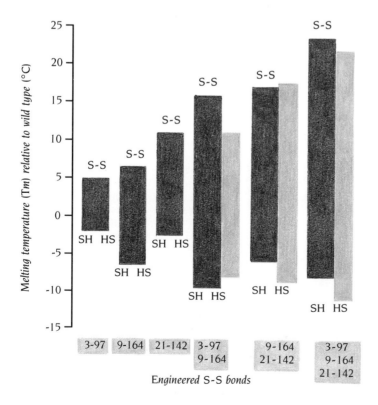

Figure 16.7 Melting temperatures, Tm, of engineered single-, double-, and triple-disulfide-containing mutants of T4 lysozyme relative to wild-type lysozyme. The red bars show the differences in Tm values of the oxidized and reduced forms of the mutant lysozymes. The green bars for the multiple-bridged proteins correspond to the sum of the differences in Tm values for the constituent single-bridged lysozymes. (Adapted from M. Matsumura et al., *Nature* 342: 291, 1989.)

Glycine and proline have opposite effects on stability

Glycine residues have more **conformational freedom** than any other amino acid, as discussed in Chapter 1. A glycine residue at a specific position in a protein has usually only one conformation in a folded structure but can have many different conformations in different unfolded structures of the same protein and thereby contribute to the diversity of unfolded conformations. Proline residues, on the other hand, have less conformational freedom in unfolded structures than any other residue since the proline side chain is fixed by a covalent bond to the main chain. Another way to decrease the number of possible unfolded structures of a protein, and hence stabilize the native structure, is, therefore, to mutate glycine residues to any other residue and to increase the number of proline residues. Such mutations can only be made at positions that neither change the conformation of the main chain in the folded structure nor introduce unfavorable contacts with neighboring side chains.

Both types of mutations have been made in T4 lysozyme. The chosen mutations were Gly 77–Ala, which caused an increase in Tm of 1°, and Ala 82 to Pro, which increased Tm by 2°. The three-dimensional structures of these mutant enzymes were also determined: the Ala 82–Pro mutant had a structure essentially identical to the wild type except for the side chain of residue 82; this strongly indicates that the effect on Tm of Ala 82–Pro is indeed due to entropy changes. Such effects are expected to be additive, so even though each mutation makes only a small contribution to increased stability, the combined effect of a number of such mutations should significantly increase a protein's stability.

Stabilizing the dipoles of α helices increases stability

In Chapter 2 we described the α helix as a dipole with a positive charge at its N terminus and a negative charge at the other end. Negative ions, such as phosphate groups in coenzymes or substrates, are usually bound to the positive ends of such helical dipoles. The α helices that are not part of a binding site frequently have a negatively charged side-chain at the N terminus or a positively charged residue at the C terminus that interacts with the dipole of the helix. Such **dipole-compensating residues** stabilize the helical forms of small synthetic peptides in solution. Do these helix-stabilizing residues also contribute to the overall stability of globular proteins? Of the 11 α helices of T4 lysozyme, 7 helices have negatively charged residues close to their N termini; 2 of

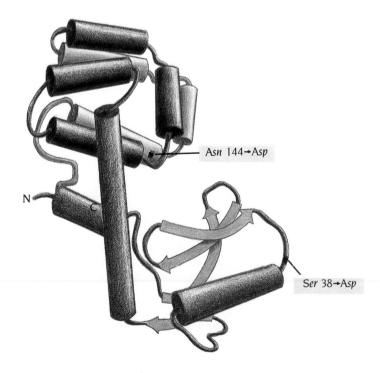

Asn 144→Asp

N

Ser 38→Asp

Figure 16.8 Diagram of the T4-lysozyme structure showing the locations of two mutations that stabilize the protein structure by providing electrostatic interactions with the dipoles of α helices. (Adapted from H. Nicholson et al., *Nature* 336: 652, 1988.)

the remaining 4 α helices were therefore chosen for engineering studies to answer this question (Figure 16.8).

Two different mutant proteins with single substitutions at the N terminus of each of these helices, Ser 38–Asp and Asn 144–Asp, were made as well as the corresponding double mutant. The single mutants both showed an increase in Tm of about 2°C; the effects are additive since the double mutant had a Tm about 4° higher than wild type. This corresponds to 1.6 kcal/mol of stabilization energy. From the x-ray structures of these mutants it is apparent that the stabilization is due to electrostatic interactions and not to specific hydrogen bonding between the substituted amino acid and the end of the helix. Alan Fersht in Cambridge, England, has shown using a different system, the small bacterial ribonuclease barnase, that a histidine residue at the C terminus of a helix stabilizes the barnase structure by about 2.1 kcal/mol. Significant stabilization of α-helical structures might, therefore, be obtained by combining several such helix-stabilizing mutations.

Mutants that fill cavities in hydrophobic cores do not stabilize T4 lysozyme

We emphasized in Chapter 2 that burying the hydrophobic side chains in the interior of the molecule, thereby shielding them from contact with solvent, is a major determinant in the folding of proteins. The surface that is buried inside a folded protein contributes directly to the stabilization energy of the molecule. Studies of destabilizing mutants in barnase, where **cavities** have been engineered into the hydrophobic core of the wild-type enzyme by mutations such as Ile to Val or Phe to Leu show that the introduction of a cavity the size of one -CH2- group destabilizes the enzyme by about 1 kcal/mol. By analogy it should be possible to stabilize a wild-type protein by making mutants that fill existing cavities in its hydrophobic core. Even though proteins have the atoms of their hydrophobic cores packed approximately as tight as atoms are packed in crystals of simple organic molecules, there are cavities in the cores of almost every protein.

T4 lysozyme has two such cavities in the hydrophobic core of its α helical domain. From a careful analysis of the side chains that form the walls of the cavities and from building models of different possible mutations, it was found that the best mutations to make would be Leu 133–Phe for one cavity and Ala 129–Val for the other. These specific mutants were chosen because the new side chains were hydrophobic and large enough to fill the cavities without making too close contacts with surrounding atoms.

The two single mutants were constructed, purified, analyzed for stability, and crystallized. They were both less stable than wild type by 0.5 to 1.0 kcal/mol. The x-ray structures of the mutants provide a rational explanation for this disappointing result. It turns out that in order to fill the cavities, the new side chains in the mutants have adopted energetically unfavorable conformations. This introduces strain in the structure, which obviously costs more in energy than is gained by the new hydrophobic interactions. Even careful model building is obviously not sufficient to predict detailed structural and energetic affects of mutations in the hydrophobic core of proteins. Apparently, the observed core structure in T4 lysozyme, and probably in most proteins, reflects a compromise between the hydrophobic effects, which will tend to maximize the core-packing density, and the strain energy that would be incurred in eliminating all packing defects. Therefore, mutations designed to fill existing cavities may be effective in some cases, but they are not likely to provide a general route to substantial improvement in protein stability.

Binding energy in molecular recognition and specificity has been analyzed by protein engineering

Binding energy between a ligand and a protein molecule is at the heart of all molecular recognition and enzyme catalysis. To design drugs and to understand

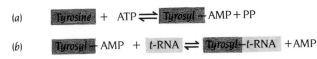

Figure 16.9 The enzyme tyrosyl-tRNA synthetase catalyzes two consecutive reactions: activation of tyrosine by formation of tyrosyl-AMP followed by transfer of the tyrosyl group to its cognate tRNA.

and modify enzyme catalysis, it is important to know the strength of the common interactions, electrostatic and hydrophobic interactions as well as hydrogen bonds between ligand and protein. Alan Fersht at Imperial College, London, pioneered the use of protein engineering to dissect the energy contribution of individual hydrogen bonds of different types to ligand binding and transition state stabilization in enzymes.

As a model for these studies, he chose the enzyme tyrosyl-tRNA synthetase (Figure 4.9–4.11), which catalyzes the addition of tyrosine to its cognate transfer RNA. The enzyme catalyzes two reactions (Figure 16.9): formation of tyrosyl-AMP from tyrosine and ATP followed by transfer of the tyrosyl group to tRNA. Two identical polypeptide chains of 419 amino acid residues form the catalytically active dimeric molecule. In the absence of tRNA the enzyme catalyzes the first reaction whereby a stable complex of the enzyme with tyrosyl-AMP accumulates. The three-dimensional structure of this complex has been determined to 2.1 Å resolution in the laboratory of David Blow at Imperial College, London.

In this complex there are many hydrogen bonds between polar groups on the enzyme and polar groups on the ligand (Figure 16.10). Many attempts previously had been made to deduce the energy contribution of hydrogen bonds to ligand binding by studying the effect on the binding strength of modifications of hydrogen-bonding groups on the ligand. Such studies gave, however, conflicting results, mainly because ligand conformation and ligand binding usually change when groups on the ligand are modified.

Alan Fersht used a different approach. He modified the enzyme instead of the ligand, by mutating one by one the amino acid residues whose side chains form hydrogen bonds to the ligand (red in Figure 16.10). Usually, but not invariably, the conformation of the protein as well as the mode of binding of ligands remained unchanged after such mutations, except that one hydrogen bond between enzyme and ligand was changed or lost. Therefore, this method has given reasonably accurate values of the contribution of hydrogen bonds to the strength of ligand binding.

Hydrogen bonds give small energy contributions to ligand binding

To understand the energy contribution of hydrogen bonds to ligand binding, we will first examine the events that occur when a hydrogen bond between a

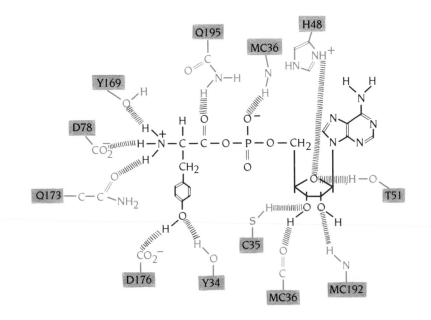

Figure 16.10 Tyrosyl-AMP forms several hydrogen bonds to side-chain and main chain (MC) groups in its complex with the enzyme tyrosyl-tRNA synthetase. Site-specific mutations that abolish some of these hydrogen bonds one at a time have been used to analyze their energy contribution to ligand binding. The ligand, tyrosyl-AMP, is red , enzyme groups are blue, and hydrogen bonds are striped. (Adapted from A. Fersht et al., *Nature* 314: 236, 1985.)

protein and a ligand is formed. We will compare the states of the groups involved in the hydrogen bond in the complex with those of free protein and free ligand in solution both for wild-type protein and for a mutant where the side chain that forms the hydrogen bond is mutated to a hydrophobic side chain.

Each hydrogen bond in the complex of wild-type protein and ligand involves two groups, one from the ligand and one from the protein. Before the complex is formed, when both protein and ligand are free in solution, both these groups form hydrogen bonds to water molecules in the solvent. On complex formation with wild-type protein these two hydrogen bonds to water are replaced by one new hydrogen bond between protein and ligand, giving a net loss of one hydrogen bond. The mutant protein, in which the side chain has been changed from polar to hydrophobic, does not form the hydrogen bond to water molecules in the solvent. Therefore, on complex formation with the mutant protein only one hydrogen bond is lost, that from the ligand to the solvent, and none is formed, giving a net loss of one hydrogen bond also in this case. Deletion of a hydrogen bond in a protein-ligand complex therefore does not lead to an overall loss of the absolute energy of a hydrogen bond and may in certain circumstances, as we will see, cause no loss or even a gain of binding energy. The amount of energy change depends on the precise nature of the interactions, in particular what type of hydrogen bond is lost, and the energy change will vary from case to case. Fersht's studies have revealed the magnitudes of these energy changes for different types of hydrogen bonds.

Hydrogen bonds involving charged groups contribute more to specificity than those between uncharged groups

The hydrogen bonds between tyrosyl-AMP and tyrosyl-tRNA synthetase that Fersht has studied are shown in Figure 16.11. The mutation Tyr 34–Phe provides an example of hydrogen bonds between uncharged groups. The hydrogen bond is formed between the side-chain -OH group of Tyr 34 of the enzyme and the -OH group of the tyrosyl part of the ligand. In the mutant enzyme this hydrogen bond is lost. The binding of the mutant protein to the ligand tyrosine is weaker than with wild-type protein but only by an amount that corresponds to an energy of 0.52 kcal/mol, which should be compared to the energy of 4.5 kcal/mol required to break a hydrogen bond between two water molecules. Similar studies of other hydrogen bonds between uncharged groups by Fersht and others have given a range of 0.5–1.5 kcal/mol for this type of hydrogen bond, provided a proper geometry is present for a hydrogen bond between ligand and protein. When there are geometric constraints preventing the formation of a proper hydrogen bond between the ligand and the wild-type protein, the energy balance can shift so that the loss of such a constrained hydrogen bond causes the ligand to bind tighter to the mutant protein. The hydrogen bond between Thr 51 and one of the ribose oxygen atoms, which is longer than a normal hydrogen bond, is such a case. The mutant Thr 51–Ala binds the ligand ATP more strongly than the wild-type enzyme by an amount corresponding to 0.44 kcal/mol, while the mutant Thr 51–Cys, which can form a more normal hydrogen bond, binds ATP even more tightly, by about 1 kcal/mol.

The energy contribution of 0.52 kcal/mol for the hydrogen bond between Tyr 34 and the -OH group of the substrate tyrosine provides only a factor of 3 to specificity. In other words, if this hydrogen bond was the only interaction that gave the enzyme specificity for Tyr compared to Phe as a substrate, it would make a mistake roughly one out of three times by forming a substrate complex with Phe instead of Tyr. As can be seen in Figure 16.11, there is another hydrogen bond to the -OH group of tyrosine from the charged group Asp 176 which provides additional and higher specificity.

Hydrogen bonds to charged groups contribute a significantly greater gain in the energy of ligand binding than those between uncharged groups. The -OH group of Tyr 169 forms a hydrogen bond to the positively charged amino group in the tyrosyl-AMP complex (Figure 16.11). Studies of the mutant Tyr 169-Phe

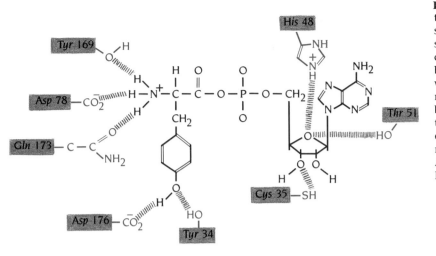

Figure 16.11 Hydrogen bonds between tyrosyl-AMP and the enzyme tyrosyl-tRNA synthetase that were used by Alan Fersht to study, by protein engineering, energy contributions of hydrogen bonds to ligand binding. The enzyme side chains shown here were mutated one at a time to hydrophobic residues, and the binding constants for ligand binding of the mutants were compared with those of the wild-type enzyme. All mutant enzymes, except those where Asp 176 was mutated, could be isolated and studied. Tyrosyl-AMP is red, enzyme groups are blue, and hydrogen bonds are striped.

showed that this hydrogen bond contributes 3.7 kcal/mol to the stability of the complex with the substrate tyrosine. A similar value has been estimated for the hydrogen bond from Asp 176 to the -OH group of the substrate tyrosine.

The energy contribution of around 4 kcal/mol for a hydrogen bond to a charged residue is worth a factor of about 10^3 in specificity. Thus hydrogen bonds between uncharged residues make a small contribution to specificity, but those involving charged residues are critical. One important driving force for high specificity in complementary interactions between a protein and a ligand is, therefore, the formation of hydrogen bonds or ionic interactions with charged groups within the interaction area of the two components.

Circularly permuted α/β barrels fold correctly

We have described how genetically altered proteins can be used to study fundamental functional properties of proteins. In these studies the effects of one or a few amino acid substitutions have been examined. We now illustrate how genetic engineering can be used to design proteins in which the amino acid sequence is changed in a much more fundamental way, by **circular permutation**.

A circular permutation can be envisaged as the result of taking a linear sequence, joining its ends to create a circle, and then cleaving the circle at another site to generate a new linear sequence that is altered only at the sites of joining and cleavage. The important question that is addressed by such experiments is whether or not the resulting rearranged protein will fold into a conformation similar to that of the original protein. The minimum requirements for such folding experiments are that the termini of the original protein are in close proximity in the wild-type structure and that the new cleavage site is in a loop region and not within the scaffold of the protein structure.

Kasper Kirschner at the Biozentrum, Basel, realized that α/β-barrel structures might be excellent candidates for such experiments (Figure 16.12). The α/β barrel is a symmetric structure with the two termini close together and with a number of loop regions at both ends of the barrel that are suitable cleavage sites. Kirschner chose for his engineering studies the gene in yeast for the enzyme PRA-isomerase, which is involved in tryptophan synthesis, as mentioned earlier in this chapter. The three-dimensional structure of this enzyme from *E. coli* was described in Chapter 4. It forms in *E. coli* one α/β-barrel domain of a double-headed enzyme (Figure 4.5). In yeast the gene for PRA isomerase is not fused to any other gene; it therefore specifies only the PRA-isomerase protein. Since the yeast enzyme has 30% amino acid sequence identity to the *E. coli* PRA-isomerase domain, it was assumed that they have very similar three-dimensional structures with similar active sites.

Figure 16.12 Schematic diagram of a "minimal" unmodified α/β-barrel structure, where the N terminus is close to the amino end of the first β strand (red) and the C terminus is close to the end of the eighth α helix (red). The β strands are numbered 1 to 8.

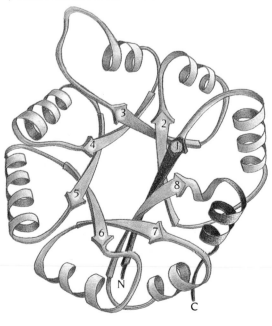

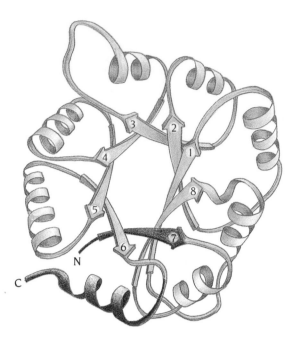

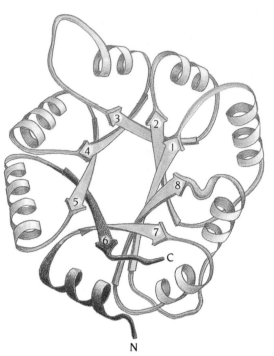

Figure 16.13 Schematic diagram of a circularly permuted α/β barrel where the original N and C termini have been joined by a short loop region (blue) and a new cleavage site has been introduced before the amino end of β strand 7. The first β strand and the last α helix are colored red. The β strands have been numbered according to their order in the unmodified protein (Figure 16.12).

Figure 16.14 A second circularly permuted α/β barrel was designed with the cleavage site at the other end of the α/β barrel, close to the active site. This redesigned protein folds into the same conformation as both the unmodified protein and the first redesigned protein (Figure 16.13).

A "minimal" α/β-barrel domain structure with eight β strands and eight α helices and no additional structure elements has its N and C termini close together in space (Figure 16.12). The *E. coli* PRA-isomerase domain has such a "minimal" α/β-barrel structure. By comparing this structure with the amino acid sequence of the yeast enzyme, Kirschner could predict that the N and C termini were also close together in the yeast enzyme. He therefore made a circular gene construct such that the corresponding protein's N and C termini were joined through a short loop region of four residues, Tyr-Asp-Pro-Ser, which would be long enough to join the two ends and polar enough for a loop region facing the solvent. The active sites of the α/β-barrel domains are formed by the loop regions that connect the carboxy end of the β strands with the α helices as described in Chapter 4.

Kirschner cut this circular gene construct at two different cleavage sites, one at each end of the α/β barrel of the protein. The first cleavage site was at the same end of the barrel as the joining loop, before the amino end of β strand 7 (Figure 16.13). The new termini are therefore far from the active site. The protein expressed by this gene could be isolated and purified and had almost identical kinetic constants, k_{cat} and K_m, as the unmodified enzyme. In other words, this circularly permuted protein folds into the same conformation as the unmodified protein.

The second cleavage site was at the other end of the barrel, after the carboxy end of β strand 6 (Figure 16.14). The two termini are now in the active-site region and would be expected to effect the catalytic function of the enzyme. Do they also affect the folding properties?

The protein could be expressed with a similar high yield to that of the first permuted protein, but, as expected, its kinetic properties were greatly changed. However, it had the same molecular weight and similar hydrodynamic and spectral properties as the wild-type protein. Furthermore, it could be unfolded in guanidine hydrochloride and subsequently refolded into a native conformation in a way similar to the wild-type protein. It seems therefore very likely that this protein also folds into the same conformation.

These protein design experiments show that neither the original amino and carboxy termini, nor the two surface loops that were cut in the process of circular permutation, are essential for correct folding of this α/β-barrel structure. Similar redesign of loop regions and termini have been done for four-helix bundle structures with similar results; loop regions are not essential for proper folding.

Protein structures can be designed from first principles

The ultimate goal of protein engineering is to design an amino acid sequence that will fold into a protein with a predetermined structure and function. Paradoxically, this goal may be easier to achieve than its inverse, the solution of the folding problem. It seems to be simpler to start from a three-dimensional structure and find one of the numerous amino acid sequences that will fold into that structure than to start from an amino acid sequence and predict its three-dimensional structure. The **design of simple domain structures** such as the four-helix bundle structure, antiparallell β barrels, and α/β barrels has met with considerable success. We will illustrate these achievements by the elegant attempts to design a four-helix bundle structure of William de Grado at E.I. du Pont de Nemours in Wilmington, Delaware, with David Eisenberg, University of California, Los Angeles.

The four-helix bundle structure, which is described in Chapter 3 (Figure 3.2), derives its stability from packing hydrophobic residues between the α helices in the bundle and arranging the helices in an antiparallell fashion so that the helix dipoles stabilize each other. Polar residues on the outside of the bundle make the domain soluble in water. The α helices are tilted about 20° to each other. This allows the packing of ridges of side chains of one helix into grooves between side chains in an adjacent helix as described in Chapter 3.

These principles of the four-helix bundle structure were applied in the initial design of the sequence of a suitable 16-residue-long helix, four of which should form a four-helix bundle. The design is based on four identical helices linked by short identical loop regions. From model building of helix packing with different hydrophobic side chains providing the packing interactions, it was found that Leu residues could be packed well against each other when the four helices were arranged in a four-helix bundle structure. Glu and Lys residues

Figure 16.15 The design of a four-helix bundle structure. (a) The amino acid sequence of the helix was chosen so that the branched hydrophobic Leu residues are concentrated on one side of the helix as can be seen from the helical wheel (Chapter 2). Alternating Glu and Lys residues face the solvent of the modeled four-helix bundle. (b) The sequence Pro-Arg-Arg provides the loop regions between the helices (c). The four-helix bundle protein is designed to start with a Met residue followed by four helices with the sequence in (a) connected by three loop regions with the sequence in (b).

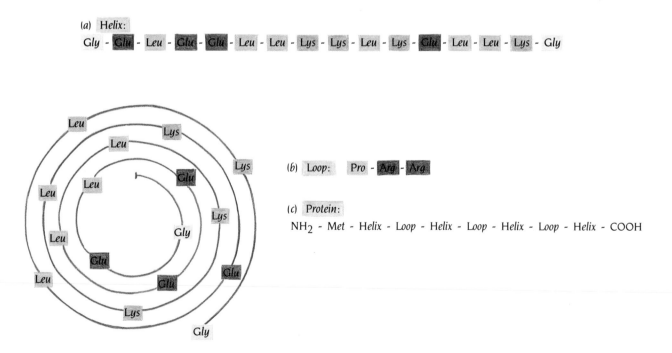

(a) Helix:

Gly - Glu - Leu - Glu - Glu - Leu - Leu - Lys - Lys - Leu - Lys - Glu - Leu - Leu - Lys - Gly

(b) Loop: Pro - Arg - Arg

(c) Protein:

NH₂ - Met - Helix - Loop - Helix - Loop - Helix - Loop - Helix - COOH

were chosen as polar, helix-stabilizing residues for the solvent-accessible exterior of the protein and were alternated one helical turn apart to help stabilize helix formation by electrostatic interactions between their side chains. Finally, Gly residues were placed at both the N and C termini of the helix to provide flexibility for coupling to the linker regions between the helices. The final sequence of the helix is shown in Figure 16.15.

To form the complete protein chain for a four-helix bundle structure, four such helical segments were joined by loop regions with the sequence Pro-Arg-Arg (Figure 16.15). Finally, a gene directing the synthesis of this complete α4 protein was synthesized, cloned, and expressed in *E. coli*. Circular dichroism spectroscopy indicates that the protein is predominantly α helical, and hydrodynamic studies show that it is monomeric and compactly folded. Furthermore, the folded protein is extraordinary stable, requiring 6M guanidine hydrochloride to unfold.

Unfortunately, the three-dimensional structure of this α4 protein is not known, so it remains to be determined whether the four helices are in fact arranged in an antiparallel bundle or in some other arrangement. Nevertheless, this example shows that a compact, folded, water-soluble protein can be designed *de novo* with predetermined secondary structure elements. No doubt we will see many more examples of such proteins that are designed from the basic principles of protein structures.

Conclusion

Proteins with homologous amino acid sequences have similar three-dimensional structures. Homologous proteins contain a core region, a scaffold of secondary structure elements, where the folds of the polypeptide chains are very similar. Loop regions that connect the building blocks of the scaffolds can vary considerably both in length and in structure. From a database of known immunoglobulin structures it has, nevertheless, been possible to predict successfully the conformation of hypervariable loop regions of antibodies of known amino acid sequence.

Methods for the prediction of secondary structure have low accuracy, but a large fraction of the errors occurs at the ends of α helices or β strands. The central regions of these secondary structure elements are often correctly predicted, but the methods do not always distinguish between α helices and β strands.

Prediction of tertiary structure from the amino acid sequence is the major unsolved problem in structural molecular biology. Current methods rely on successful prediction of secondary structure. Prediction in the form of classifying an amino acid sequence to have a high probability of folding into an α-, β-, or α/β-type structure can be achieved. The α subunit of the enzyme tryptophan synthase was correctly predicted to have an α/β-barrel structure.

Folded and unfolded proteins are almost equally stable, the difference being about 5–15 kcal/mol. Protein engineering, via site-directed mutagenesis, can be used to make proteins more stable. Correct engineering of disulfide bridges and helix-dipole stabilization increases the stability of proteins. Removal of glycine residues and introduction of proline residues can also make the folded conformation more stable than the unfolded protein.

Protein engineering has been used to analyze the contribution of hydrogen bonds to the stability and specificity of ligand binding to proteins. Specificity in ligand binding is caused to some small extent by hydrogen bonds between uncharged residues but is best mediated by charged residues.

Circularly permuted α/β barrels have been designed by genetic engineering. These design experiments show that neither the original amino or carboxy termini nor some of the surface loops are essential for correct folding of the α/β barrel.

Protein structures can be designed from first principles. It is simpler to start from a three-dimensional structure and find an amino acid sequence that will fold into that structure than to start from a sequence and predict its three-dimensional structure. A compact, globular, water-soluble protein comprising four α helices has been designed and expressed from a synthetic gene of predetermined sequence.

Selected readings

General

Alber, T. Mutational effects on protein stability. *Annu. Rev. Biochem.* 58: 765–798, 1989.

Baldwin, R.L. How does protein folding get started? *Trends Biochem. Sci.* 14: 291–294, 1989.

Blundell, T.L., et al. Knowledge-based prediction of protein structures and the design of novel molecules. *Nature* 326: 347–352, 1987.

DeGrado, W.F.; Wasserman,Z.R.; Lear, J.D. Protein design, a minimalist approach. *Science* 243: 622–628, 1989.

Fasman, G.D. Protein conformational prediction. *Trends Biochem. Sci.* 14: 295–299, 1989.

Fasman, G.D., ed. Prediction of protein structure and the principles of protein conformation. New York: Plenum, 1989.

Fersht, A.R. Protein engineering. *Prot. Eng.* 1: 7–16, 1986.

Fersht, A.R. The hydrogen bond in molecular recognition. *Trends Biochem. Sci.* 12: 301–304, 1987.

Fersht, A.R.; Leatherbarrow, R.J.; Wells, T.N.C. Binding energy and catalysis: a lesson from protein engineering of the tyrosyl-tRNA synthetase. *Trends Biochem. Sci.* 11: 321–325, 1986.

Fletterick, R.; Zoller, M., eds. Protein structure and design. Current communications in molecular biology. Cold Spring Harbor, NY: Cold Spring Harbor Laboratory, 1986.

Ikehara, M., ed. Protein engineering. Protein design in basic research, medicine and industry. New York: Springer, 1990.

Kaiser, E.T. Design principles in the construction of biologically active peptides. *Trends Biochem. Sci.* 12: 305–309, 1987.

Kolata, G. Trying to crack the second half of the genetic code. *Science* 233: 1037–1039, 1986.

Oxender, D.; Fox, C., eds. Protein engineering. New York: Liss, 1987.

Richardson, J.S.; Richardson, D.C. The *de novo* design of protein structures. *Trends Biochem. Sci.* 14: 304–309, 1989.

Schulz, G. A critical evaluation of methods for prediction of protein secondary structures. *Annu. Rev. Biophys. Biophys. Chem.* 17: 1–21, 1988.

Taylor, W. Pattern-matching methods in protein sequence comparison and structure prediction. *Prot. Eng.* 2: 77–86, 1988.

Thornton, J. The shape of things to come. *Nature* 335: 10–11, 1988.

Thornton, J.M.; Gardner, S.P. Protein motifs and data-base searching. *Trends Biochem. Sci.* 14: 300–304, 1989.

von Heijne, G. Sequence analysis in molecular biology, treasure trove or trivial pursuit. San Diego: Academic Press, 1987.

Specific structures

Argos, P. A sensitive procedure to compare amino acid sequences. *J. Mol. Biol.* 193: 385–396, 1987.

Bowie, J.U. Deciphering the message in protein sequences: tolerance to amino acid substitutions. *Science* 247: 1306–1310, 1990.

Brange, J., et al. Monomeric insulins obtained by protein engineering and their medical implications. *Nature* 333: 679–682, 1988.

Brick, P.; Bhat, T.N.; Blow, D.M. Structure of tyrosyl-tRNA synthetase refined at 2.3 Å resolution. Interaction of the enzyme with the tyrosyl adenylate intermediate. *J. Mol. Biol.* 208: 83–98, 1988.

Chothia, C.; Lesk, A. The relation between the divergence of sequence and structure in proteins. *EMBO J.* 5: 823–826, 1986.

Chothia, C., et al. The predicted structure of immuno-globulin D 1.3 and its comparison with the crystal structure. *Science* 233: 755–758, 1986.

Chothia, C., et al. Conformations of immunoglobulin hypervariable regions. *Nature* 342: 877–883, 1989.

Chou, P.Y.; Fasman, G.D. Prediction of protein conformation. *Biochemistry* 13: 222–245, 1974.

Cohen, C.; Parry, D.A.D. α-helical coiled coils and bundles: how to design an α-helical protein. *Prot.: Struct. Funct. Gen.* 7: 1–15, 1990.

Cohen, F.E., et al. Secondary structure assignment for α/β proteins by a combinatorial approach. *Biochemistry* 22: 4894–4904, 1983.

Cohen, F.E., et al. Turn prediction in proteins using a pattern-matching approach. *Biochemistry* 25: 266–275, 1986.

Crawford, I.P.; Niermann, T.; Kirschner, K. Prediction of secondary structure by evolutionary comparison: application to the a subunit of tryptophan synthase. *Prot.: Struct. Funct. Gen.* 2: 118–129, 1987.

DeGrado, W.F.; Regan, L.; Ho, S.P. The design of a four-helix bundle protein. *Cold Spring Harbor Symp. Quant. Biol.* 52: 521–526, 1987.

Faber, H.R.; Matthews, B.W. A mutant T4 lysozyme displays five different crystal conformations. *Nature* 348: 263–266, 1990.

Fersht, A.R. Dissection of the structure and activity of the tyrosyl-tRNA synthetase by site-directed mutagenesis. *Biochemistry* 26: 8031–8037, 1987.

Fersht, A.R.; Leatherbarrow, R.J.; Wells, T.N.C. Quantitative analysis of structure-activity relationships in engineered proteins by linear free-energy relationships. *Nature* 322: 284–286, 1986.

Fersht, A.R., et al. Hydrogen bonding and biological specificity analyzed by protein engineering. *Nature* 314: 235–238, 1985.

Garnier, J.; Osguthorpe, D.J.; Robson, B. Analysis of the accuracy and implications of simple methods for predicting the secondary structure of globular proteins. *J. Mol. Biol.* 120: 97–120, 1978.

Goldenberg, D.P. Circularly permuted proteins. *Prot. Eng.* 2: 493–495, 1989.

Goraj, K.; Renard, A.; Martial, J.A. Synthesis, purification and initial structural characterization of octarellin, a *de novo* polypeptide modeled on the α/β-barrel proteins. *Prot. Eng.* 3: 259–266, 1990.

Gribskov, M.; McLaschlan, A.D.; Eisenberg, D.E. Profile analysis: detection of distantly related proteins. *Proc. Natl. Acad. Sci. USA* 84: 4355–4358, 1987.

Handel, T. *De novo* design of an α/β-barrel protein. *Prot. Eng.* 3: 233–234, 1990.

Hurle, M.R., et al. Prediction of the tertiary structure of the α subunit of tryptophan synthase. *Prot.: Struct. Funct. Gen.* 2: 210–224, 1987.

Hyde, C.C., et al. Three-dimensional structure of the tryptophan synthase $\alpha_2\beta_2$ multienzyme complex from *Salmonella typhimurium*. *J. Biol. Chem.* 263: 17857–17871, 1988.

Karpusas, M., et al. Hydrophobic packing in T4 lysozyme probed by cavity-filling mutants. *Proc. Natl. Acad. Sci. USA* 86: 8237–8241, 1989.

Kellis, J.T., et al. Contribution of hydrophobic interactions to protein stability. *Nature* 333: 784–786, 1988.

Lear, J.D.; Wasserman, Z.R.; DeGrado, W.F. Synthetic amphiphilic peptide models for protein ion channels. *Science* 240: 1177–1181, 1988.

Lim, V.I. Algorithms for prediction of α-helical and β-structural regions in globular proteins. *J. Mol. Biol.* 88: 873–894, 1974.

Luger, K.; Szadkowski, H.; Kirschner, K. An 8-fold βα-barrel protein with redundant folding possibilities. *Prot. Eng.* 3: 249–258, 1990.

Luger, K., et al. Correct folding of circularly permuted variants of a βα-barrel enzyme *in vivo*. *Science* 243: 206–210, 1989.

Matsumura, M.; Signor, G.; Matthews, B.W. Substantial increase of protein stability by multiple disulfide bonds. *Nature* 342: 291–293, 1989.

Matthews, B.W.; Nicholson, H.; Becktel, W.J. Enhanced protein thermostability from site-directed mutations that decrease the entropy of unfolding. *Proc. Natl. Acad. Sci. USA* 84: 6663–6667, 1987.

Matthews, B.W.; Remington, S.J. The three-dimensional structure of the lysozyme from bacteriophage T4. *Proc. Natl. Acad. Sci. USA* 71: 4178–4182, 1974.

Mutter, M., et al. The construction of new proteins: V. A template-assembled synthetic protein (TASP) containing both a 4-helix bundle and β-barrel-like structure. *Prot.: Struct. Funct. Gen.* 5: 13–21 1989.

Nicholson, H.; Becktel, W.J.; Matthews, B.W. Enhanced protein thermostability from designed mutations that interact with α-helix dipoles. *Nature* 336: 651–656, 1988.

Regan, L.; DeGrado, W.F. Characterization of a helical protein designed from first principles. *Science* 241: 976–978, 1988.

Richardson, J.S.; Richardson, D.C. Amino acid preferences for specific locations at the ends of α helices. *Science* 240: 1648–1652, 1988.

Weaver, L.H.; Matthews, B.W. Structure of bacteriophage T4 lysozyme refined at 1.7 Å resolution. *J. Mol. Biol.* 193: 189–199, 1987.

Wells, T.N.C.; Fersht, A.R. Hydrogen bonding in enzymatic catalysis analyzed by protein engineering. *Nature* 316: 656–657, 1985.

Wetzel, R. Harnessing disulfide bonds using protein engineering. *Trends Biochem. Sci.* 12: 478–482, 1987.

Wilkinson, A.J., et al. A large increase in enzyme-substrate affinity by protein engineering. *Nature* 307: 187–188, 1984.

Winter, G., et al. Redesigning enzyme structure by site-directed mutagenesis: tyrosyl-tRNA synthetase and ATP binding. *Nature* 299: 756–758, 1982.

Zvelebil, M.J., et al. Prediction of protein secondary structure and active sites using the alignment of homologous sequences. *J. Mol. Biol.* 195: 957–961, 1987.

Determination of Protein Structures

<big>**17**</big>

The structures described in this book have been determined by physical methods: most of them by **x-ray crystallography**, some of the smaller ones by methods involving **NMR**. We conclude the book with a short description of these techniques. It is not our aim to convert biologists into x-ray crystallographers and NMR spectroscopists; a complete explanation of the physical basis of these techniques and of the methods as currently practiced would fill more than one textbook. Our purpose is rather to convey the essence of the principles and procedures involved, so as to provide a general understanding of what is entailed in solving protein structures by these means. We will see how deriving a three-dimensional protein structure from x-ray or NMR data depends not only on the quality of the data themselves, but also on biochemical and sometimes genetic information that are essential to their interpretation.

Several different techniques are used to study the structure of protein molecules

Different techniques give different and complementary information about protein structure. The primary structure is obtained by biochemical methods, either by direct determination of the amino acid sequence from the protein or indirectly, but more rapidly, from the nucleotide sequence of the corresponding

Figure 17.1 A crystal is built up from many billions of small identical units, or unit cells. These unit cells are packed against each other in three dimensions much as identical boxes are packed and stored in a warehouse. The unit cell may contain one or more than one molecule. Although the number of molecules per unit cell is always the same for all the unit cells of a single crystal, it may vary between different crystal forms of the same protein. The diagram shows in two dimensions several identical unit cells, each containing two objects packed against each other. The two objects within each unit cell are related by twofold symmetry to illustrate that each unit cell in a protein crystal can contain several molecules that are related by symmetry to each other. (The pattern is adapted from a Japanese stencil of unknown origin from the nineteenth century.)

(a)

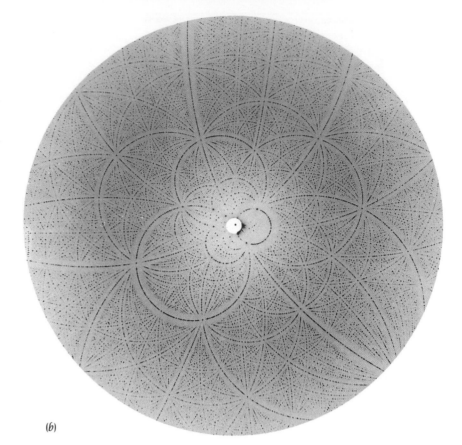

(b)

gene or cDNA. The quaternary structure of large proteins or aggregates such as virus particles, ribosomes, or gap junctions can be determined by electron microscopy. In general, this method gives structural information at very low resolution, with no details, although, if one can obtain ordered two-dimensional arrays of the object, the noise in the electron microscopic image can be reduced enough to reveal the shape of individual subunits or, in rare cases, even determine the path of the polypeptide chain within a protein molecule.

To obtain the secondary and tertiary structure, which requires detailed information about the arrangement of atoms within a protein, the main method so far has been x-ray crystallography. In recent years NMR methods have been developed to obtain three-dimensional models of small protein molecules, and NMR will probably become increasingly useful as it is further developed.

Figure 17.2 Well-ordered protein crystals diffract x-rays and produce diffraction patterns that can be recorded on film. The crystal shown in (a) is from the enzyme RuBisCo from spinach and the photograph in (b) is a recording (Laue photograph) of the diffraction pattern of a similar crystal of the same enzyme. The diffraction pattern was obtained using polychromatic radiation from a synchrotron source in the wavelength region 0.5 to 2.0 Å. More than 100,000 diffracted beams have been recorded on this film during an exposure of the crystal to x-rays for less than one second. (The Laue photograph was recorded by Janos Haijdu, Oxford, and Inger Andersson, Uppsala, at the synchrotron radiation source in Daresbury, England.)

Protein crystals are difficult to grow

The first prerequisite for solving the three-dimensional structure of a protein by x-ray crystallography is a well-ordered **crystal** that will diffract x-rays strongly. The crystallographic method depends, as we will see, upon directing a beam of x-rays onto a regular, repeating array of many identical molecules (Figure 17.1) so that the x-rays are diffracted from it in a pattern from which the structure of an individual molecule can be retrieved. Well-ordered crystals (Figure 17.2) are difficult to grow because globular protein molecules are large, spherical, or ellipsoidal objects with irregular surfaces, and it is impossible to pack them into a crystal without forming large holes or **channels** between the individual molecules. These channels, which usually occupy more than half the volume of the crystal, are filled with disordered solvent molecules (Figure 17.3). The protein molecules are in contact with each other at only a few small regions, and even in these regions many interactions are indirect, through one or several layers of solvent molecules. This is one reason why structures of proteins determined by x-ray crystallography are the same as those for the proteins in solution.

270

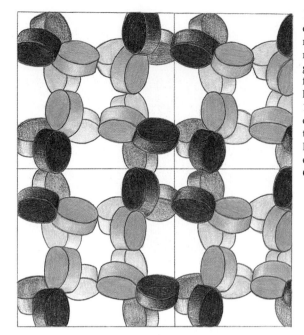

Figure 17.3 Protein crystals contain large channels and holes filled with solvent molecules as shown in this diagram of the molecular packing in crystals of the enzyme glycolate oxidase. The subunits (colored disks) form octamers of molecular weight around 300 kD with a hole in the middle of each of about 15 Å diameter. Between the molecules there are channels (white) of around 70 Å diameter through the crystal. (Courtesy of Ylva Lindqvist, who determined the structure of this enzyme to 2.0 Å resolution in the laboratory of Carl Branden, Uppsala.)

Crystallization is usually quite difficult to achieve, and crystal growth can be slow; in some cases it may require months for sufficiently large crystals (~0.5 mm) to grow from microcrystals. The formation of crystals is also critically dependent on a number of different parameters, including pH, temperature, protein concentration, the nature of the solvent and precipitant as well as the presence of added ions or ligands to the protein. Many crystallization experiments are therefore required to screen all these parameters for the few combinations that might give crystals suitable for x-ray diffraction analysis. Recently **crystallization robots** have been built to automate and speed up the tedious work of reproducibly setting up large numbers of crystallization experiments. Much interest is also currently focused on **gravity** as an obstacle to the formation of well-ordered crystals, and a number of costly protein crystallization experiments have been performed in space, under microgravity conditions. However, it still remains to be shown that these conditions have more than a marginal effect on protein crystallization.

A pure and homogeneous protein sample is crucial for successful crystallization, and recombinant DNA techniques have been a major breakthrough in this regard. Proteins obtained from cloned genes in efficient expression vectors can be purified quickly to homogeneity in large quantities in a few purification steps. As a rule of thumb, a protein to be crystallized should ideally be more than 97% pure according to standard criteria of homogeneity. Crystals form when molecules are precipitated very slowly from **supersaturated solutions**. The most frequently used procedure for making protein crystals is the **hanging-drop** method (Figure 17.4), in which a drop of protein solution is brought very gradually to supersaturation by loss of water from the droplet to the larger reservoir that contains salt or polyethylene glycol solution.

Since there are so few direct packing interactions between protein molecules in a crystal, small changes in, for example, the pH of the solution can cause the molecules to pack in different ways to produce different crystal forms. The structures of some protein molecules such as lysozyme and myoglobin have been determined in different crystal forms and found to be essentially similar, except for a few side chains involved in packing interactions. Because they are so few, these interactions between protein molecules in a crystal do not change the overall structure of the protein. However, different crystal forms can be more or less well-ordered and hence give diffraction patterns of different quality. As a general rule, the more closely the protein molecules pack, and consequently the less water the crystals contain, the better is the diffraction pattern because the molecules are better ordered in the crystal.

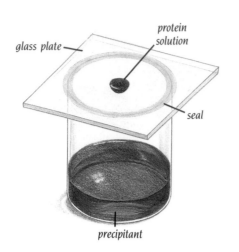

(a)

glass plate

protein solution

seal

precipitant

(b)

Figure 17.4 The hanging-drop method of protein crystallization. (a) About 10 μl of a 10 mg/ml protein solution in a buffer with added precipitant such as ammonium sulfate in a concentration below the saturation point is put on a thin glass plate that is sealed upside down on the top of a small container. In the container there is about 1 ml of the precipitant solution (ammonium sulfate in buffer) with a concentration of precipitant above the saturation point. Equilibrium between the drop and the container is slowly reached through vapor diffusion, and the salt concentration in the drop is increased by loss of water to the reservoir. If other conditions such as pH and temperature are right, protein crystals will occur in the drop. (b) Crystals of recombinant enzyme RuBisCo from *Anacystis nidulans* formed by the hanging-drop method. (Courtesy of Janet Newman, Uppsala, who produced these crystals.)

X-ray sources are either monochromatic or polychromatic

X-rays are electromagnetic radiation at short wavelengths, emitted when electrons jump from a higher to a lower energy state. In conventional sources in the laboratory, x-rays are produced by high-voltage tubes in which a metal plate, the anode, is bombarded with accelerating electrons and thereby caused to emit x-rays of a specific wavelength, so-called **monochromatic x-rays**. The high voltage rapidly heats up the metal plate, which therefore has to be cooled. Efficient cooling is achieved by so-called **rotating anode** x-ray generators, where the metal plate revolves during the experiment so that different parts are heated up. Rotating anode x-ray generators are the conventional equipment used in most protein crystallography laboratories.

More powerful x-ray beams can be produced in **synchrotron storage rings** where electrons (or positrons) travel close to the speed of light. These particles emit very strong radiation at all wavelengths from short gamma rays to visible light. When used as an x-ray source, only radiation within a window of suitable wavelengths is channeled from the storage ring. **Polychromatic x-ray** beams are produced by having a broad window that allows through x-ray radiation with wavelengths of 0.2–2.0 Å. Such beams were used to record the **Laue-diffraction** picture shown in Figure 17.2b. A very narrow window produces monochromatic radiation that is still several orders of magnitude more intense

Figure 17.5 Schematic view of a diffraction experiment. (a) A narrow beam of x-rays (red) is taken out from the x-ray source through a collimating device. When the primary beam hits the crystal, most of it passes straight through, but some is diffracted by the crystal. These diffracted beams, which leave the crystal in many different directions, are recorded on a detector, either a piece of x-ray film or an area detector. (b) A diffraction pattern from a crystal of the enzyme RuBisCo using monochromatic radiation (compare with Figure 17.2b, the pattern using polychromatic radiation). The crystal was rotated one degree while this pattern was recorded.

(b)

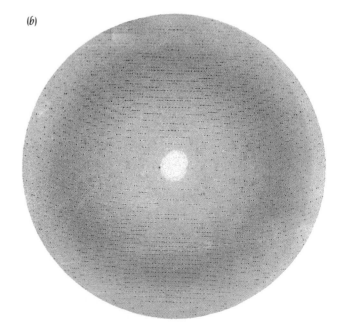

(a)

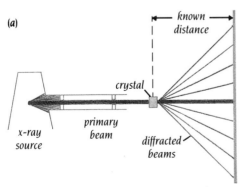

known distance

crystal

x-ray source

primary beam

diffracted beams

detector

than the beam from conventional rotating anode x-ray sources. Such strong beams allow very short exposure times in diffraction experiments and are used to collect data from complex structures or when only small crystals are available so that the diffraction patterns produced from conventional x-ray sources would be too faint. Most of the virus structures described in Chapter 11, as well as the HLA structure discussed in Chapter 12, were determined from diffraction data obtained with monochromatic beams from synchrotron sources. The HLA crystals used were very thin (0.02 mm) plates about 0.5 mm square.

X-ray data are recorded either on films or by electronic detectors

In diffraction experiments a narrow and parallel beam of x-rays is taken out from the x-ray source and directed onto the crystal to produce **diffracted beams** (Figure 17.5a). The incident primary beam causes damage to both protein and solvent molecules. This produces free radicals that in turn damage other molecules in the crystal. In addition, heat is generated, especially from synchrotron radiation, and eventually the primary beam burns through the crystal. The crystal is, therefore, usually cooled to prolong its lifetime. The primary beam must strike the crystal from many different directions to produce all possible diffraction spots, and so the crystal is rotated in the beam during the experiment. Rotating the crystal is much easier than rotating the x-ray source, especially when it is a synchrotron.

The diffracted spots are recorded either on a film, the classical method (Figure 17.5b), or by an **electronic detector**. The exposed film has to be measured and digitized by a scanning device, whereas electronic detectors feed the signals they detect directly in a digitized form into a computer. Electronic **area detectors**, which can be regarded as an electronic film, have recently become available and have significantly reduced the time required to collect and measure diffraction data. To determine the structure of a protein, as we will see, it is necessary to compare x-ray data from native crystals of the protein with those from crystals in which different atoms of the protein are complexed with **heavy metals**. Moreover, to elucidate a protein's function x-ray data must also be collected from complexes with different types of bound ligands. In total, therefore, several hundred thousand diffraction spots are usually collected and measured for each protein.

The rules for diffraction are given by Bragg's law

When the primary beam from an x-ray source strikes the crystal, most of the x-rays travel straight through it. Some, however, interact with the electrons on each atom and cause them to oscillate. The oscillating electrons serve as a new source of x-rays, which are emitted in almost all directions. We refer to this rather loosely as **scattering**. When atoms and hence their electrons are arranged in a regular three-dimensional array, as in a crystal, the x-rays emitted from the oscillating electrons interfere with one another. In most cases, these x-rays, colliding from different directions, cancel each other out; those from certain directions, however, will add together to produce diffracted beams of radiation that can be recorded as a pattern on a photographic plate or detector (Figure 17.6a).

How is the diffraction pattern obtained in an x-ray experiment such as that shown in Figure 17.5b related to the crystal that caused the diffraction? This question was addressed in the early days of x-ray crystallography by Sir Lawrence Bragg of Cambridge University, who showed that diffraction by a crystal can be regarded as the reflection of the primary beam by sets of parallel planes, rather like a set of mirrors, through the unit cells of the crystal (Figure 17.6b and c).

X-rays that are reflected from adjacent planes travel different distances (Figure 17.6c), and Bragg showed that diffraction only occurs when the

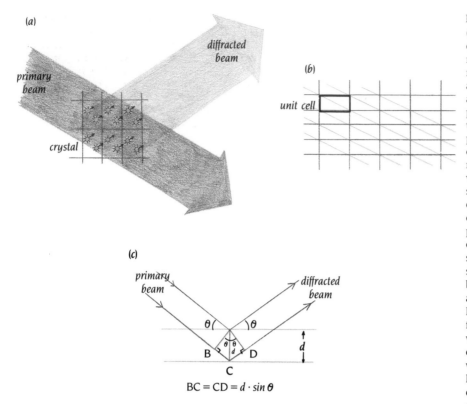

(a)

diffracted beam

primary beam

crystal

(b)

unit cell

(c)

primary beam

diffracted beam

θ   θ

B   θ θ   D
      d

d

C

BC = CD = d · sin θ

Figure 17.6 Diffraction of x-rays by a crystal. (a) When a beam of x-rays (red) shines on a crystal all atoms (green) in the crystal scatter x-rays in all directions. Most of these scattered x-rays cancel out, but in certain directions (blue arrow) they reinforce each other and add up to a diffracted beam. (b) Different sets of parallel planes can be arranged through the crystal so that each corner of all unit cells is on one of the planes of the set. The diagram shows in two dimensions three simple sets of parallel lines: red, blue, and green. A similar effect is seen when driving past a plantation of regularly spaced trees. One sees the trees arranged in different sets of parallel rows. (c) X-ray diffraction can be regarded as reflection of the primary beam from sets of parallel planes in the crystal. Two such planes are shown (green), separated by a distance d. The primary beam strikes the planes at an angle θ and the reflected beam leaves at the same angle, the reflection angle. X-rays (red) that are reflected from the lower plane have traveled farther than those from the upper plane by a distance BC + CD, which is equal to 2d sin θ. Reflection can only occur when this distance is equal to the wavelength λ of the x-ray beam and Bragg's law—2d sin θ = λ—gives the conditions for diffraction. To determine the size of the unit cell, the crystal is oriented in the beam so that reflection is obtained from the specific set of planes in which any two adjacent planes are separated by the length of one of the unit cell axes. This distance, d, is then equal to λ/(2 sin θ). The wavelength, λ, of the beam is known since we use monochromatic radiation. The reflection angle, θ, can be calculated from the position of the diffracted spot on the film, using the relation derived in Figure 17.7, where the crystal to film distance can be easily measured. The crystal is then reoriented, and the procedure is repeated for the other two axes of the unit cell.

difference in distance is equal to the wavelength of the x-ray beam. This distance is dependent on the reflection angle, which is equal to the angle between the primary beam and the planes (Figure 17.6c).

The relationship between the reflection angle, θ, the distance between the planes, d, and the wavelength, λ, is given by **Bragg's law** $2d \sin \theta = \lambda$. This relation can be used to determine the size of the unit cell (see legend to Figure 17.6c and Figure 17.7). Briefly, the position on the film of the diffraction data relates each spot to a specific set of planes through the crystal. By using Bragg's law, these positions can be used to determine the size of the unit cell.

Phase determination is the major crystallographic problem

Each atom in a crystal scatters x-rays in all directions, and only those that positively interfere with one another, according to Bragg's law, give rise to diffracted beams (Figure 17.6a) that can be recorded as a distinct **diffraction**

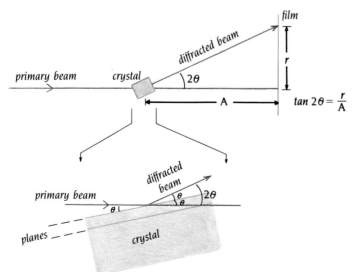

film

r

diffracted beam

primary beam   crystal   2θ

A

tan 2θ = r/A

primary beam
θ

diffracted beam
θ
θ   2θ

planes

crystal

Figure 17.7 The reflection angle, θ, for a diffracted beam can be calculated from the distance (r) between the diffracted spot on a film and the position where the primary beam hits the film. From the geometry shown in the diagram the tangent of the angle 2θ = r/A. A is the distance between crystal and film that can be measured on the experimental equipment, while r can be measured on the film. Hence θ can be calculated. The angle between the primary beam and the diffracted beam is 2θ as can be seen on the enlarged insert at the bottom. It shows that this angle is equal to the angle between the primary beam and the reflecting plane plus the reflection angle, both of which are equal to θ.

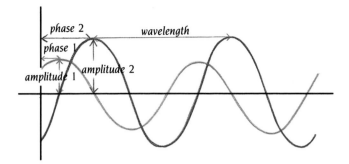

Figure 17.8 Two diffracted beams (purple and orange), each of which is defined by three properties: amplitude, which is a measure of the strength of the beam and which is proportional to the intensity of the recorded spot; phase, which is related to its interference, positive or negative, with other beams; and wavelength, which is set by the x-ray source for monochromatic radiation.

spot above background. Each diffraction spot is the result of interference of all x-rays with the same diffraction angle emerging from all atoms. For a typical protein crystal, myoglobin, each of the about 20,000 diffracted beams that have been measured contains scattered x-rays from each of the around 1500 atoms in the molecule. To extract information about individual atoms from such a system requires considerable computation. The mathematical tool that is used to handle such problems is called the **Fourier transform**, invented by the French mathematician Jean Baptiste Joseph Fourier while he served as a bureaucrat in the government of Napoleon Bonaparte.

Each diffracted beam, which is recorded as a spot on the film, is defined by three properties (Figure 17.8): the **amplitude**, which we can measure from the intensity of the spot; the **wavelength**, which is set by the x-ray source; and the **phase**, which is lost in x-ray experiments. We need to know all three properties for all of the diffracted beams to determine the position of the atoms giving rise to the diffracted beams. How do we find the phases of the diffracted beams? This is the so-called phase problem in x-ray crystallography.

In small-molecule crystallography the phase problem was solved by so-called direct methods (recognized by the award of a Nobel Prize in chemistry to Jerome Karle, U.S. Naval Research Laboratory, Washington, D.C., and Herbert Hauptman, the Medical Foundation, Buffalo). For larger molecules, protein crystallographers have stayed at the laboratory bench using a method pioneered by Max Perutz and John Kendrew and their co-workers to circumvent the phase problem. This method, called **multiple isomorphous replacement** (MIR), requires the introduction of new x-ray scatterers into the unit cell of the crystal. These additions should be heavy atoms (so that they make a significant contribution to the diffraction pattern); there should not be too many of them (so that their positions can be located); and they should not change the structure of the molecule or of the crystal cell—in other words, the crystals should be isomorphous. In practice, isomorphous replacement is usually done by diffusing different heavy-metal complexes into the channels of preformed protein crystals. With luck the protein molecules expose into these solvent channels side chains, such as SH groups, that are able to bind heavy metals. It is also possible to replace endogenous light metals in metalloproteins with heavier ones, e.g., zinc by mercury or calcium by samarium.

Since such heavy metals contain many more electrons than the light atoms, H, N, C, O, and S, of the protein, they scatter x-rays more strongly. All diffracted beams would therefore increase in intensity after heavy-metal substitution if all interference were positive. In fact, however, some interference is negative; consequently, following heavy-metal substitution, some spots measurably increase in intensity, others decrease, and many show no detectable difference.

How do we find phase differences between diffracted spots from intensity changes following heavy-metal substitution? We first use the intensity differences to deduce the positions of the heavy atoms in the crystal unit cell. Fourier summations of these intensity differences give maps of the vectors between the heavy atoms, the so-called **Patterson maps** (Figure 17.9). From these vector maps it is relatively easy to deduce the atomic arrangement of the heavy atoms, so long as there are not too many of them. From the positions of the heavy metals in the unit cell, one can calculate the amplitudes and phases of their

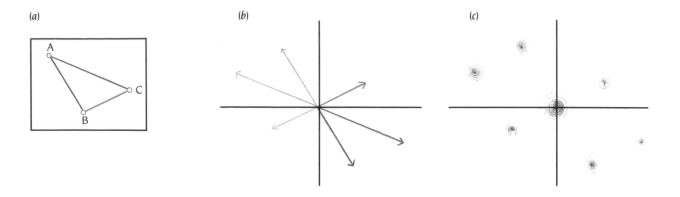

contribution to the diffracted beams of the protein crystals containing heavy metals.

How is that knowledge used to find the phase of the contribution from the protein in the absence of the heavy-metal atoms? We know the phase and amplitude of the heavy metals and the amplitude of the protein alone. In addition, we know the amplitude of protein plus heavy metals (i.e., protein heavy-metal complex); thus we know one phase and three amplitudes. From this we can calculate whether the interference of the x-rays scattered by the heavy metals and protein is constructive or destructive (Figure 17.10). The extent of positive or negative interference plus knowledge of the phase of the heavy metal together give an estimate of the phase of the protein. Unfortunately, the problem is underdetermined so that two different phase angles are equally good solutions. To distinguish between these two possible solutions, a second heavy-metal complex must be used, which also gives two possible phase angles. Only one of these will have the same value as one of the two previous phase angles; it therefore represents the correct phase angle. In practice, more than two different heavy-metal complexes are needed to give a reasonably good phase determination for all reflections. Each individual phase estimate contains experimental errors arising from errors in the measured amplitudes; furthermore, for many reflections, the intensity differences are too small to measure after one particular isomorphous replacement, and others must be tried.

Figure 17.9 Fourier summations of the intensity differences between diffracted spots from crystals of the protein alone and protein plus heavy metals give vector maps between the heavy atoms. Three atoms—A, B, and C—are at specific positions in the unit cell in (a). They give vectors A–B, A–C, and B–C, which are drawn from a common origin in (b) in dark colors. They also give the same vectors in the opposite directions as shown in light colors. The experimentally observed vector map is shown in (c) with a large peak at the origin corresponding to zero vectors between an atom and itself. It is straightforward to deduce the map in (c) from the atomic arrangement in (a). It is more difficult to do the reverse, to deduce the atomic arrangement in (a) from the vector map in (c), especially if there are many atoms in the unit cell that give rise to a large number of peaks in the vector map. For example, with 10 atoms in the unit cell there are 90 different vectors between the atoms.

Building a model involves subjective interpretation of the data

The amplitudes and the phases of the diffraction data from the protein crystals are used to calculate an **electron-density map** of the repeating unit of the crystal. This map then has to be interpreted as a polypeptide chain with a particular amino acid sequence. The interpretation of the electron-density map is complicated by several limitations of the data. First of all, the map itself contains errors, mainly due to errors in the phase angles. In addition, the quality of the map depends on the **resolution** of the diffraction data, which in turn depends on how well-ordered the crystals are. This directly influences the image that can be produced. The resolution is measured in Å units; the smaller this number is, the higher the resolution and therefore the greater the amount of detail that can be seen (Figure 17.11).

Figure 17.10 The diffracted waves from the protein part (red) and from the heavy metals (green) interfere with each other in crystals of a heavy-atom derivative. If this interference is positive as illustrated in (a), the intensity of the spot from the heavy-atom derivative (blue) crystal will be stronger than that of the protein (red) alone (larger amplitude). If the interference is negative as in (b), the reverse is true (smaller amplitude).

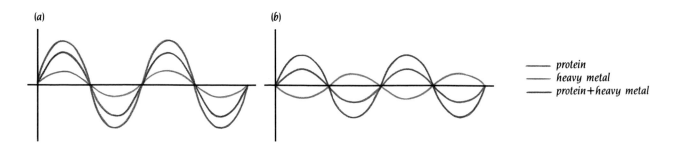

—— protein
—— heavy metal
—— protein+heavy metal

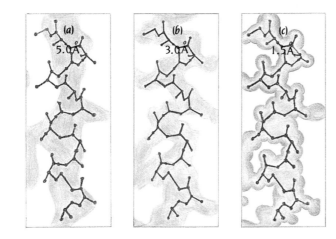

Figure 17.11 Electron-density maps at different resolution show more detail at higher resolution. At low (a) resolution (5.0 Å) individual groups of atoms are not resolved, and only the rodlike feature of an α helix can be deduced. At medium (b) resolution (3.0 Å) the path of the polypeptide chain can be traced, and at high (c) resolution (1.5 Å) individual atoms start to become resolved. Relevant parts of the protein chain (red) are superimposed on the electron densities (grey). The diagrams show one α helix from a small plant protein, crambin. (Adapted from W.A. Hendrickson in Protein Engineering, eds. D.L. Oxender and C.F. Fox. P. 11. New York: Liss, 1987.)

From a map at low resolution (5 Å or higher) one can obtain the shape of the molecule and sometimes identify α-helical regions as rods of electron density. At medium resolution (around 3 Å) it is usually possible to trace the path of the polypeptide chain and to fit a known amino acid sequence into the map. At this resolution it should be possible to distinguish the density of an alanine side chain from that of a leucine, whereas at 4 Å resolution there is little side-chain detail. Gross features of functionally important aspects of a structure usually can be deduced at 3 Å resolution, including the identification of active-site residues. At 2 Å resolution details are sufficiently well resolved in the map to decide between a leucine and an isoleucine side chain, and at 1 Å resolution one sees atoms as discrete balls of density. However, the structures of only a few small proteins have been determined to such high resolution.

Building the initial **model** is a trial-and-error process. First, one has to decide how the polypeptide chain weaves its way through the electron-density map. The resulting chain trace constitutes a hypothesis, by which one tries to match the density of the side chains to the known sequence of the polypeptide. This sounds easy, but it is not; a map showing continuous density from N terminus to C terminus is rare. More usually one produces a number of matches between the electron density and discontinuous regions of the sequence that may initially account for only a small fraction of the molecule and may be internally inconsistent. When a reasonable chain trace has finally been obtained, an initial model is built to give the best fit of the atoms to the electron density (Figure 17.12). Today **computer graphics** are exploited both for chain tracing and for model building to present the data and manipulate the models.

Errors in the initial model are removed by refinement

The initial model will contain errors. Provided the protein crystals diffract to high enough resolution (better than ~ 2.5 Å), most of the errors can be removed by crystallographic refinement of the model. In this process the model is changed to minimize the difference between the experimentally observed diffraction amplitudes and those calculated for a hypothetical crystal containing the model instead of the real molecule. This difference is expressed as an **R factor**, residual disagreement, which is 0.0 for exact agreement and around 0.59 for total disagreement.

In general, the R factor is between 0.15 and 0.20 for a well-determined protein structure. The residual difference rarely is due to large errors in the model of the protein molecule, but rather it is an inevitable consequence of errors and imperfections in the data. These derive from various sources, including slight variations in conformation of the protein molecules and inaccurate corrections both for the presence of solvent and for differences in the orientation of the microcrystals from which the crystal is built. This means that the final model represents an average of molecules that are slightly different both in conformation and orientation, and not surprisingly the model never corresponds precisely to the actual crystal.

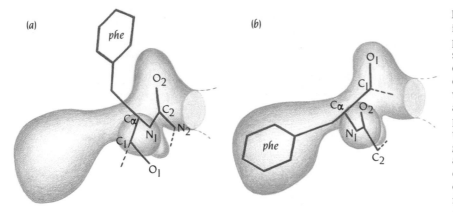

(a) (b)

Figure 17.12 The electron-density map is interpreted by fitting into it pieces of a polypeptide chain with known stereochemistry such as peptide groups and phenyl rings. The electron density (blue) above a certain level is displayed on a graphics screen in combination with a part of the polypeptide chain (red) in an arbitrary orientation (a). The units of the polypeptide chain can then be rotated and translated relative to the electron density until a good fit is obtained (b). Notice that individual atoms are not resolved in such electron densities, there are instead lumps of density corresponding to groups of atoms. (Adapted from Alwyn Jones in Methods in Enzymology, eds. H.W. Wyckoff, C.H. Hirs, and S.N. Timasheff. Vol. 115, part B, p. 162. New York: Academic Press, 1985.)

In refined structures at high resolution (around 2 Å) there are usually no major errors in the orientation of individual residues, and the estimated errors in atomic positions are around 0.1–0.2 Å provided the amino acid sequence is known. Hydrogen bonds both within the protein and to bound ligands can be identified with a high degree of confidence.

At medium resolution (around 3 Å) it is possible to make serious errors in the interpretation of the electron-density map, and there are, unfortunately, a growing number of them in the literature. Errors usually arise because elements of secondary structure are wrongly connected by the loop regions. α helices and β strands in the interior of the protein are rigid in structure and well defined in the electron-density map. The loop regions, however, are usually more flexible, and therefore the corresponding electron density is less well defined. It is easy to make errors in such regions in the preliminary interpretations of electron-density maps at medium resolution. These errors are usually caught and corrected before publication since such models will not refine properly and are likely to be incompatible with existing biochemical data.

However, some models containing serious errors have been published. They all have been based on data to only medium resolution together with insufficient phase information, which gives large errors in the electron density. It should, therefore, be kept in mind that unrefined structures with R values higher than 0.30 at medium resolution may contain errors, although the overwhelming majority of such published structures have survived subsequent refinement at high resolution.

Amino acid sequence is essential for x-ray structure determination

Most x-ray structures are determined to a resolution between 1.7 Å and 3.0 Å, and it is not possible to determine the structure of an unknown amino acid sequence from a map with a resolution in this range. Electron-density maps with this resolution range have to be interpreted by fitting the known amino acid sequences into regions of electron density in which individual atoms are not resolved. Chemically different groups, such as O, NH_2, or CH_3, are indistinguishable in an electron-density map at 1.7–3.0 Å resolution. This is because x-rays are scattered only by electrons, and these groups contain the same or similar numbers of electrons around the central atom.

To give some specific examples, electron densities corresponding to the side chains of Thr (T) and Val (V) look the same, as do the Asp (D)–Asn (N) and Glu (E)–Gln (Q) pairs. It is also usually very difficult to distinguish a His (H) from a Phe (F) side chain, and sometimes a Glu (E) or an Asp (D) can be misinterpreted as a His (H). For instance, the structure of the enzyme carboxypeptidase, which is described in Chapter 4, was initially determined to 2.0 Å resolution before the complete amino acid sequence was known and one of the zinc ligands, His (H) 196, was wrongly interpreted as Lys (K).

In addition to these uncertainties, many side chains on the surface of a protein molecule lack a fixed structure and can adopt several different conformations. As a result, no electron density is observed for these side chains except for the first atom. One consequence is that amino acids with flexible side chains would be interpreted as Ala (A), even if they were Arg (R), Lys (K), or other residues with large side chains.

When Tom Steitz at Yale University determined the structure of the enzyme hexokinase from yeast, the amino acid sequence was not known. He refined the structure to 2.2 Å resolution and tried to determine the amino acid sequence by x-ray methods. Steitz found that 40% of the amino acids in the polypeptide chain could not be guessed at, mainly because they were located on the outside of the molecule and the electron density faded after the first or second atom of the side chains. The remaining 60% of the residues contributed sufficiently to the electron-density map to allow Steitz to attempt to identify them. Some years later the gene was cloned, and its nucleotide sequence showed that only half of these residues, 30% of the complete chain, had been correctly identified. It is quite obvious from this example that an amino acid sequence is indispensible for accurate x-ray structure determination. Thus recombinant DNA techniques have had a double impact on x-ray structural work. When a protein is cloned and overexpressed for structural studies, the amino acid sequence, necessary for the x-ray work, is also quickly obtained via the nucleotide sequence. Recombinant DNA techniques give us not only abundant supplies of rare proteins, but also their amino acid sequence as a bonus. It should be kept in mind, however, that there also may be errors in published sequences; in fact, protein x-ray structures as well as NMR structures have several times identified and corrected such errors.

Recent technological advances have greatly influenced protein crystallography

In the early days of protein crystallography the determination of a protein structure was laborious and time consuming. The diffracted beams were obtained from weak x-ray sources and recorded on films that had to be manually scanned and measured. The available computers were far from adequate for the problem, with a computing power roughly equal to present-day pocket calculators. Computer graphics were not available, and models of the protein had to be built manually from pieces of steel rod (see frontispiece illustration). To determine the structure of even a small protein molecule, therefore, required many years of work and entailed time-consuming bottlenecks at almost every stage.

The situation is radically different today. The diffraction pattern can now be recorded on electronic area detectors coupled to powerful microcomputers that immediately interpret and process the recorded signals. Data collection that only a few years ago required many months of work is now done in a few days. If the in-house x-ray source is too weak for the problem, there are synchrotron sources available in several centers around the world that provide x-ray beams that are brighter by several orders of magnitude. Powerful computers in the laboratory provide the crystallographer with immediate access to almost all the computing power he or she needs. The electron-density maps are interpreted, and models of the protein molecules are built by the crystallographer sitting in front of a computer graphics screen. He or she is greatly aided by sophisticated software that involves semiautomatic methods for the model building using knowledge from databases of previously determined and refined protein structures.

These technical advances have greatly facilitated the use of crystallography for protein structure determination. One significant problem, however, still remains, namely, obtaining crystals that diffract to high resolution. Some protein molecules give excellent crystals after the first few trials, others may require several months of screening for the proper crystallization conditions, and many have so far resisted all attempts to crystallize them, for example, the *lac* repressor. Fortunately, it is now possible to determine the structure of small protein molecules in solution by nuclear magnetic resonance (NMR) methods.

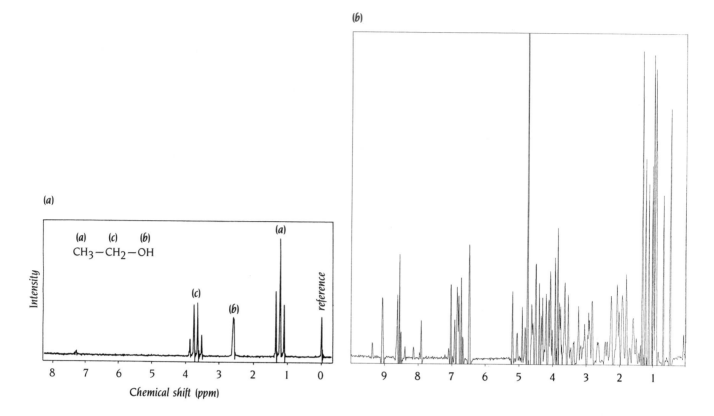

Figure 17.13 One-dimensional NMR spectra. (a) ^{1}H-NMR spectrum of ethanol. The NMR signals (chemical shifts) for all the hydrogen atoms in this small molecule are clearly separated from each other. In this spectrum the signal from the CH_3 protons is split into three peaks and that from the CH_2 protons into four peaks close to each other, due to the experimental conditions. (b) ^{1}H-NMR spectrum of a small protein, the C-terminal domain of the enzyme cellulase, comprising 36 amino acid residues. The NMR signals from many individual hydrogen atoms overlap and peaks are obtained that comprise signals from many hydrogen atoms. (Courtesy of Per Kraulis, Uppsala, from data published in Kraulis et al., *Biochemistry* 28: 7241, 1989.)

NMR methods use the magnetic properties of atomic nuclei

Certain atomic nuclei, such as ^{1}H, ^{13}C, ^{15}N, and ^{31}P have a magnetic moment or **spin**. The chemical environment of such nuclei can be probed by **nuclear magnetic resonance, NMR**, and this technique can be exploited to give information on the distances between atoms in a molecule. These distances can then be used to derive a three-dimensional model of the molecule. Most structure determinations of protein molecules by NMR have used the spin of ^{1}H, since hydrogen atoms are abundant in proteins.

When protein molecules are placed in a strong **magnetic field**, the spin of their hydrogen atoms aligns along the field. This equilibrium alignment can be changed to an excited state by applying **radio frequency** (RF) pulses to the sample. When the nuclei of the protein molecule revert to their equilibrium state, they emit RF radiation that can be measured. The exact frequency of the emitted radiation from each nucleus depends on the molecular environment of the nucleus and is different for each atom, unless they are chemically equivalent and have the same molecular environment (Figure 17.13a). These different frequencies are obtained relative to a reference signal and are called **chemical shifts**. The nature, duration, and combination of applied RF pulses can be varied enormously, and different molecular properties of the sample can be probed by selecting the appropriate combination of pulses.

In principle, it is possible to obtain a unique signal (chemical shift) for each hydrogen atom in a protein molecule, except those that are chemically equivalent, for example, the protons on the CH_3 side chain of an alanine residue. In practice, however, such one-dimensional NMR spectra of protein molecules (Figure 17.13b) contain overlapping signals from many hydrogen atoms because the differences in chemical shifts are often smaller than the resolving power of the experiment. In recent years this problem has been bypassed by designing experimental conditions that yield a **two-dimensional NMR spectrum**, the results of which are usually plotted in a diagram as shown in Figure 17.14.

The diagonal in such a diagram corresponds to a normal one-dimensional NMR spectrum. The peaks off the diagonal result from interactions between hydrogen atoms that are close to each other in space. By varying the nature of the applied RF pulses these off-diagonal peaks can reveal different types of

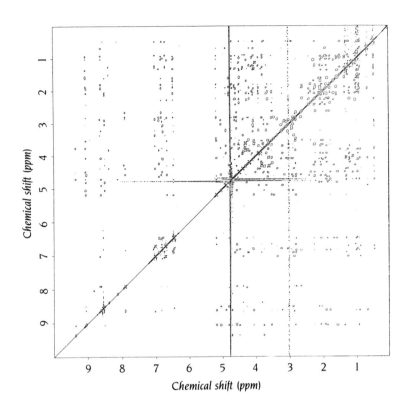

Figure 17.14 Two-dimensional NMR spectrum of the C-terminal domain of cellulase. The peaks along the diagonal correspond to the spectrum shown in Figure 17.13b. The off-diagonal peaks in this NOE spectrum represent interactions between hydrogen atoms that are closer than 5 Å to each other in space. From such a spectrum one can obtain information on both the secondary and tertiary structures of the protein. (Courtesy of Per Kraulis, Uppsala.)

interactions. A **COSY** (*correlation spectroscopy*) experiment gives peaks between hydrogen atoms that are covalently connected through one or two other atoms, for example, the hydrogen atoms attached to the nitrogen and C_α atoms within the same amino acid residue (Figure 17.15a). An **NOE** (*nuclear Overhauser effect*) spectrum, on the other hand, gives peaks between pairs of hydrogen atoms that are close together in space even if they are from amino acid residues that are quite distant in the primary sequence (Figure 17.15b).

Two-dimensional NMR spectra of proteins are interpreted by the method of sequential assignment

Obviously, two-dimensional NOE spectra, by specifying which groups are close together in space, contain three-dimensional information about the protein molecule. It is far from trivial, however, to assign the observed peaks in the spectra to hydrogen atoms in specific residues along the polypeptide chain because the order of peaks along the diagonal has no simple relation to the order of amino acids along the polypeptide chain. This problem has in principle been solved in the laboratory of Kurt Wüthrich in the E.T.H., Zürich, where the method of **sequential assignment** was invented.

Sequential assignment is based on the differences in the number of hydrogen atoms and their covalent connectivity in the different amino acid residues. Each type of amino acid has a specific set of covalently connected hydrogen atoms that will give a specific combination of **cross-peaks**, a "fin-

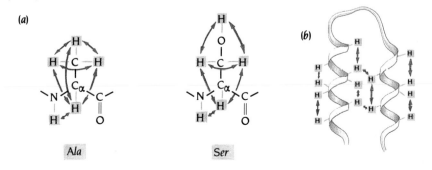

(a)

Ala

Ser

(b)

Figure 17.15 (a) COSY NMR experiments give signals that correspond to hydrogen atoms that are covalently connected through one or two other atoms. Since hydrogen atoms in two adjacent residues are covalently connected through at least three other atoms (for instance, HCα-C'-NH), all COSY signals reveal interactions within the same amino acid residue. These interactions are different for different types of side chains. The NMR signals therefore give a "fingerprint" of each amino acid. The diagram illustrates fingerprints (red) of residues Ala and Ser. (b) NOE NMR experiments give signals that correspond to hydrogen atoms that are close together in space (less than 5 Å), even though they may be far apart in the amino acid sequence. Both secondary and tertiary structures of small protein molecules can be derived from a collection of such signals, which define distance constraints between a number of hydrogen atoms along the polypeptide chain.

gerprint," in a COSY spectrum (Figure 17.15a). From the COSY spectrum it is therefore possible to identify the H atoms that belong to each amino acid residue and, in addition, determine the nature of the side chain of that residue. However, the order of these fingerprints along the diagonal has no relation to the amino acid sequence of the protein. For example, when the fingerprint in one specific region of the COSY spectrum of the *lac*-repressor segment was assigned to a Ser residue, it was not known whether this fingerprint corresponded to Ser 16, Ser 28, or Ser 31 in the amino acid sequence.

The sequence-specific assignment, however, can be made from NOE spectra (Figure 17.14 and 17.15b) that record signals from H atoms that are close together in space. In addition to the interactions between H atoms that are far apart in the sequence, these spectra also record interactions between H atoms from sequentially adjacent residues, specifically, interactions from the H atom attached to the main chain N of residue number $i + 1$ to H atoms bonded to N, C_α, and C_β of residue number i (Figure 17.16a). These signals in the NOE spectra therefore in principle make it possible to determine which fingerprint in the COSY spectrum comes from a residue adjacent to the one previously identified. For example, in the case of the *lac*-repressor fragment the specific Ser residue that was identified from the COSY spectrum was shown in the NOE spectrum to interact with a His residue, which in turn interacted with a Val residue.

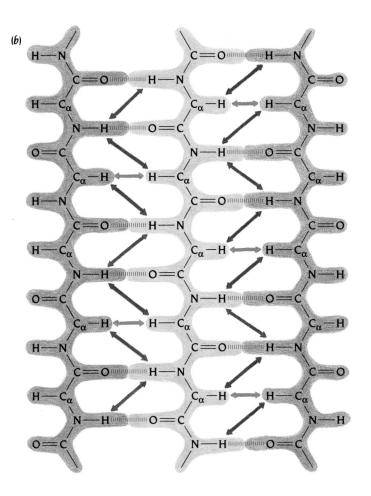

(a)

Figure 17.16 (a) Adjacent residues in the amino acid sequence of a protein can be identified from NOE spectra. The H atom attached to residue $i + 1$ (orange) is close to and interacts with (purple arrows) the H atoms attached to N, C_α, and C_β of residue i (light green). These interactions give cross-peaks in the NOE spectrum that identify adjacent residues and are used for sequence-specific assignment of the amino acid fingerprints derived from a COSY spectrum. (b) Regions of secondary structure in a protein have specific interactions between hydrogen atoms in sequentially nonadjacent residues that give a characteristic pattern of cross-peaks in an NOE spectrum. In antiparallel β-sheet regions there are interactions between C_α-H atoms of one strand (green) to C_α-H atoms (pink) and N-H atoms (purple) of adjacent strands (orange). Likewise there are interactions between N-H atoms of one strand to N-H atoms (red) as well as C_α-H atoms (purple) of adjacent strands. The corresponding pattern of cross-peaks in an NOE spectrum identifies the residues that form the antiparallel β sheet. Parallel β sheets and α helices are identified in a similar way.

Comparison with the known amino acid sequence revealed that the tripeptide Ser-His-Val occurred only once, for residues 28–30.

In practice, it is difficult to make unique assignments for longer pieces than di- or tri-peptides, since NOE signals also occur between residues close together in space but far apart in the sequence. Therefore, the peptide segments that have been uniquely identified by NMR are usually matched with corresponding segments in the independently determined amino acid sequence of the protein. Thus knowledge of the amino acid sequence is just as essential for the correct interpretation of NMR spectra as it is for the interpretation of electron-density maps in x-ray crystallography. Whereas x-ray crystallography directly gives an image of a three-dimensional model of the protein molecule, NMR spectroscopy identifies H atoms in the protein that are close together in space, and this information is then used to derive, indirectly, a three-dimensional model of the protein.

Distance constraints are used to derive possible structures of a protein molecule

The final result of the sequence-specific assignment of NMR signals, preferably done using interactive computer graphics, is a list of **distance constraints** from specific hydrogen atoms in one residue to hydrogen atoms in a second residue. The list contains a large number of such distances, which are usually divided into three intervals within the region 1.8 Å to 5 Å, depending on the intensity of the NOE peak. This list immediately identifies the secondary structure elements of the protein molecule because both α helices and β sheets have very specific sets of interactions of less than 5 Å between their hydrogen atoms (Figure 17.16b). It is also possible to derive models of the three-dimensional structure of the protein molecule. However, usually a set of possible structures rather than a unique structure (Figure 17.17) is obtained, each of the possible structures obeying the distance constraints equally well. The sets of possible structures, which are frequently seen in NMR articles, do not, therefore, represent different actual conformations of a protein molecule present in solution. Rather they are simply the different structures that are compatible with data obtained by current methods. The primary source of this ambiguity is an insufficient number of measured distance constraints. Because of this ambiguity, the accuracy of an NMR structure is not constant over the whole molecule and is also difficult to quantify.

In addition to the problem of ambiguity, there are other limitations to the use of NMR methods for the determination of protein structures. The most severe concerns the size of the protein molecules whose structures can be determined. Currently, the upper limit is molecules with molecular weights of around 15kD. Even though the methods are rapidly being improved, there seems little hope, today, that this limit can be changed by more than a factor of 2 to 30kD. Furthermore, the method requires highly concentrated protein solutions, on the order of 1–2 mM, with the additional requirement that the protein molecules must not aggregate at these concentrations. In addition, the pH of the solution should be lower than about 6 for proton NMR experiments. The exchanges of the NH protons in the main chain become so fast at higher pH that it is very difficult to observe them with NMR, and the signals from these hydrogen atoms are essential for the sequential assignment procedure.

These limitations notwithstanding, about 60 protein structures all with molecular weights of less than 15kD have been determined by NMR methods at the time of writing this book. How well do these NMR-derived structures agree with those determined by x-ray methods? There are currently six different globular proteins for which similar structures have been independently obtained by the two methods: bovine pancreatic trypsin inhibitor (Figure 2.14a), carboxypeptidase inhibitor, barley serine proteinase inhibitor, an α-amylase inhibitor called tendamistat, plastocyanin (Figure 2:11c), and thioredoxin from *E. coli* (Figure 2:7). The general conclusion from these comparisons is that the structures are nearly identical. The minor differences that exist are of the same

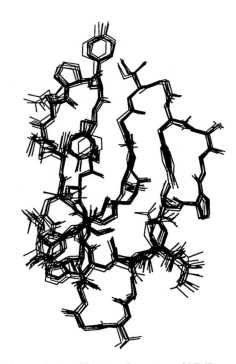

Figure 17.17 The two-dimensional NMR spectrum shown in Figure 17.14 was used to derive a number of distance constraints for different hydrogen atoms along the polypeptide chain of the C-terminal domain of cellulase. The diagram shows 10 superimposed structures that all satisfy the distance constraints equally well. These structures are all quite similar since a large number of constraints were experimentally obtained. (Courtesy of P. Kraulis, Uppsala, from data published in P. Kraulis et al., *Biochemistry* 28: 7241, 1989.)

order of magnitude as usually seen between x-ray structures of unrelated crystal forms of the same protein or determinations made under different experimental conditions. In other words, they are mostly small differences in loop regions of the main chain and different conformations of exposed side chains.

The situation is different for two other examples—namely, the peptide hormone glucagon and a small peptide, metallothionein, which binds seven cadmium or zinc atoms. Here large discrepancies were found between the structures determined by x-ray diffraction and NMR methods. The differences in the case of glucagon can be attributed to genuine conformational variability under different experimental conditions, whereas the disagreement in the metallothionein case was later shown to be due to an incorrectly determined x-ray structure. A recent reexamination of the x-ray data of metallothionein gave a structure very similar to the NMR structure.

NMR and x-ray crystallography are in many respects complementary. X-ray crystallography deals with the structure of proteins in the crystalline state, while NMR determines the structure in solution. The time scales of the measurements are different: NMR is more suitable for investigation of various dynamic processes such as those during folding, while x-ray crystallography is more suitable for characterization of protein surfaces and the water structure around the protein. X-ray crystallography remains the only method available to determine the structure of large protein molecules, whereas NMR is the method of choice for small protein molecules that might be difficult to crystallize.

Biochemical studies and molecular structure give complementary functional information

Our current knowledge of the relation between structure and function of protein molecules is insufficient to deduce the function of a protein from its structure alone, although, as we have seen, structural homology with proteins of known function can sometimes allow this. It is necessary to combine biochemical studies with structural information. Biochemical and cell biological studies can tell us if a protein is a receptor, a transport molecule, or an enzyme and, in addition, which ligands can bind to it, as well as the functional effects of such ligand binding. Studies of the three-dimensional structure of complexes between specific ligands and the protein will then give detailed information on how the active site is constructed and which amino acid residues are involved in ligand binding. Examples that we have described include protein–DNA interaction in Chapters 7, 8, and 9, sugar binding to a sugar transport protein in Chapter 4, and binding of inhibitors to enzymes that cleave peptide bonds in Chapter 15.

The specific role of each amino acid residue for the function of the protein can be tested by making specific mutations of the residue in question and examining the properties of the mutant protein. By combining in this way functional studies in solution, site-directed mutagenesis by recombinant DNA techniques, and three-dimensional structure determination, we are now in a position to gain fresh insights into the way protein molecules work.

Conclusion

The three-dimensional structure of protein molecules can be experimentally determined by two different methods, x-ray crystallography and NMR. The interaction of x-rays with electrons in molecules arranged in a crystal is used to obtain an electron-density map of the molecule, which can be interpreted in terms of an atomic model. Recent technical advances, such as powerful computers including graphics work stations, electronic area detectors, and very strong x-ray sources from synchrotron radiation, have greatly facilitated the use of x-ray crystallography.

Crystallization of proteins can be difficult to achieve and usually requires many different experiments varying a number of parameters, such as pH, temperature, protein concentration, and the nature of solvent and precipitant. Protein crystals contain large channels and holes filled with solvents, which can be used for diffusion of heavy metals into the crystals. The addition of heavy metals is necessary for the phase determination of the diffracted beams.

X-ray structures are determined at different levels of resolution. At low resolution only the shape of the molecule is obtained, whereas at high resolution most atomic positions can be determined to a high degree of accuracy. At medium resolution the fold of the polypeptide chain is usually correctly revealed as well as the approximate positions of the side chains, including those at the active site. The quality of the final three-dimensional model of the protein depends on the resolution of the x-ray data and on the degree of refinement. In a highly refined structure with R value less than 0.20 at a resolution around 2.0 Å the estimated errors in atomic positions are around 0.1 Å to 0.2 Å, provided the amino acid sequence is known.

In the NMR method the magnetic-spin properties of atomic nuclei within a molecule are used to obtain a list of distance constraints between atoms in the molecule, from which a three-dimensional structure of the protein molecule can be obtained. The method does not require protein crystals and can be used on protein molecules in concentrated solutions. It is, however, restricted in its use to small protein molecules.

Selected readings

Bernstein, F.C., et al. The protein data bank: a computer-based archival file for macromolecular structures. *J. Mol. Biol.* 112: 535–542, 1977.

Blundell, T.L.; Johnson, L.N. Protein Crystallography. London: Academic Press, 1976.

Branden, C.-I. ; Jones, A. Between objectivity and subjectivity. *Nature* 343: 687–689, 1990.

Clore, G.M.; Gronenborn, A.M. Determination of three-dimensional structures of protiens and nucleic acids in solution by nuclear magetic resonance spectroscopy. *CRC Crit. Rev. Biochem.* 24: 479–564, 1989.

Eisenberg, D.; Hill, C.P. Protein crystallography: more surprises ahead. *Trends Biochem Sci.* 14: 260–264, 1989.

Hendrickson, W. X-ray diffraction. In Protein Engineering. Eds. D. Oxender; C.F. Fox. New York: Liss, 1986.

McPherson, A. The Preparation and Analysis of Protein Crystals. New York: Wiley, 1982.

Wright, P. What can two-dimensional NMR tell us about proteins? *Trends Biochem. Sci.* 14: 255–260, 1989.

Wyckoff, H.W.; Hirs, C.H.W.; Timasheff, S.N., eds. Diffraction methods for biological macromolecules. Vols. 114–115: Methods in Enzymology. Orlando, FL: Academic Press, 1985.

Wütrich, K. NMR of Proteins and Nucleic Acids. New York: Wiley, 1986.

Wütrich, K. Protein structure determination in solution by nuclear magnetic resonance spectroscopy. *Science* 243: 45–50, 1989.

Index

287

Bacteriophage 434
  Cro-DNA complex, **93–99**, 95FF
  operator regions
    OR1–OR3, OL1–OL3, 94–95, 94F
    recognition signal, 96–97, 97F
  repressor and Cro-protein role, 88–89, 88F
  repressor-DNA complex, **93–99**, 93FF
Bacteriophage lambda
  see also Cro, phage lambda; Repressor, phage lambda
  operator region, 89, 89F
  repressor and Cro, 87–93, 88FF
Bacteriophage MS2, 173–174, 173FF
Bacteriophage P22
  arc repressor, 108
  repressor and Cro-protein role, 88–89, 88F
  repressor-DNA complex, 93–94, 93F
Bacteriophage T4, 257
Bacteriophage T7, DNA polymerase, 136–137, 136F
Bacteriophages
  see also individual bacteriophages
  lysogenic, 87–88
    genetic on-off switch, 88–89, 88F
  T-even, 162F
Bacteriopheophytin, 205
Bacteriorhodopsin, structure to 3 Å, 203, 203F
Bal I, 83, 83F
Banner, David, 33
Bence-Jones protein, 184
Bent helical axis, in DNA-protein complexes, 96, 96F
Beta (β) barrel. See antiparallel beta (β) barrels
β2 macroglobulin, 191, 195F
β-α-β motifs, 24–25, 25F
  in barrels and open sheets, 44, 44F
  domains and, 27, 27F, 29
  left-handed of subtilisin, 243, 243F
  of NAD-binding domains, 145, 145FF
β crystallin, gene organization, 70
β domains, 28
β-lactoglobulin, function, structure, similarity to RBP, 63–64
β meander, 152
β sheets, 12, **15–17**, 16FF
  DNA-binding motif, 108–109, 108FF
  Greek key motif in antiparallel, 24, 24F
  hairpin β motif of, 23, 24F
  mixed
    in α/β proteins, active sites, 53–54, 54F
  open twisted
    of Klenow fragment, large domain, 132, 132FF

of NAD-binding domains, 145, 145F
open twisted mixed
  of Klenow fragment, small domain, 133–134, 134F
packing among Ig domains, 187, 187FF
parallel, antiparallel, mixed, 15–16, 16FF
pleats in, 15, 16F
topology diagrams of, 21, 21F
twist of, 16, 17F
β strands, **15–17**, 16FF
  in α/β barrels, 45, 45F
  in barrels and sheets, 43–44, 43FF
  β-α-β motif of parallel, 24–25, 25F
  hairpin β motif of, 23, 24F
  hairpin loops in antiparallel, 19, 19F
  membrane-spanning, of porin, 214
  packing interaction with α helices, 46, 46T
  prediction, methods, 251–252
  Ramachadran plot, 9F
  schematic symbol for, 19, 20F
β structures, antiparallel. See Antiparallel β structures
β1-αA-β2 motif
  of FAD-binding regions, 151–152, 152F
  in NAD-binding regions, 149F, 150–151
Bilayered sheet, of membranes, 201
Biliverdin, 62–63, 63F
  as algal photoreceptor, 63
Biliverdin-binding protein, 62–63, 63F
Binding energy
  of hydrogen bonds, 262–263, 263F
  of ligand/protein, analysis by protein engineering, 260–263, 261F, 263F
Binuclear zinc cluster
  in DNA-binding domain, 119–120, 120F
  in Gal 4, 120, 120F
Bjorkman, Pamela, 194
Blow, David, 52, 236, 261,
Blundell, Tom 68, 69
Bragg, Sir Lawrence, 11, 273
Bragg's law, 274, 274F
Branden, Carl, 17F, 48F, 51, 153

C$_H$. See Constant domains (immunoglobulin)
Calcium-binding motif, 21–23, 22F, 23T, 27F
Calmodulin, calcium-binding motif, 23, 23T
Cancer cells, oncogenes and, 219
  transforming ras oncogenes in, 223, 226

CAP
  amino acid sequence of helix-turn-helix motif, 102, 102F, 103T
  effects, 106, 106F
  structure, 106–107, 107FF
CAP family, modular construction, functions, 137–138
Capsids
  defined, 161
  uses of knowledge of, 162
Capsids, spherical viruses, 70, **162–175**, 162FF
  see also Icosahedron and individual viruses
  jelly roll barrel structure, 169–171, 170F
  phage MS2 structure, 173–174, 173FF
  plant and animal virus similarities, 169–171, 170F
  quasi-equivalent packing, 166–167, 166FF
  specificity and economy, 162–163
  symmetry, 162
    icosahedral, 163, 163F
  triangulation numbers, T, 165–166, 166F
Carboxy group, of amino acids, 4, 4F
Carboxypeptidase, 53–54, 54F
Carboxypeptidase inhibitor, 283
Carotenoids, 62
Carter, P., 243
Caspar, Don, 165
Catabolite activating protein. See CAP
Catalysis, rate of, 231–233
Catalytic triad
  catalysis without, 243–244, 244F
  of chymotrypsin, 238, 237FF
  defined, 235, 235F
  evolution of, 238
  of subtilisin, 241–242, 242F
    mutants, 243–244, 244F
Cavities, engineering of, 260
CDR loops
  see also Hypervariable regions and individual CDRs
  conformation, models of, 186
  lysozyme binding to antibody, 190–191, 191FF
CDR1, in topology of Ig, 185, 185F
CDR2
  conformation of, 186
  hapten binding without, 189
  in topology of Ig, 184–186, 185F
CDR3
  conformation prediction, 250, 250F
  encoding in D segment, 191
  in lysozyme-antibody binding, 190–191, 191F
  in topology of Ig, 185, 185F, 186

homology to EF-Tu and cH-ras p21, 223, 225

as molecular amplifiers, 222–223, 223F

physiological processes mediated by, 222, 222T

gag proteins, zinc finger motif of, 115F, 120, 121F

Gal 4, 119–120, 120F

Gamma (γ) crystallins, **67–70**, 68FF

  domains, amino acid sequence homology, 70

    structural similarity, 69–70

  evolution, 70

  schematic and topological diagram, 68, 68F, 69F

GAP, 227

GAPDH. *See* Glyceraldehyde-3-phosphate dehydrogenase

Gap junctions, 203

GCN 4, 124FF, 125–126

Gene duplication

  chymotrypsin superfamily evolution by, 238

  Ig evolution by, 182

Gene expression, regulation

  see also transcription factors

  allosteric effectors, 104

  in eucaryotes, 113–114, 114F

  in lysogenic phages, 87–89, 88FF

  negative feedback, *trp* repressor, 104–106, 104FF

  positive control, CAP, 106–108, 106FF

Gene fusion

  double α/β barrels and, 47

  evolution of NAD-binding domains, 148

  of FMN-binding domain and cytochrome, 152–153, 153F, 155, 155F

Genes, for immunoglobulins, 182–183, 182F

Genetic on-off switch, in lysogenic bacteriophages, 88–89, 88F

Gilbert, Walter, 70

Globin fold, **35–38**, 36F

  in evolution, 37–38

  packing of, 36–37, 37F

Globins, conservation of structure in evolution, 38

  see also Hemoglobin; Myoglobin

Globular protein

  hydrophobic and hydrophilic sequences, 209

  hydrophobic core, 12

  membrane pore, change to, 212–213, 213F

Glucagon, 284

Glucocorticoid receptor, DNA-binding domain, 115F, 118–119, 118FF

Glutamate, in NAD-binding domains, 149F, 150–151

Glutamic acid, model, structural formula, 6P

Glutamine

  model, structural formula, 6P

  phage 434 operator region and, 97, 97F

Glutathione reductase

  amino acid sequence of β1-αA-β2 motif, 149F

  domains, biochemistry, 150F, 151

  structure (3-D), 151–152, 152F

Glyceraldehyde-3-phosphate dehydrogenase (GAPDH), 142–151, 143T, 143FF, 253F

  amino acid sequence of β1-αA-β2 motif, 149F

  as class B member, 147–148, 147F

  structure (3-D), 144FF, 145

Glycine, 5, 7P

  conformational freedom, 259

  conformations, 8, 9F

  in helix-turn-helix motif, 102, 102F, 104

  in NAD-binding domains, 149F, 150–151

  specificity of serine proteinases and, 238–239, 239F

  stability of lysozyme and, 259

Glycolate, 154, 145F

Glycolate oxidase, 154–155, 154F

  α/β barrel of, 45, 45F, 45T

  crystals, 271F

  evolution, 155, 155F

Glycolytic enzymes, α/β structures of, 43

Goldman, A., 210, 210T

Gravity, protein crystallization and, 271

Greek key motifs, 24, 24F

  in antiparallel β structures, **66–70**, 67FF

  of chymotrypsin, 236FF, 237

  complex, arrangements of, 28, 28F

  of immunoglobulin fold, 184FF

Growth factors. *See* Epidermal growth factor; EGF receptor; Insulinlike growth factors

GTP

  binding GTP-binding proteins, 225–226, 225F

  G-protein activation and, 222–223, 223F

GTP-binding proteins

  see also individual proteins

  cell transformation and mutations in, 223, 226

  consensus sequences, 225

GTP-GDP exchange catalyst, 227

GTPase Activating Protein (GAP), 227

Guanine, in DNA, hydrogen bonds in, 81FF

H subunit, of photosynthetic reaction center, 205–206, 206F

Hairpin β motifs, 23, 24F

  see also Antiparallel β hairpin motif

  arrangements of complex, 28, 28F

Hairpin loops, in β strands, 19, 19F

Halobacterium halobium, 203

Hanging-drop method, of protein crystallization, 271, 272F

Haptens, 189–190, 189F

Hardman, Karl, 19

Harrison, Stephen, 94, 166, 175

Hartmut, Michel, 203, 204

Hauptman, Herbert, 275

Heavy (H) chain (immunoglobulins), 180F, 181

  association of domains in, 186–188, 187F

  generation of, 182–183

Heavy chain, class I MHC molecule, 194

Heavy metals, in x-ray diffraction, 273, 275–276, 276F

Helical wheel, of α helices, 15, 15F

Helix-loop-helix. *See* Helix-turn-helix motif

Helix-turn-helix motif

  amino acid sequences of, 102–103, 102F, 103T

  of CAP, 106–108, 107FF

  in colicin A, 213, 213F

  of DNA-binding proteins, 22–23, 22F, 22T, **90–108**, 90FF, 100T, 103T

  eucaryotic, 103T, 104

  of 434 repressor and Cro dimer, 95, 95F

  of homeodomain proteins, 115, **121–124**, 122FF

  invariant residues, 123

  of *lac* repressor, 138

  in lambda Cro, 90, 90F,

  of lambda repressor, 91, 91F, 93

  prediction, approaches for, 99–100, 100T, 101FF, 103–104, 103T

  recognition helix of, 92–94, 92FF

  stereochemical constraints, 101–102, 102F, 104

  trp repressor, 104FF, 105

Hemagglutinin, 64, 70, 72–74, 73FF, 201

  infection by influenza virus and, 74

  as membrane fusogen, 74

  receptor binding site, 74, 74FF

  subunit structure, 73, 73F

  synthesis, 72–73

  trimer, 73, 74F

M subunit, of photosynthetic reaction center
α helices of, 205–206, 206F
hydropathy plot, 211F
pigment binding to, 206–208, 207F
McCormick, Frank, 227
Magnesium atoms
in nucleotide triphosphate hydrolyzing enzymes, 225, 225F
photosynthetic reaction center, in 211
Magnetic field, NMR and, 280
Main chain
in four-helix bundle, 34, 34F
of serine proteinases/substrate complex, 235, 235F
Major groove, of DNA, 80–81, 81F
Major histocompatibility complex. See MHC antigen; MHC molecules
Malaria, sickle-cell anemia and, 39–40, 40F
Malate dehydrogenase (MDH), 142–143, 143T, 143F
Manduca sexta, biliverdin-binding protein, 62–63
Mat a1, 103T, 104
Mathews, Scott, 35F, 153
Matthews, Brian, 90–93, 102, 257–258
Melting temperatures
of lysozyme engineered with S-S, 258, 258F
of proteins, 256
Membrane fusogen, hemagglutinin as, 74
Membrane lipids, interaction with proteins, 212
Membrane proteins, 201–213, 201FF, 210T, 212T
see also Transmembrane α helices and individual proteins
channel formation by colicin A, 212–213, 213F
functions, 201
high-resolution EM of, 203, 203F
solubilization, purification, crystallization, 202, 202F
types, 201
x-ray diffraction study of, 203–204, 204F
Membranes, 201, 201FF
insertion of colicin A into, 212–213, 213F
Mengo virus, 161, 167, 171
Menten, Maud, 232
Mesophiles, 256
met repressor, 108–109, 108FF
Metallothionein, 284
Methionine, 108
model, structural formula, 7P
MHC antigen,·class I
antigen binding site, 196–197, 196FF

genes, 194
polymorphism, 196, 196F
structure, **194–197**, 195FF
T-cell activation and, 194
virus, foreign protein interaction with, 196
MHC molecules, 179F, 180
Michaelis, Leonor, 232
Michaelis constant, K_m, 232, 232F
of engineered trypsin mutants, 239–240, 240T, 240F
Michaelis-Menten equation, 232, 232F
Minor groove, of DNA, 80–81, 81F
MIR. See Multiple isomorphous replacement
Mitochondria, 142
Model building
of antigen binding sites, 250, 250F
of Cro-DNA complex, 92–93, 92F
from x-ray diffraction data, 277–279, 278F
Modules, NAD-dependent dehydrogenase construction from, 144, 144FF
Modules, protein construction from, 27F, 137–138, 137F
Molecular disease, 39
Molecular weight, estimates from mobility, 204
Monochromatic x-rays, 272, 274F
Monoclonal antibodies, 190
Monod, Jacques, 104
Mononucleotide-binding motif, 145, 145F
Motifs
see also Domains and individual motifs
β-α-β, 24–25, 25F
domains and, 27–29, 27F
EF hand, 22–23, 22F, 22T
exon, correlation with, 70
four-helix bundle, 205, 206, 206FF
Greek key, 24, 24F, 66–70, 67FF
hairpin β, 23, 24F
helix-loop-helix, 22–23, 22F, 22T
jelly roll, 70–74, 71FF
prediction approaches, 99–100, 100T, 101FF, 103–104, 103T
of secondary structure, 21–25, 22FF
simple to complex, 28, 28F
in superbarrel of neuraminidase, 64–65, 64FF, 67F
Muirhead, Hilary, 47
Multiple isomorphous replacement, 275–276, 276F
Muscarinic receptor, 222T, 228, 228F
Mutants
lysozyme, 257–260
of tyrosyl-tRNA synthetase, 262–263, 263F

Mutations
conservation of protein structure and, 38
engineered, of trypsin, 239–241, 240F
in rhinoviruses, 173
for sickle-cell hemoglobin, 39–40, 39FF
single, protein function and, 247
to stabilize proteins, 257–260, 257FF
for study of binding energy, 261–263, 261F, 263F
myc, 125
Myeloma, Ig isolation and, 184
Myeloma proteins, hapten binding, 190
Myoglobin
α helix of, 13F
crystal forms and structure, 271
globin fold and active site, 35
schematic diagrams, 20F
structure, early results, 11, 11F
Myohemerythrin
active site, 35, 35F
four-helix bundle, 34, 35F

NAD, 141–142, 141F
binding NAD-binding domain, 146–147, 146F
NAD-binding domains
β1-αA-β2 motif, conserved structure, 149F, 150–151
evolution, 148
location in polypeptide chain, 144, 145F
NAD binding, 146–147, 146F
prediction of binding region, 149F, 150–151
stereochemical constraints, 148–151, 149F
structures (3-D), 144–145, 144FF
NAD-binding motif, prediction from amino acid sequence, 148–151, 149F
NAD-dependent dehydrogenases, **142–152**, 143T, 144FF
active site, 144, 144F, 146–147, 147F
chemical reactions catalyzed, 143, 143F
class A and class B, 147–148, 147F
functional domains, 144
homologies, 143
modular construction, 143–144, 144FF
NAD-binding domains. See NAD-binding domains
Negative feedback, in tryptophan synthesis, 104–105
Neuraminidase, **64–66**, 64FF, 201
active site, 65–66, 67F
antigen-antibody complex of, 191

function of viral, 64
schematic diagram of folding motif,
    64, 64F
subunits, 65–66, 65FF
superbarrel structure, 65, 65FF, 75
Neurospora crassa, trifunctional
    enzyme in, 47
Nicotinamide, 141–142, 141F
Nicotinamide adenine dinucleotide.
    See NAD
NMR
    advantages, 284, 285
    ambiguity and limitations, 283–285
    for homeodomain structure, 122–123
    interpretation of data, 281–283,
        281FF
    principle of methods, 280–281,
        280FF
    secondary and tertiary protein
        structures using, 270, **280–285**,
        280FF
    transcription factor structures using,
        115, 116, 117F, 118, 120–121,
        120FF, 126
NMR spectra (1-D), 280, 280F
NMR spectra (2-D), 280–281, 281F
    interpretation via sequential assign-
        ment, 281–283, 281FF
NOE NMR experiments, 281–283,
    281FF
Nopaline synthase, 149F
North, Tony, 63
Nuclear magnetic resonance. *See* NMR
5'-3' Nuclease activity, in DNA
    synthesis, 130, 130F
Nucleosides, 141F
Nucleotide-binding proteins
    evolution, 142
    NAD-dependent dehydrogenases,
        142–157, 143T, 144FF
Nucleotide binding sites, on hexoki-
    nase and DNA polymerase, 134,
    134F
Nucleotide polymerases, 129
    see also specific polymerases
Nucleotides, 141–142, 141F
Nyborg, Jens, 224

Okazaki fragments, 129FF, 130
Oncogene products, 218
    cell transformation and, 219
    growth-factor and protein kinase
        homologies, 219, 219F
    leucine zipper motif, 125–126, 124FF
Operator regions
    conformational changes on protein
        binding, 95–96, 96F, 98–99, 98FF
    for *met* repressor, 109
    of phage 434, 94–95, 94F
        recognition signal, 96–97, 97F

of phage lambda, palindromic
    sequences, 89, 89F
of phages lambda, 434, P22, control
    of, 88–89, 88F
Overwound, in DNA-protein com-
    plexes, 96, 96F, 98
Oxidases, cofactors for, 142F
Oxidation-reduction (redox) processes,
    141–142, 141F
Oxidative phosphorylation, 142
Oxyanion hole
    of chymotrypsin, 237, 237FF
    defined, 235, 235F
    of subtilisin, 242, 242F
        mutants, 244

P2 family, 62
p21. *See* ras p21
Pabo, Carl, 91, 122, 123
Palindromic nature, of phage lambda
    OR, 89, 89F
Pancreatic trypsin inhibitor, 283
Papain, 183
Parallel β sheets, 15–16, 17F
Parvalbumin, calcium-binding motif
    of, 22–23, 22F, 23T
Pattern recognition, structure predic-
    tion from, 248
Patterson maps, in x-ray diffraction,
    275–276, 276F
Pattus, Franc, 213
Pauling, Linus, 12, 39, 231
PDGF (platelet-derived growth factor),
    220F, 221
PDGF receptor, 220F, 221
Pepsin, 183
Peptide bonds
    formation, 4, 4F
    hydrolysis by serine proteinases, 234,
        234F
Peptide plane, 8, 8F
Peptide units, 8, 8F
Periplasmic sugar-binding proteins,
    138
Perutz, Max, 12, 39F, 275
Phase, of diffracted beam, 275–276,
    275F
Phenylalanine
    model, structural formula, 6P
    in RBP-retinol binding site, 61F
Pheophytin
    in photosynthesis, 208
    in photosynthetic reaction center,
        207, 207F
Phi (φ), 8, 8F
Phi (φ), psi (ψ) angle pairs
    in α helices, 12
    in β strands, 15
    Ramachadran plot, 8, 9F
Phillips, David, 20F, 63

Phillips, Simon, 108
Phosphate groups, α helices, binding
    to, 13, 14F
Phosphoglycerate mutase, 50F
PhosphoRibosyl Anthranilate (PRA)
    isomerase, 47, 47F
    biosynthesis of tryptophan and, 253
    circular permutation of, 263–265,
        263FF
Phosphorylation, protein kinases. *See*
    Serine kinase; Tyrosine kinase
Phosphorylcholine, as hapten, 189,
    189F
Photons, absorption in photosynthe-
    sis, 208–209
Photoreceptor, biliverdin in
    phycobilisomes, 63
Photorespiration, 154
Photosynthesis
    amplification, 204, 205
    basic process, 208
Photosynthetic pigments, in photo-
    synthetic reaction center, 206–
    208, 207F
Photosynthetic reaction center, 202,
    203–208, 204FF
    see also individual subunits and
        pigments
    amino acid sequences, 211, 212T
    electron transfer through, 208–209
    hydropathy plots, 211, 211F
    photon
        absorption, 208
        efficiency, 209
    pigment molecules, 204–205, 205F
        location, 206–208, 207FF
    subunits, 204
        structure, 205–206, 206FF
    twofold symmetry axis, 206, 207F,
        208
Photosynthetic vesicles, of
    Rhodopseudomonal bacteria, 204
Phycobilisomes, 63
Picorna viruses, 161, **167–173**, 168F
    antigen binding sites, 168
    jelly roll structure, 169–171, 170F,
        172F
    sequence homologies, 171
    structural proteins (VP1 to VP4), 168
    subunit arrangement in capsids,
        168–169, 169F
    virion assembly, 168
Pieris brassicae, biliverdin-binding
    protein, 62
Plasminogen, domains, 27F
Plastocyanin, 283
    topology diagram, 21F
Platelet-derived growth factor, 220F,
    221
Pleated β sheets, 16

structural domains, function, active site, 51–53, 52FF

Ullrich, Axel, 221
Ultraviolet radiation, lysis-lysogeny switch and, 87–88, 89
Uniquinone, 205, 205F
Unwin, Nigel, 203
Up-and-down β barrel
  β-lactoglobulin, 63–64
  biliverdin-binding protein, 62–63, 63F
  hydrophobic and hydrophilic residues, 62, 62T
  retinol-binding protein as, 60–62, 60FF, 62T
  schematic and topological diagrams, 60, 60F
  superfamilies, 62
Up-and-down β sheets, **60–66**, 60FF, 62T
  see also Up-and-down β barrels
  neuraminidase, **64–66**, 64FF
  of phage MS2, 173–174, 173FF
Up-and-down β strands, in HLA-A2 α1 and α2, 195, 195F
Upstream promoter element, 113, 114, 114F
Urokinase, domains, 27F

VH. See Variable domain
V_{max}, 232, 232F
Valegård, Karin, 173
Valine
  α/β packing and, 46, 46T
  model, structural formula, 6P
Variable domain (immunoglobulin)
  association and packing within, 187–188, 187FF
  β-sheet structure of, 184, 184FF
  defined, 180FF, 181–182
  hypervariable region, loop structure, 185–186, 185F
  switch peptide and orientation of, 193–194, 193F
Virus infection, antigens and MHC in, 196

Viruses
  see also Capsids; Capsids, spherical viruses; and individual viruses
  enveloped, **174–175**, 174FF
  infections by, 161
  picorna. See Picorna viruses
  self-sufficient, 165
  simplest, 161, 162F
  sizes, shapes, compositions, 161, 162F
  smallest known, 164
  spherical, capsids of, **162–175**, 162FF
    evolutionary relationships, 171
    T = 3
      capsid structure, 166–167, 166FF
      Sindbis nucleocapsid, 174, 174F
    T = 3 plant, **166–167**, 166FF
      jelly roll structure, 169–171, 170F, 172F
    T = 4, capsid structure, 166, 166F
      Sindbis spike, 174, 174F
Vitamin A. See Retinol
Vitamin D3 receptor, 118, 118F
VP0–VP4, 168–169, 169F, 171
VP1 subunits, antiviral drug design and, 172–173, 172F

"W" regions, of HLA-A2 α1 and α2, 195, 195F
Watson, Herman, 50F
Watson, James, 11, 79
Watson/Crick base pairing, in DNA replication, 129FF, 130
Wavelength, of diffracted beam, 275, 275F
Wells, J.A., 243
Wiley, Don, 73, 194
Wilkins, Maurice, 79
WIN 51711, 172
Wright, Peter, 116
Wüthrich, Kurt, 122, 123, 281

X-ray crystallography
  see also X-ray diffraction
  advantages and limitations, 284, 285
  diffraction pattern, reading of, 273–276, 274FF

of 434 repressor–DNA complexes, 94–95, 94FF
  homeodomain structure using, 122–123
  of myoglobin, 11, 11F
  phase determination, 275–276, 275F
  secondary and tertiary protein structures using, 269FF, **270–279**, 284–285
X-ray diffraction
  see also X-ray crystallography
  Bragg's law, 273–274, 274F
  electron-density maps and interpretation, 276–278, 277FF
  model building, 277–279, 278F
  of photosynthetic reaction center, 203–204, 204F
  protein crystallization, 270–271, 270FF
  recent technological advances, 279
  resolution, 276–277, 277F
  x-ray data recording, 272F, 273
  x-ray sources, 272–273
Xenopus laevis, 115, 116, 117F
Xfin protein, 116–117, 118F

Yanofsky, Charles, 106

Zinc atom
  binuclear clusters, 120, 120F
  in carboxypeptidase, 53–54, 54F
  in LADH, 144F
  in zinc fingers, 115F, 116–117, 117FF, 119, 120–121
Zinc finger motifs
  amino acid sequences of, 115F
  binuclear zinc cluster (Gal 4), 119–120, 120F
  classic [*Xfin* protein], 116–118, 117F
  DNA binding
    classic family, 118
    hormone-receptor family, 119, 119F
    TFIII, 116, 116F
  families (3), 115F, 116
  of hormone-receptor family, 115F, 118–119, 118FF
  retroviral gag proteins, 120, 121F
  of transcription factors, **115–121**, 115FF